Quality Measures and the Design of Telecommunications Systems

Quality Measures and the Design of Telecommunications Systems

John Fennick

Artech House

Library of Congress Cataloging-in-Publication Data

Fennick, John
  Quality measures and the design of telecommunication systems /
John Fennick.
    p. cm.
    Includes bibliographies and index.
    ISBN 0-89006-258-7
    1. Telecommunication systems — Design and construction.
    2. Telecommunication systems — Testing. I. Title.
TK5102.5.F45 1988
621.38 — dc19   88-6318

International Standard Book Number: 0-89006-258-7
Library of Congress Catalog Card Number: 88-6318

10   9   8   7   6   5   4   3   2   1

To Sally

Contents

Foreword

One of the benefits of retirement is having time available for reflection. No doubt most retirees do reflect on a lifetime of work and wonder how it all has gone by so quickly. Most of them probably do little more than that. John Fennick, on the other hand, has taken this experience and realized his lifetime of work was a little more involved than it need be. He appreciated the aphorism laid down by Lord Kelvin: "When you can measure what you are speaking about and express it in numbers, you know something about it . . .," but as he looked back on it all, he wondered why another person should struggle through such a maze of measurements. During his long career at Bell Telephone Laboratories he had developed a map of the tortuous path through the field of establishing a measure of transmission quality. Not only that, but he knew some shortcuts. Here in this book he takes the neophyte by the hand and leads the way to all the historical measurements, one by one, highlighting the evolutionary nature of the "structures" until one arrives at today's "citadel." At that place, Fennick goes into great detail so that future designers understand the parameters, the limits, the theoretical calculations, and the testing methods entering into measurements of transmission quality.

D. L. Favin

Preface

Telecommunication is concerned with the transport of information over a distance. In engineering terms, this is described as the delivery of a signal from an origin or source, through some media, to a destination or sink. Engineering for telecommunication is concerned with the manipulation of the signal or media to achieve a set of ranked goals. These may include speed, efficiency, integrity of the signal delivered, and of course cost. The engineering process can be considered as that of striking an acceptable trade-off among a set of such goals that are inevitably conflicting. The process results in a product that, over time, earns a reputation regarding its quality. This is the final judgment on the success of the process.

To evaluate trade-offs, the parameters in all equations must be known. This means they must be measured, and measurement is the first subject of this book. Engineers routinely make or use numerous kinds of measurements in the pursuit of design goals. When the goals are high, every measurement should be examined and interpreted for its relevance to the problem at hand. This requires a thorough understanding of what the measurement is and how it was made. Measurement and measurement technique are, therefore, fundamental to the engineering process. In spite of this, or perhaps because of it, in transmission engineering, it has been arduous for the newcomer to even learn the terminology. Understanding transmission measurements appears a never-ending struggle. The important transmission measures and the accompanying vocabulary, which often adds to the confusion, are presented at a level high enough to grasp concepts, and then explored in depth when deemed necessary.

Chapter 1 reviews the nature of transmission parameters and their measurement in telecommunication systems. In the first portion, each one is defined, its impact on quality is discussed, and typical values encountered in real systems presented. For a number of parameters, measurements cannot be made without affecting what is being measured and the measurement details must be known for results to be understood. In other cases, the name given to the measurement is ambiguous or even misleading. These are examined in detail in the second portion of the chapter.

Prominent in nearly every set of design goals is cost. It is almost always regarded as large so there is constant effort to reduce it, and one way is to allow increased degradation of the delivered signal. Eventually, the very difficult question, "What is the penalty in quality for this cost saving?", must be addressed. The question is difficult because of its subjective nature; none of the usual measurements apply.

A great deal of work has been done to derive quantitative measures of subjective quality. They can be used in answering the question above, but are not widely known. The most useful of these are grouped under the name *transmission grade-of-service,* or simply GoS. They enable a measure of quality to be included in the list of ranked goals in the same manner as cost or physical size. Relative penalty to the user becomes a part of the trade-off equations.

Quantifying an opinion and relating it through objective measures to cost is a subject likely to be new to engineers. Chapter 2 delves into the fundamental work for the basic measure, loudness. Chapter 3 expands the results to engineering measures, and shows how these and others are used in subjective testing to relate objective measures to opinion.

In Chapter 4 a description of the grade-of-service model for quality of speech communication and a discussion of a number of objective measures for data transmission is given. Though data transmission (telegraphy) is older than telephony, subjective measures for its quality are essentially nonexistent. That will be apparent from the relative space devoted to the two subjects.

The objective measures of Chapter 1 and the quality measures developed in Chapters 2, 3, and 4 come together when a telephone is used as an interface with a network. Chapters 5 and 6 provide a brief summary of important aspects of voice terminals, the network interface, and characteristics of a number of different kinds of networks popular in North America.

The last chapter comprises an example of the application of the tools developed earlier to the design of a general purpose telephone using linear transducers. Chapter 7 points to compromises made, what types of users are favored by the design choices, and how to determine the penalties to others.

The subject of transmission is replete with specialized terms and acronyms. An effort has been made to identify those not well known in other branches of engineering, and to define terms and acronyms when first used. No one can prejudge what is "well known," so if the reader is not clear as to just what is meant, please refer to the glossary, where a definition should be available.

The motivation for this text was a desire to bring together a body of knowledge scattered throughout the literature and to relate the diverse subjects in a sensible manner. I want to express my deep regard and appreciation for the critical eye and unending patience of my friend and colleague, Mr. David L. Favin. His continual review has been a major factor in this effort to provide an easier introduction to transmission measurements and quality.

Chapter 1
Transmission Parameters

Quality of transmission has two components: the first comprises the physical phenomena that cause a received signal to be an imperfect replication of the transmitted signal; the second is the user's perception of the received signal. In this chapter, we summarize and review significant elements of the first component, called *transmission parameters,* to explain their effect on quality.

The parameters are split into three groups: analog, data, and digital. Analog parameters include impairments that are significant to the delivery of a message comprising a continuous waveform, such as in speech transmission. Data parameters include impairments of most interest when the delivered message is in a discrete form, such as a telegram or a computer file. Digital parameters refer to impairments that may affect the transport of either continuous or discrete information, but are unique to digital transmission facilities (i.e., facilities that encode all information in a stream of pulses). Assignment of a parameter to the analog or data group is based on the premise that it is usually of more concern in the transmission of that particular class of message than in the other. (Noise is arbitrarily placed in the analog group.)

In several cases, it is not possible to obtain an exact measure of a parameter in a real transmission system. Therefore, the later sections of the chapter describe how the parameters are estimated or characterized, and alert the reader to cautions that must be exercised in interpreting and using data concerning these parameters.

1.1 CAUTIONS CONCERNING MEASURED QUANTITIES

One of the first concerns in the design of a transmission system is the signal power that must be delivered to the end user.[1] This determines how loud a voice message will sound, or the required sensitivity of a recording or display device at the signal sink. Because of the importance of this quantity, the results of many transmission measurements are expressed in units of power, but for ease and convenience are made with a voltage sensitive type of measuring instrument. Power is a function of both voltage and impedance and expressing a voltage measurement in power terms implies that the impedance across which the voltage was measured is known. In practice, the impedance is seldom known accurately and almost never measured. This is probably true for a number of reasons: first, impedance is relatively difficult to measure in a working communication system; second, calculations to determine true power from voltage and impedance measurements involve the use of complex arithmetic and so are not desirable on a routine basis; third, in most cases, impedances can be adequately controlled by design, manufacture, and installation procedures, and are not normally subject to significant changes in time. These facts suggest that it is practical to assume that the impedances encountered at designated test points are fixed, and so a simple voltage measurement provides an adequate estimate of the associated power.

If we work with logarithms of numbers rather than numbers, some calculations are simplified. Multiplication reduces to addition for example. This practice is so common in electrical engineering that a particular form of logarithmic expression is given the name decibel (dB). A decibel is one tenth of a bel, a unit named after Alexander Graham Bell. A decibel is defined in the following way.

Let two powers, p_1 and p_2, be expressed in common units such as watts. Their ratio is a dimensionless quality, and

$$X = 10 \log_{10}(p_1/p_2) \text{ dB} \tag{1.1}$$

The decibel is therefore a description of one power *relative* to another. Absolute amounts of power can be described if p_2 in this equation is replaced with a standard value. In electrical engineering for telecommunications a common standard for p_2 is one milliwatt, 0.001 watts, and the corresponding logarithmic relation is given a special name, dB referred to one milliwatt, abbreviated dBm:

$$P = 10 \log_{10}(p_1/1.0) \text{ dBm}, \quad p_1 \text{ in milliwatts} \tag{1.2}$$

Although the decibel is strictly defined as a power ratio, it is also used to describe the ratio of two voltages. For mathematical consistency, the multiplier preceding the logarithm must be 20 rather than 10 because power is a function of the square of voltage:

$$Y = 20 \log_{10}(v_1/v_2) \text{ dB} \quad \text{(of volts)} \tag{1.3}$$

Absolute voltage levels are conveyed by use of one volt for v_2. In addition, the one volt is taken to mean one volt rms unless otherwise specified. The corresponding special name is dB volts, abbreviated dBV:

$$Z = 20 \log_{10}(v_1/1.0) \text{ dBV}, \quad v_1 \text{ in volts} \tag{1.4}$$

In most telecommunication networks in North America, electrical interfaces are designed to have an impedance of 600 or 900 ohms (Ω), resistive, the choice being determined by where the interface appears in the network. Roughly speaking, 600 Ω is used in those parts of a network dedicated to long distance service, the *toll*[2] portion, and 900 Ω is used in the short distance, local, or *exchange* portion.[3] The scales of voltmeters designed for power measurements are labeled in units of power and the instruments are marked to indicate the impedance used in their calibration. Because the toll and exchange portions of a network are well defined, it is assumed that the instruments will be used only at interfaces (test points) that present an impedance that is correct for the instrument. This working assumption is easily forgotten and can lead to confusion and apparent inconsistencies in measurements. For example, telecommunication test equipment is often provided with both 600- and 900-Ω meter calibrations. Many central offices or switching centers contain equipment serving both the 900-Ω exchange area and the 600-Ω toll area. When manual measurements are made, it is the responsibility of the tester to adjust the equipment to the correct impedance. Errors in power readings of nearly 2 dB can result if this is not done. Two decibels is a small enough error to go easily unnoticed at the time of a measurement because it is within the range of expected variations of magnitudes of several parameters. This is especially true in noise measurements where normal variations may be significantly larger than this. The error may not be suspected until some time later when the data are used in an analysis or report. This may lead to expensive and time-consuming repeat measurements or may be reason to discard blocks of data.

As indicated earlier, power is a measure of great interest in transmission. However, in work closely associated with acoustic levels, as in voice terminals, the meaningful measure is voltage, or dBV; voltage referred to one volt (rms). This is because the ear responds to acoustic pressure rather than power, and electrical voltage is mathematically similar to acoustic pressure. For example, electrical power is proportional to the square of voltage, and acoustical power is proportional to the square of pressure. Thus, designers of terminal equipment are accustomed to working with dBV, while network people are accustomed to the use of dBm. If we are involved with interface devices or otherwise concerned with levels in both domains, great care must be exercised to determine or specify all impedances and define all quantities explicitly in diagrams and other documents. Amusing but time-wasting conversations may take place between engineers whose experience is in the different areas.

1.2 ANALOG PARAMETERS

The parameters chosen here to be classed as analog are listed in Table 1.1 and discussed in order in the following sections.

Table 1.1
Analog Parameters

Parameter	*Section*
Transmission level point	1.2.1
Noise	1.2.2
Loss	1.2.3
Echo and delay	1.2.4
Return loss	1.2.5
Singing margin	1.2.6
Balance	1.2.7
Interference	1.2.8
Crosstalk	1.2.9

1.2.1 Transmission Level Point

Transmission level point (TLP) is a concept used to control signal power or *transmission level* (TL) in a network. It is best explained through examples.

The examples used in this discussion relate to a particular type of network architecture known as *hierarchical* that, until recently, was used throughout North America. This architecture incorporates a rank ordering of switching machines and prioritized routing of traffic.[4] Hierarchical networks may have any number (greater than one) of ranks of switching offices; the one considered here has five. The concepts are valid, however, for any number of ranks, and may be used in analog networks of any architecture.

Signal power and TLP only have real meaning in an analog network because the signal referred to is the information signal, power in modulation sidebands, for example, as opposed to carrier power. In digital systems, all the power is in pulses of some type, and the information is contained in the sequencing or other characteristic of those pulses. Thus, signal power has no meaning in a digital system. The concept is still used, however, for purposes of interfacing and compatibility with analog systems in mixed networks.

The power levels of signals in an analog network must be carefully controlled. Not only must adequate power be delivered to the destination, but an excessive decrease anywhere within the network can result in poor signal-to-noise ratio, and an excessive increase can result in severe distortion of the waveform by overloading various devices. Level control is administered through the concept of transmission level.

At any point in a network, the TL can be expressed as a ratio, in dB, of the signal power there to the power of the same signal at some other point referred to as a *zero transmission level point* (0TLP). That ratio, X, is simply the net gain if positive, or loss if negative, from the reference point to the point at which the TL or power is specified.[5] Any point in the network may, therefore, be described as having a TLP of X dB referred to a reference point. *If the power of a signal is known at the 0TLP, its power at any other point will be above or below that value by a number of dB equal to the stated TLP.* The TLP concept is, therefore, a simple and convenient way of keeping track of losses and signal levels in networks. Note, however, that a TLP at a point in a network has meaning only for signals originating, or defined on an absolute basis, at the point at which the 0TLP was taken to be. The derived TLP may change if another reference point or source is considered.

It is important to note here that the discussion above was about the TL concept in a network. It is used in establishing and controlling the level of signals as they enter and flow through a network. Another use of TL is in the control of loss in components of networks. In this case, TLPs are rigidly defined. This application and how the two relate are described below, but first we need the notion of a standard level test tone.

A standard level test tone is a signal that is adjusted to have a level of 0 dBm at a 0TLP, and so its measured level, in dBm, at any point is numerically equal to the TLP at that point. Thus, the TLP at any point in a network, relative to a reference point, may be determined by measuring (at that point) the level of a standard test tone applied at the reference point. Sources of standard tones are also referred to as milliwatt supplies because 0 dBm is one milliwatt of power.

The building blocks of connections (transmission paths) established through a network are called *trunks*. Trunks serve as connecting links between switching offices and introduce a controlled amount of loss. The TLP concept is used in the administration and testing of trunk loss values. To illustrate this usage, it is helpful to first explain a typical trunk testing arrangement. A diagram defining some basic terms and illustrating a testing arrangement is provided in Figure 1.1.

The figure shows portions of the first (or last) pair of offices encountered in a long distance connection through the network. A class-5 (the switch rank number) office is shown at the left. Class-5 offices connect subscribers to the network and so are also known as *serving central offices*. A trunk is shown connecting the class-5 office to a class-4 office. A class-4 office represents the lowest level of so-called toll switching machines because local, or nontoll connections, have no need for the services of switches of rank higher than five. A switching office may often be considered simply as one switch that connects two trunks but, in fact, a number of switches are required for correct trunk selection and other purposes. This leads to a nomenclature to define specific points within the path through an office. The first switch encountered is referred to as the *incoming switch,* and to be even more precise, the term *incoming side of the incoming switch* may be used. Similarly, the last switch used in the intraoffice path is the *outgoing switch,* and one may speak of the outgoing side of the outgoing switch. Incoming sides of the incoming switches are identified by asterisks in the figure.

The outgoing sides of switches are defined to be at a certain TLP as shown in the figure. At class-5 offices, the TLP is designated as zero because, from a network view, that is where signals originate. At class-4 and other higher-ranked analog toll offices, the outgoing switch TLP is defined to be 2 dB less (−2 TLP). The smaller values are consistent with the fact that signal levels in the higher ranked portions of the network are expected to be somewhat attenuated by normal network losses, but the reason for the choice of −2 is not clear. There is some merit to arguments that relate these signal levels to loss design plans for networks, discussed in Chapter 6, but they are not conclusive. A typical value of loss on an analog trunk from a class-5 to a class-4 office is about 3 dB in the North American (analog) network, not 2 as the −2 TLP for a class-4 office implies. The −2 dB may be historically significant. With the TL of the outgoing switch defined, and because this switch appears as a source to the trunk and to the remainder of a connection, the term transmitting TL is also used in reference to such a point.

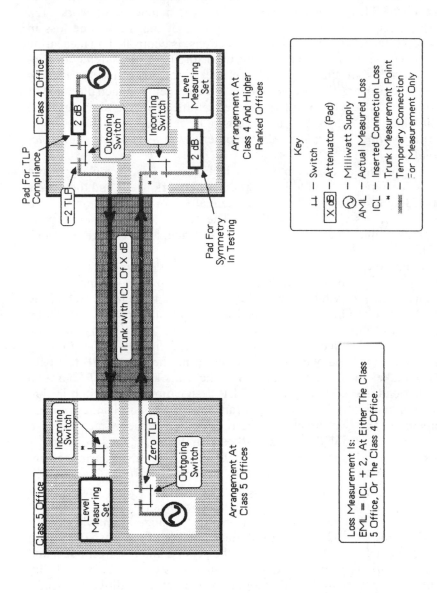

Fig. 1.1 A Trunk Loss Measuring Arrangement

The incoming switch receives signals from an outgoing switch and the receiving TL may be measured in the same manner as network TLs, and the receiving TL depends upon the loss of the trunk. The loss in trunks and, therefore, the loss on connections, is controlled by measuring trunk-receiving TLs. The equipment used for these measurements consists of a milliwatt supply and a level measuring set connected (during tests) to the trunk as illustrated; one set of equipment for each direction of transmission. The ends of the trunk are identified by several names. The end at which the supply is connected may be referred to as the *originating, transmitting, or sending end;* the other as the *terminating, or receiving end.*

The testing arrangement shown in Figure 1.1 is exemplary; many different designs are used to suit particular network needs and the details of providing test access. The use of attenuators, or pads as shown in the figure at the class-4 office, is variable. The attenuators shown are typical for a trunk connecting a class-4 to a class-5 office, a *toll connecting trunk* (TCT).

Trunk loss is defined as the *inserted connection loss* (ICL) (see the Glossary for a complete definition) of a 1000-Hz tone traversing the path from the outgoing switch of one office through the outgoing switch of the next.[6] Switching offices are equipped with milliwatt supplies that can be connected to trunks at the outgoing switch for use in measuring the trunk loss. At a class-5 office, the supply is connected directly to the trunk; at other offices it is usually connected to the trunk through a 2-dB test pad. The pad makes the test source appear to the trunk as a −2-TLP source to be numerically consistent with the TLP designation for a switch in that office. For symmetry in the testing arrangements and consistency of measurements in the two directions, the received power level meter is also connected to the trunk through a 2-dB pad. In the figure, the level meter is shown connected to the incoming switch, and the net testing arrangement is consistent with the loss definition: ". . . through the outgoing switch." In practice, the actual point of connection depends upon the type of switch used. Different types of switching machines provide different access points. The level measured in any test configuration will always differ from the true ICL of the trunk for a number of reasons. The length of office cabling cannot be identical in the service and testing modes, and the use of various pieces of equipment such as signaling devices (trunk circuits) that contribute small amounts of loss may be different. Minor deviations such as those due to cable length differences are allowed for in tolerances associated with measured values. Larger deviations are included in design calculations made to determine what the measured loss value should be. The calculated value is called the *expected measured loss* (EML). The loss that is measured in a test is called the *actual measured loss* (AML) and should agree with the prescribed EML, but not necessarily with the ICL.

Trunk maintenance loss values are stated in terms of the level that should be measured at the receiving end of the trunk when the test source is connected to the sending end. For toll-connecting trunks (those between class-4 and class-5 offices such as illustrated), these loss values are equal to the ICL + 2 dB because of the pads at the class-4 office sending end. Trunks between high-level offices, class-4 and above, are known as intertoll trunks and have EMLs equal to ICL + 4 because measurements include two 2-dB pads, one at each office. These two equations, ICL + 2 and ICL + 4, are generic and have variations appropriate to the type of switching machine and office equipment used, as mentioned in the previous paragraph.

It is important to note again that trunk TLPs are permanently defined at the sending end, and that receiving TLPs are not subject to redefinition, as are network TLPs, because the reference point is fixed. Trunk TLPs can be used to estimate signal power, with reference to a zero level, at some point in a connection in the following manner. To determine the expected power, the TLPs of the offices from the tone source to the point of interest must be known and included in the calculation. The loss, X, from a test tone source to the point of interest is found by the expression:

$$X = \text{(the sum of the AMLs or EMLs for the trunks)}$$
$$+ \text{(the sum of the intervening TLPs)}$$
$$- 2 \cdot \text{(the number of intertoll trunks)} \tag{1.5}$$

The last term accounts for the inclusion of two 2-dB pads in the EMLs of the intertoll trunks. The expected power of the standard test tone would be X dB below one milliwatt, and the network TLP would be $-X$ at that point. (Note that TLP values within one network are negative numbers.) For reasons explained in the discussion of loss plans in Chapter 6, trunks cannot have a net gain. (Certain test access points within equipment comprising a trunk may have positive TLs.)

It is common practice to specify desired levels for signals other than the standard test tone that may be applied to a network in terms of the level at the originating class-5 office, the 0TLP. Signals from data modems, for example, were required, by regulation, to arrive at the class-5 office over the subscriber's loop at -10 dBm and, therefore, to be 10 dB below the standard test tone or -10 dBm0. (The notation x dBm0 is used to indicate a power level of x dB relative to a 0TLP.) These signals should then appear throughout the network at -10 dBm0. If the TLP at a point in the network has been estimated, as explained earlier, to be $-X$, the data signal power there should be $(-10 - X)$ dBm.

These concepts are useful in network design for signal-to-noise ratio estimates and other power sensitive measures. Figure 1.2 illustrates the use of these concepts. The measuring apparatus and the trunk-to-trunk connections are indicated as being simultaneously on the same physical switch for illustration only. Calculations shown are made from left to right to determine the connection TLP (top row of numbers) at the extreme right with respect to the outgoing switch of the class-5 office. The office TLPs at the outgoing sides of the switches are defined as indicated, consistent with the rank of the offices. The inserted connection losses of the three trunks are 2, 1, and 3 dB. Testing arrangements use one pad of 2 dB for the toll-connecting trunk, and two pads, one at the tone source and one at the power level meter, for each of the intertoll trunks. The EMLs then become 4, 5, and 7 dB, and these numbers represent the loss values that should be found if the trunks were measured. A trunk measurement is from a source on the left to the meter in the nearest office to the right. Using (1.5), the sum of the EMLs and TLPs less twice the number of intertoll trunks, the connection loss from the 0TLP to the point of interest, is found to be 6 dB, so the connection TLP is -6.

In North America, the frequency of the oscillators used in the standard test tone sources is either 1000, 10004, or 1020 Hz. The value 1000 was universal until pulse code modulation (PCM) facilities were introduced and measurement problems encountered. Most of the energy in a signal that is an exact submultiple of the 8-kHz sampling frequency used in PCM is converted to harmonics. This causes the measured level of the test tone to be time variable. Shifting the test tone slightly, by 4 or 20 Hz, avoids the problem.

Internationally an 800-Hz tone is often used. Because the standard test tone frequencies are also commonly used for reference gains or losses in equipment specifications, the possibility of compatibility problems arises with a difference of 200 Hz between the North American and the international standard frequencies. Any measured loss differences arising from the different test tone frequencies are insignificant because both these frequencies lie near the flat center portion of typical loss-frequency characteristics of trunks.

Another choice of frequency is suggested by the CCITT[7] and used by some administrations in specifying subscriber loop losses. The loss on cable increases in proportion to the square root of frequency, a definition

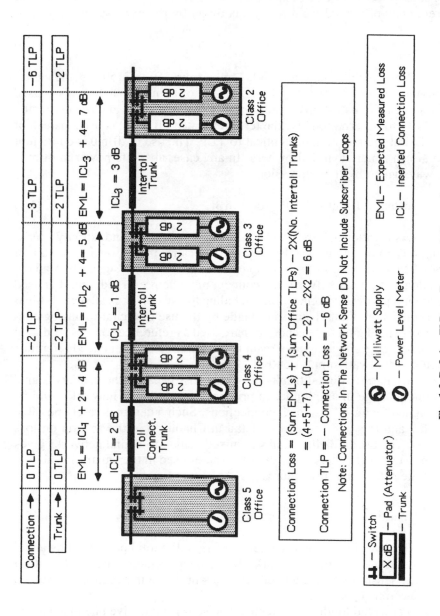

Fig. 1.2 Defining TLPs on Trunks and Connections

of mid-voice-band can be made on this basis. C, the square root center frequency of a band from f_1 to f_2, is found by the expression:

$$C = f_1 + [(f_1^{1/2} + f_2^{1/2})/2]^2 \tag{1.6}$$

For example, taking f_1 and f_2 to be from 250 to 3150 Hz, the center frequency, on a square root basis, is 1543 or about 1500 Hz. This is used by some administrations in subscriber loop documents and requirements.

Also, internationally, the term reference dB (dBr) is used instead of TLP, but the meaning is identical to TLP. The use of dBm0 is consistent in North America and elsewhere. In any case, absolute power in dBm is always given by the expression:

$$dBm = dBm0 + TLP \quad (or + dBr) \tag{1.7}$$

As mentioned earlier, TLP is an analog concept and, strictly speaking, has no meaning on a digital facility because the power in a stream of bits bears no relation to the amplitude of the encoded signal. In a mixed analog and digital network the concept must be preserved, and it is used even if a particular connection, including the switches, is all digital. Measurements on digital trunks are made by the use of digital-to-analog converters and digital tone generators are used as reference sources. A digital tone source emits a digital stream representative of an encoded reference source (milliwatt supply) output.

By nature, digital trunks have zero loss because the encoded signal cannot change in amplitude. In principle, an all-digital network could be designed for zero loss on all connections. Such a design is rarely possible today and this is discussed in detail in Chapter 6 in the section on loss plans. The use of digital trunks in mixed analog-to-digital networks gives rise to test-pad values other than the values used in the illustration above, but the calculation procedures are the same.

1.2.2 Noise

The word noise as used here, refers to the total power of unwanted signals at a point in a network. Noise measurements are made with a voltage sensitive device and either a 600- or 900-Ω impedance is assumed (see Section 1.1).

Perceived loudness of acoustic noise is a subjective phenomena and a function of level and frequency. The dependency of loudness on relative power level is not linear in dB but this is ignored in noise measurements. It is taken into account in quality evaluations, as described in Chapter 4.

Noise measurements do attempt to account for the frequency dependency, however, through the use of special weighting networks or filters. The filter shapes are determined through subjective tests using samples of many kinds of noise found on telephone circuits. The intent of the shaping is to make the noise meter yield the same measure for noises of differing frequency content if the subjective effects of the noises were determined to be equivalent. Thus, if a high-pitched noise is more annoying than a low-pitched noise of the same acoustic pressure, the weighting must attenuate the low-pitched (electrical) signal so the meter display will correlate with its effective value.

In North America, the shape of the filter currently used is known as *C-MESSAGE WEIGHTING* (C-MESS) ([2]). A drawing of the C-MESS weighting characteristic is shown in Figure 1.3. Note that 60 Hz, the frequency of power-line interference, is suppressed by more than 50 dB. This seems to imply that power-line interference, or hum, is much less annoying than white noise (in a telephone conversation), but the subjective tests were conducted using telephone receivers and the noise was measured at the input to the receiver. Typical telephone receiver frequency characteristics begin to roll off sharply at about 300 Hz, so hum is attenuated considerably before being converted to acoustic pressure. Thus, the frequency shaping of a typical telephone receiver is embedded in the C-MESS characteristic.

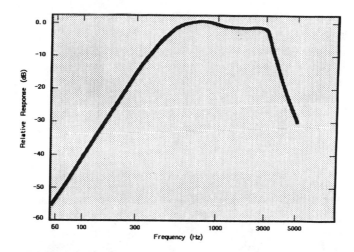

Fig. 1.3 C-MESSAGE WEIGHTING Characteristic
Source: Reprinted from ANSI/IEEE Std. 743-1984, copyright 1984 by the Institute of Electrical Engineers, Inc., by permission of the Standards Department

A number of other weighting characteristics are available for special applications. Two of the more common characteristics in North America are 3-kHz FLAT and Program. The first characteristic is intended for use if 60 Hz and its first several harmonics are of interest; the second is intended for measuring noise on special wideband (up to 15 kHz) broadcast radio channels between studios and transmitters for example.

The international CCITT weighting is called *psophometric*. The shape is similar to that of C-MESS, and the CCITT has stated that measurements may be interchanged with no conversion factor attributable to the shape of the weighting.

In North America, noise measurements, expressed in decibels, refer to an absolute power that is also subjectively based. The reference is −90 dBm, and is known as *reference noise* (rn). The −90-dBm corresponds to a noise power, measured at the input to a standard reference telephone, that produces an acoustic pressure deemed barely audible in the presence of normal received speech, by a listener in a quiet room. North American noise measurements are, therefore, expressed in decibels above reference noise, with C-MESS weighting, abbreviated dBrnC. For example, average noise on a telephone connection within a local area is about 18 dBrnC.

Outside North America it is common to specify the noise power in picowatts or in decibels above (or below) one milliwatt instead of decibels above reference noise. Noise values are, therefore, expressed in picowatts psophometric (pWp) or as dBm psophometric (dBmp). If p is noise power in watts (1000 mW), relations between these metrics are as follows: First,

$$X = 10 \log(p/10^{-3}) \text{ dB}$$

referred to one milliwatt (dBm), or

$$X = 10 \log(p) + 30 \text{ dBm} \tag{1.8}$$

and

$$Y = 10 \log(p/10^{-12}) \text{ dB}$$

referred to −90 dBm, or

$$Y = 10 \log(p) + 120 \text{ dBrn} \tag{1.9}$$

Then, for example, to express a typical weighted noise power of 100 pWp in other ways:

$$100 \text{ pW} = 10^{-10} \text{ W}$$

Then, using (1.8):

$$10 \log(10^{-10}) + 30 = -100 + 30 = -70 \text{ dBmp}$$

and, by (1.9):

$$10 \log(10^{-10}) + 120 = -100 + 120 = 20 \text{ dBrnC}$$

Note that the weightings play no role in the numerics of the conversions. Because reference noise is -90 dBm, dBrn and dBmp differ simply by 90, thus,

$$\text{dBmp} = \text{dBrn} - 90 \quad \text{(weighted noise conversion)} \tag{1.10}$$

From data (1983) reported in [3], noise on toll connecting trunks, referred to as exchange area access in the reference, has a median value of approximately 19 dBrnC (-71 dBmp) with a standard deviation on the order of 5 to 7 dB. Noise on connections in the switched North American network (1982), reported in [4], exhibited a mean value of 25 to 31 dBrnC (-65 to -59 dBmp), depending on the length of the connection. The standard deviation of noise on connections was approximately 3 dB.

1.2.3 Loss

In network engineering applications, loss is used as an abbreviated form of the more exact ICL. It is a measure of the power lost when a given device is inserted between the particular source and load under consideration. Because loss measurements are made at a single frequency, 800 or 1000 Hz, they have meaning only at that frequency. The total loss on connections is determined and controlled primarily *via* the losses of component trunks.

Losses for trunks are specified for installation and maintenance testing, as discussed in Section 1.2.1, in terms of EML, and measurements are stated as AML. These quantities include the 2-dB pads associated with trunks. Naturally, small differences, on the order of tenths of a decibel, are expected between EML and AML due to office cabling as mentioned in Section 1.2.1, temperature changes, and drift due to aging. On average, the small differences are usually negligible in terms of quality, and periodic testing and maintenance routines are designed to keep them that way. These concepts, as illustrated in Figure 1.2, were discussed earlier. Choosing loss values for trunks can be a complicated task and two approaches are discussed in Chapter 6 in the section on network loss plans.

For consideration of transmission quality, the total loss of the connection between user locations is the quantity of interest and it is usually obtained as the sum of three components: The loss of the loop from the transmitting end to the local switching machine or class-5 office, plus the loss of the network connection between that office and the one serving the receive location, plus the loss of the receiving loop. If one office is serving both the transmitting and receiving loop, a local call, 0.5 dB is added to the loss of the two loops to account for the loss in the switch. Losses on loops are considered in detail in Chapter 5 in the section on optimizing grade-of-service.

Magnitudes of losses to be expected in network connections can be obtained from data in [4]. Expected loss is shown there to range approximately from 6.5 to 7.5 dB with a standard deviation of approximately 1.3. These values are for connections that require at least two central offices in the path and so do not apply to local calls as defined here.

1.2.4 Echo and Delay

These two parameters are closely related. *Echo,* or more precisely, talker echo for this discussion, is the name applied to the delayed and unwanted sound of a person's own speech heard in the receiver of a telephone in use. Some level of sound of this nature is always present although the talker may not be aware of it. It comes from power reflected back toward its source at junctions of equipment with dissimilar impedances.

Talker echo may be of little interest, or the dominating impairment on a connection, depending upon its magnitude and the time delay between the speaker's utterance of a sound, and when the echo is heard. This time is called the round trip delay. If the round trip delay is close to zero (< 1 ms), the echo is called *sidetone* and a small amount is desirable for ease of conversation. If some sidetone is not present, the telephone appears to be *dead* and the user may be uncertain that it is working. An echo of small magnitude delayed up to a few milliseconds usually will go unnoticed because of masking by intentional sidetone introduced in the station set. As the round trip delay increases beyond that range, the echo becomes increasingly annoying and conversation increasingly difficult. If delays on the order of 300 to 400 ms are encountered, and the amplitude of the echo is not adequately controlled, speech of more than a few words at a time becomes impossible for most people. Delay time is determined by the effective velocities of transmission in the transmission media. This varies widely as shown in Table 1.2 where typical values for some common media are listed. The table includes delays associated with repeaters and terminal equipment in typical installations.

Table 1.2
Mean Velocities of Propagation

Medium	Velocity kmi/s
VF cable	5–30
Optical fiber	125
Coaxial cable	155
Microwave radio	155
Satellite	186

Repeaters (amplifiers in the satellite) and ground terminal equipment add a delay of a few milliseconds to that incurred in transit on satellite systems. Nonsatellite North American network transcontinental calls may have round trip delays of up to 45 ms ([4]), but for many of them it will be somewhat shorter. For comparison, connections that utilize satellite facilities encounter round trip delays on the order of 520 ms. Some relative round trip delays encountered in portions of typical telecommunication connections are illustrated in Figure 1.4. For satellite facilities, the delay is dependent upon the viewing angles of the satellite from the ground stations (see Figure 1.4). However, the total possible variation due to this delay is only about two percent.

Because of its adverse effects, the control of echo in networks is of prime concern. Two techniques are commonly employed. The first is to introduce loss in the trunks comprising the connection. Because the echo travels a portion of the connection in each of two directions, from the talker to a reflection point and then back to the talker, it undergoes more loss (at least twice as much) than the desired signal. Long connections

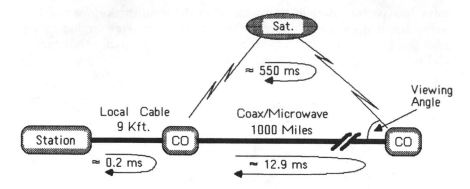

Fig. 1.4 Some Nominal Round-Trip Delay Times
Source: Telecommunications Transmission Engineering, Volume 1, Principles, Second Edition, American Telegraph and Telephone Company, 1980.
Reprinted with permission of AT&T—All rights reserved

have greater delay and so must use more loss to adequately attenuate the echo. Remember that longer delays cause more annoying echoes and so they must be reduced more in level. Detailed engineering rules have been established that determine necessary trunk design loss as a function of length. One set of widely used rules, known as *via net loss* (VNL) *design* is described in detail in [5] and discussed in Chapter 6. For now, it is sufficient to be aware that such rules exist. When the round trip delay exceeds 40 ms, the loss required for echo control becomes so large that the speech heard by the listener would be too low in level. At this point, the trunks are designed to have 0 ICL, and echo suppressors or echo cancelers are added to them. Echo suppressors are loss switches that effectively open the return path of a trunk when speech is detected in the forward direction. Echo cancelers use adaptive filters to create a replica of the potential echo and subtract it from that echo, thus canceling it. Either method can achieve echo suppression of up to about 40 dB. The subjective effects of echo have been studied in great detail and are very significant in the calculation of grade-of-service as shown in Chapters 4 and 7.

1.2.5 Return Loss

Any electrical junction can be treated as a source driving a load; the voltage developed across the load may be considered to be the sum of two vectors: $(1/2)V_{source}$, which would be developed if the source and load impedances were identical, and the other due to any impedance difference, the mismatch, between the source and load. The magnitude of the second vector is a function of the size of mismatch. The difference of these two vectors is referred to as the reflected voltage. If the source impedance is Z_s, and the load impedance Z_L, a coefficient of reflection, ρ, is given by the expression:

$$\rho = (Z_L - Z_s)/(Z_L + Z_s) \quad \text{(reflection coefficient)} \tag{1.11}$$

The quantity ρ is a complex number with a magnitude in the range 0 to 1 and an angle of value 0 to 2π. The reflected voltage is equal to $\rho \cdot V_i$, where V_i is voltage delivered in the matched case, one half of the source voltage. Note from (1.11) that only if $Z_L = Z_s$ is the reflection coefficient, and so reflected voltage, equal to zero.

Reflected voltages added to the sidetone in a receiver on a vector basis, appear as echoes and, if not adequately controlled, can be the basis of sustained oscillations in a circuit. The latter condition is referred to as *singing*. In general, reflections are undesirable, so the amount by which they are suppressed is of interest. This suppression is called *return loss* (RL), expressed in decibels and calculated as

$$\text{RL} = -20 \log_{10}|\rho| \quad \text{(return loss)} \tag{1.12}$$

Return loss ranges from 0 dB, for a shorted or open load, to infinity, for the matched case. If the magnitude of the impedance difference, $|Z_L - Z_s|$, is near 10 percent of the source magnitude, the return loss is approximately 26 dB; a 5 percent difference equates to approximately 32 dB.

The condition for maximum return loss, or zero reflected voltage, should not be confused with the condition for maximum power transfer from source to load which requires that the source and load have impedances that are complex conjugates of each other. Maximum power transfer does not imply zero reflected voltage. If the load impedance is the complex conjugate of the source impedance and written in rectangular notation as $a + jb$, the source impedance is $a - jb$. Then, for a voltage source, V, the current, i, is found to be $V/2a$. The voltage across the load, V_L, is given by the relation:

$$V_L = V(1/2)(1 + jb/a) \tag{1.13}$$

and the reflection coefficient, ρ, is jb/a. Note that the multiplier $1/2$, together with the bracketed term, define two components of V_L: one half of the source voltage and one half of $\rho \cdot V_{\text{source}}$, as claimed above. The voltage-current product, $e \cdot i$, at the load is $(1/4)(a + jb)/a^2$ or $(1/4a)(1 + jb/a)$. Only the real part of this is used by the load (the $e \cdot i \cdot \cos \theta$ term familiar to power engineers). The imaginary or reactive component is returned or reflected back to the source.

For matched (equal) complex impedances, the power delivered to the load, p_1, is

$$p_1 = V_{\text{source}}(1/4)a/(a^2 + b^2) \tag{1.14}$$

and may be far from the possible maximum. In the maximum power transfer case with complex impedances, the voltage reflected must appear across, or be absorbed by the source. This is the undesirable echo situation in communication systems.

Because any impedance other than a pure resistance is a function of frequency, return loss is also a function of frequency. This gives rise to two distinct transmission impairments. First, if echoes have no net attenuation, sustained oscillations may arise, and the circuit is said to be in a singing condition. The minimum return loss at any point in the frequency band of interest, is a measure of protection against this condition and so is called the *singing margin*. This is a separate topic and is discussed further in the next section.

If singing margin is adequate to prevent oscillation, a second impairment, *echo return loss* (ERL), must be considered. Echoes may be subjectively annoying and the annoyance is, not surprisingly, frequency dependent and a function of the total power in the band, not just that at one frequency. The measure of annoyance is a frequency weighted power measurement of the echo and is called echo return loss. In North America, a common frequency weighting characteristic for echo evaluation is that of a simple bandpass filter with 3-dB points at 560 and 1965 Hz ([6]). A second measure of echo return loss, developed in grade-of-service work is *weighted echo path loss* (WEPL). WEPL is defined as

$$\text{WEPL} = -20 \log \left[(1/3200) \int_{200}^{3400} 10^{-\text{EPL}(f)/20} \, df \right] \tag{1.15}$$

where $\text{EPL}(f)$ is the return loss associated with the entire round trip echo path as a function of frequency. WEPL is just an average voltage measure.

Other administrations generally use the weighting defined in CCITT Rec. G.122 found in [7]. This is similar to that for WEPL but is a power sum weighted by a characteristic with a negative slope of 3 dB per octave. It is computed as

$$\text{echo loss} = 3.85 - 10 \log \left(\int_{300}^{3400} A(f)/f \, df \right)$$

where $A(f) = 10^{-L(f)/10}$, and $L(f)$ is the *echo* loss measured at specified locations in a network where it is expected to be a minimum. Such locations are considered in Chapter 6 where various types of networks are described in some detail.

The magnitude of echo return loss, or any of the other subjective measures of return loss should generally be at least 6 to 8 dB, or more if the echo delay is significant. Its magnitude is controlled by how well impedances are matched at junctions ($|\rho|$, the magnitude of (1.11)) and by the loss in the transmission path. Echo return loss is, therefore, a major factor in establishing loss values for trunks in networks. Two methods used to establish loss values are described in Chapter 6. Return loss, in conjunction with delay, also enters quantitatively into grade-of-service calculations as shown in Chapter 4.

1.2.6 Singing Margin

Whenever an arbitrary transmission path forming a closed loop exists, the potential for sustained oscillations, or singing, is created. Communication systems contain many such loops. Every connection and every four wire trunk terminating *via* hybrids[8] on two-wire switches, is such a loop. Speech emanating from the receiver of a telephone handset must, in turn, be picked up by the handset transmitter and returned to the receiver through the sidetone path. This loop does not oscillate because of the large acoustic attenuation normally present between the receiver and transmitter. On the other hand, speaker phone design is complicated with special circuitry to open this loop because the acoustic attenuation may be inadequate in that arrangement. Feedback theory dictates that if the net closed loop gain is less than 0 dB, the system will be stable.[9] In practice, at least 2B of net loss, referred to as singing margin, is considered necessary to allow for variations in the numerous components involved in the loops. The CCITT recommends that terminations for international circuits maintain a minimum single frequency return loss of at least 6 dB ($|\rho| \leq 0.5$) for singing protection ([7], Rec. G.122).

In the United States, a weighted return loss measurement is used as a measure of protection against singing. Impedances tend to be less controlled near the edges of the frequency response of channels and so minimum return losses are expected to be encountered at either the high or low ends of the band. With this in mind, weighting characteristics have been devised that permit measuring reflected energy in relatively narrow bands centered near the channel edges. These are known as *singing return loss low and high* (SRL Low and SRL High). The weightings are those of bandpass filters with 3-dB points at 260 and 500 Hz, with high and low rolloff of 18 dB per octave for SRL Low, and at 2200 and 3400 Hz, with rolloff of 30 dB per octave for SRL High (abstracted from [6]).

Data reported in [3] show that the distributions of echo return losses measured at the four-wire to two-wire, intertoll-to-toll connecting, interface in the North American network have mean values of approximately 21 to 28 dB with standard deviations on the order of 5 dB. The smaller values are encountered on toll connecting trunks comprising cable facilities, the higher values on trunks using carrier facilities. It must be pointed out, however, that these measurements were made with the toll connecting trunk terminated in a standard compromise impedance at the class-5 office, rather than the subscriber loop and telephone as in a real connection. The impedances presented by loops and telephones cover a significantly large area in the complex plane (see the section on the exchange area in Chapter 6), and cause the relatively good distributions of return losses reported in [3] to become much more dispersed and include much smaller values. Calculations based on impedance measurements made in exchange areas may reveal minimum single frequency return loss values of 2 dB or less.

1.2.7 Balance

The desired current in a pair of wires used as a transmission medium is known as the metallic current because its path lies completely within the metal conductors. The current is supplied by a source at one end of the pair. It flows in a loop path to a load at the other end and returns to the source. Because there is no such thing as a perfect insulator, all terrestrial devices have finite impedances from their components to the earth, and this impedance can be part of an electrical circuit. Two such impedances, from two distinct devices, are connected by virtue of the common earth, which also has a finite impedance value. The three impedances in series create what is commonly called the *ground path* between the two devices. If the impedances to ground, as measured from either terminal of a two terminal device are identical, the device is said to be balanced to ground. A high degree of balance (nearly identical impedances) is often desirable. To see why this is so, consider the circuit of Figure 1.5. A source with an impedance, Z_s, is shown driving a transmitting line comprising a pair of identical wires in a section of cable that connects to a load impedance, Z_L, on the right. The distributed impedances to ground are illustrated as effectively connecting to some point within the source and load, defined by the parameters a and b. The a and b define Z_s and Z_L as sums of two parts,

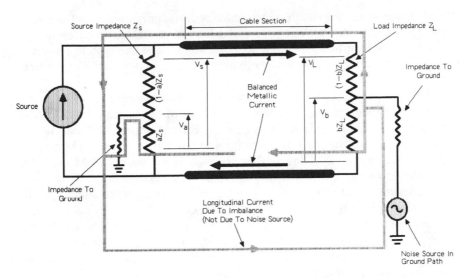

Fig. 1.5 Illustration of Source Imbalance and Resultant Longitudinal Current

i.e., $Z_s = aZ_s + (1 - aZ_s)$, $0 \leq a \leq 1$. The source or load will be unbalanced to ground if either a or b is not equal to 0.5.

The current from the source through the load is indicated by the wide dark arrows. It is called the *metallic current* because it is flowing in the metal conductors of the pair of wires in the cable. If the impedances to ground were infinite, the current to the load would be identical to that from the load, a situation described by the word *balanced* associated with the metallic current. Voltages V_s and V_L appear across Z_s and Z_L which act as voltage dividers with taps at the ground impedance points defined by a and b. Most often, V_s is not equal to V_L so if a and b are not identically 0.5, then V_a is not equal to V_b and their difference will cause a current (different in the upper and lower portions of the cable), indicated by the gray lines, to flow through the ground path. This is called a longitudinal current because the direction of flow is the same in both wire conductors. Because this current arises from voltages incident to the flow of the metallic current, the circuit is said to exhibit metallic-to-longitudinal conversion. The circuit itself is said to be unbalanced to ground because a is not equal to b. These facts by themselves are of no great concern to the operation of the circuit except for a minor power loss.

What is important is that the ground path contains noise sources. One such noise source is illustrated near the junction of the load impedance to ground and ground in the drawing. Current driven by this source flows in the same path as that just described, the gray line. A net nonzero noise voltage then develops across the load because of the difference in current from its ground impedance to each side, and the system output is noisy. The circuit is now said to exhibit longitudinal-to-metallic conversion.

For this example, the unbalance conditions were placed in the source and load; in practice they may occur in the source, load, or the transmission medium. In a many-pair cable, components of the ground paths for each of the pairs are common, and metallic currents in one pair may appear as sources in another through these conversions between metallic and longitudinal currents, giving rise to crosstalk.

An important source of noise introduced through unbalanced conditions stems from the fact that metallic protective sheaths on cables commonly have large induced currents in them from nearby power distribution systems, radio transmitters, *etcetera*. These are also referred to as longitudinal currents because of their singular direction compared to the bidirectional metallic currents in the pairs. Longitudinal-to-metallic conversion can give rise to significant noise or interference from these currents.

The two kinds of conversion, metallic-to-longitudinal and longitudinal-to-metallic, define two types of balance measurements with corresponding names. The names are often shortened to just metallic balance, referring to the first type of conversion, and longitudinal balance for the second type. Testing arrangements are similar to the circuit of Figure 1.5.

If the cable section is removed, then $V_s = V_L$ and, if the noise source illustrated is zero, and Z_s is an ideally balanced source, metallic-to-longitudinal balance, B_m, is defined by

$$B_m = -20 \log[(V_a - V_b)/V_L] \text{ dB} \quad \text{(metallic balance)} \qquad (1.16)$$

Then, with the source current removed and the noise generator adjusted to a voltage, V_n,[10] the longitudinal-to-metallic balance, B_L, is defined by

$$B_L = -20 \log(V_L/V_n) \text{ dB} \quad \text{(longitudinal balance)} \qquad (1.17)$$

The two measures are usually not the same due to the distributed nature of the impedances to ground, which can be considered as short transmitting lines and need not be symmetrical when viewed from either end.

The magnitudes of balance frequently encountered range from approximately 40 to 55 dB for miscellaneous transmitting devices, 50 to 70 dB for common test equipment, and up to 100 dB or more for precision equipment and for pairs in high quality communication cable. Great care to maintain such balance is exercised in the manufacture of cable because of its potential for exposure to strong electromagnetic fields when placed in service. Measures of balance are not considered directly in grade-of-service work, the effects of imperfect balance simply represent another contribution to the total noise.

1.2.8 Interference

In general, the term interference can be used to describe an amount of any undesired energy in the circuit of interest. Phenomena such as noise from power sources and crosstalk are important subsets of interference. In the design of transmitting devices it is essential to protect the device from external sources of interference, and to minimize effects within the device that might appear as sources of interference to others. Therefore, two kinds of interference measurements are defined: The first kind, in-band measurements, determine susceptibility of a device to external sources of interference. The term, in-band, is somewhat misleading because it includes interference from modulated *radio frequency* (RF) sources. These can induce RF currents that are demodulated by nonlinear elements in the device and then produce audible *in-band* interference. The second kind, called out-of-band interference, measures any interference sources within a device that may couple in some manner to other devices. In-band interference is often abbreviated to simply *in-band* to indicate that measurements of the interference are made within the pass band of the system. Similarly, *out-of-band* is often just used in the second case and has the complementary implication. These definitions are diagrammed in Figure 1.6. Potential in-band disturbances to other devices are usually not measured explicitly but are partially covered by crosstalk requirements in large systems. Requirements for both in-band and out-of-band interference may extend to tens or hundreds of mHz and in-band interference specifications usually call for satisfactory operation in RF field intensities of about two volts per meter. For some systems, out-of-band requirements may identify sources intrinsic to the technology used and call for special tests. Measurement of radiation at multiples of the system clock frequency in digital devices is an example.

1.2.9 Crosstalk

Crosstalk is said to exist when portions of the transmission from one or more other telephone connections can be heard on the desired one. It may be intelligible or unintelligible. The first kind is most common in base band transmitting systems over cables. The mechanisms causing this were described in Section 1.2.7 in the discussion of balance. If the source

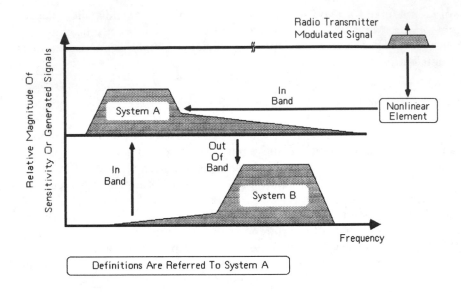

Fig. 1.6 Concepts of In-Band and Out-of-Band Interference

of the crosstalk is located at the same end of the cable or trunk or connection as the listener, it is referred to as *near-end crosstalk*. This would be the case if the source in Figure 1.5 were a talker, and the listener represented a load on an adjacent cable pair in which the desired direction of transmission was reversed as illustrated in Figure 1.7. Talker 1 causes a ground current through a metallic-to-longitudinal unbalance at his location, and listener 2 hears talker 1 because of the longitudinal-to-metallic conversion at the number 2 location. If the source and listener are at different ends, it is referred to as *far-end crosstalk*. Near-end crosstalk is usually the controlling design parameter simply because the level of the potential interferer is most often greatest at the near end. In multiplexing systems, inadequate or faulty modulation or filtering can give rise to intelligible or unintelligible crosstalk. It is unintelligible if the audible frequencies are inverted across the voice band, a phenomenon that can arise in single sideband transmission, popular in communications systems. Situations can arise in which a number of other transmissions are heard simultaneously. Some refer to this event as hearing garble but it is also properly referred to as unintelligible crosstalk. In cables, a transmission on one pair may be coupled first to one or more other pairs which then couple to yet another, *etcetera*. The multiple couplings are referred

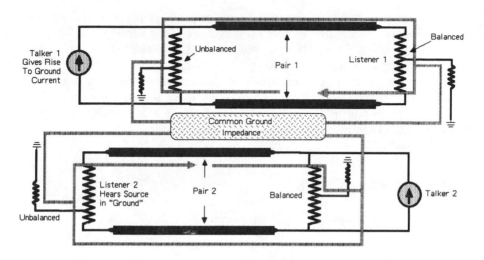

Fig. 1.7 Illustration of Near-End Crosstalk

to as *secondary, tertiary, etcetera* crosstalk paths. Similar and even more complex situations occur in carrier systems in which many conversations are frequency multiplexed into a broadband channel. In these cases crosstalk can arise from extraneous modulation products generated if a system becomes overloaded, or by inadequate suppression of modulation sidebands that are not required for normal operation. Overload conditions may be caused by failures in automatic regulators used in these systems or by inadvertent application of test tones with excessively high levels.

As suggested by the last paragraph, in any transmitting or switching system with a capacity of more than a small number of channels, the number of possible crosstalk coupling paths is extremely large. Any attempt to identify each of them and determine its loss would be impractical. Crosstalk is, therefore, a statistical parameter. Crosstalk objectives recognize this and are often stated statistically. For example, "the (estimated) mean crosstalk coupling loss shall be at least X dB, and 95 percent (estimated) of all crosstalk losses shall be at least Y dB," the intent being to bound one side of the distribution of coupling losses. Testing for crosstalk to meet requirements on a system should be treated, as the requirements suggest, on a sampling basis. At the equipment design stage, the minimum possible crosstalk coupling loss may be identified and measures taken to assure that this minimum is well within stated limits, thus providing some confidence that other path losses will also be acceptable.

Some low level of crosstalk is nearly always present and, if not intelligible or identifiable as *voices,* it is considered simply as a component of the noise. Crosstalk does not enter into grade-of-service calculations as a separate quantity. Intelligible crosstalk is a fault condition that demands remedial action and low-level unintelligible crosstalk is simply more noise. In high-capacity transmitting systems, low-level unintelligible crosstalk from multiple intermodulation products is impossible to eliminate. It results in a facility design parameter known as *intermodulation noise.*

1.3 DATA PARAMETERS

Transmission parameters considered under this heading are listed in Table 1.3.

Table 1.3
Data Parameters

Parameter	*Section*
Attenuation distortion	1.3.1
Phase distortion	1.3.2
Frequency offset	1.3.3
Gain, phase, and delay hits	1.3.4
Impulse noise	1.3.5
Nonlinear distortion	1.3.6
Tracking error	1.3.7
Phase jitter	1.3.8
Signal-to-noise ratio	1.3.9
Single frequency interference	1.3.10
Eye patterns	1.3.11
Peak-to-average ratio	1.3.12

1.3.1 Attenuation Distortion

Attenuation distortion is departure of the frequency characteristic of a channel from a desired shape, often that of an ideal bandpass filter. This statement immediately presents a measurement problem. Even when it is convenient to measure a complete frequency response, it is not obvious how the task of quantifying departures of the measured curve from the desired one should be accomplished. A second problem exists in evaluating the effects of such departures on data transmission.

E. D. Sunde proposed a method that makes use of the Fourier transform of the departures from an ideal characteristic to translate them into effective peak and rms intersymbol interference for base band pulse transmission ([8, pp. 16–22]). These measures of interference can be used to predict error probabilities for any given detector. The method could perhaps be generalized to include data transmission using modulated carriers, but it does not seem to have attracted a great deal of attention. Possible reasons for this are the need to derive an expression describing the departures before the technique can be applied, or simply the lack of adequate data on real channels to explore applications. The paper was published in 1954 when not much was known about the actual detailed frequency responses of telecommunication channels. Because of these two problems, description and evaluation of departures from ideal, requirements for control of this impairment are actually empirical rules of thumb built around easily measured quantities.

When data transmission over voice channels was first attempted, very little was known about the transmission characteristics of those channels other than the loss, as measured at 1 kHz, and expected C-MESS noise levels. Extensive surveys were undertaken to fill the void. Analysis of those data showed that a simple and useful approximation to the frequency responses found could be generated with four straight lines. (Refer to Figure 1.8(a).) The data are normalized to the value at 1 kHz. Two lines perpendicular to the frequency axis are drawn where the response is 10 dB below that at 1 kHz. These define a 10-dB bandwidth, and the frequencies identified are called the *lower and upper cutoff frequencies*. A horizontal line, at the 1-kHz value, from the low-frequency cutoff to 1 kHz defines the *flat* portion of the curve. The fourth and last line is drawn from the flat response at 1 kHz to the upper-frequency cutoff and passes through the measured value at 2.8 kHz. This usually downward-sloping line determines the *slope* of the channel, defined as the value of the loss (in decibels), at 2.8 kHz relative to that at 1 kHz.

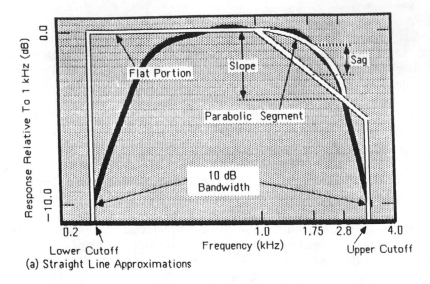

(a) Straight Line Approximations

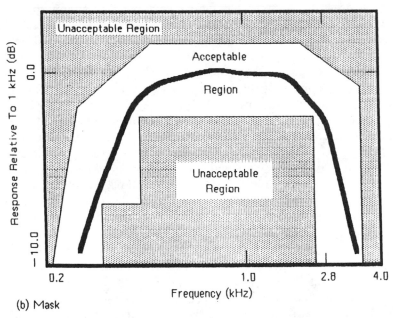

(b) Mask

Fig. 1.8 Methods of Describing Attenuation Distortion

As higher bit rates and different modulation techniques were incorporated in data modems, the *best* frequency at which to define the slope was moved around over the range from about 2.5 to 3.2 kHz. Best, in this context, implies a best correlation of the slope measure with the effects of the actual channel upper-end rolloff on observed data error rates.

In communication channels, the frequency characteristic rolloff at frequencies below 1 kHz is usually not significant for moderate data rates. Attenuation distortion is thus often specified by the 10-dB bandwidth and slope, with the defining slope frequency stated, i.e., 2.8 kHz, or just the slope, in cases where the frequency is implicit. More refined recommendations may include a *low-end* slope from the 1-kHz value to the lower-frequency cutoff, or be shown as a mask on a loss-frequency plot. Such masks outline an area, centered vertically about a 1-kHz value taken as 0 dB, with upper and lower boundaries drawn with straight lines similar to the construct just described. These are intended to be more flexible, and possibly more definitive, by accepting any real response curve which fits within the enclosing boundaries. A mask type of requirement is illustrated in Figure 1.8(b). Permitted slopes or acceptable regions are determined empirically through extensive testing of modems on transmission-line simulators, and may be augmented by theoretical analyses for some modulation schemes.

Some methods have been devised that substitute a parabolic segment for the straight-line defining slope. The parabola is made to pass through the slope intercepts with the measured characteristic at 1 and 2.8 kHz, and the measured characteristic at 1.75 kHz. This permits another parameter called *sag* to be specified. It is defined as the difference in decibels, between the straight-line defining slope and the actual response at 1.75 kHz:

sag = (response at **1.75** kHz) − (slope value at **1.75** kHz)

Thus, sag is positive when the parabolic segment is concave downward. On real channels, sag is more often positive than negative. The sag measure is shown in Figure 1.8(a). Slope and sag together define the three points through which the parabola pass, and so define the parabola, which may then be used as a second-order approximation to the higher-end frequency characteristic. Thus, slope and sag include all the information that might be derived by use of such an approximation in, for example, Sunde's method to obtain estimates of equivalent intersymbol interference. They are also helpful in the design and use of transmission-line simulators for modem testing as discussed in Chapter 4. Requirements for the maximum slope may range from 1 to 10 dB, depending on the types of service or data rate to be supported on the channel.

Table VI of [4] provides a summary of the relative frequency response of connections between class-5 offices in the North American network as measured in 1983. From these data the median 2.8-kHz slope is 0.3 dB, the 90% case is 5.8 dB, and the 99% case is 10 dB. The reader is reminded that the contribution to slope contributed by the two subscriber loops at the ends of the connections may add significantly to these values.

1.3.2 Phase Distortion

The consequences of departure from ideal phase characteristics in real filters can be illustrated as follows:

Let a periodic signal, $f(t)$, be represented by its Fourier series as

$$f(t) = \frac{C_0}{2} + \sum_{n=1}^{n=\infty} C_n \cos(n\omega_0 t + \theta_n) \tag{1.18}$$

In traversing a transmission system, the phases of the component terms will be modified by the phase characteristic, $\phi(\omega)$, producing a signal:

$$f(x) = \frac{C_0}{2} + \sum_{n=1}^{n=\infty} C_n \cos[\phi(n\omega_0)x + \theta_n] \tag{1.19}$$

If, for every n:

$$\phi(n\omega_0)x + \theta_n = n\omega_0(t + \tau) + \theta_n + 2\pi m \tag{1.20}$$

where $m = 0, 1, 2, \ldots$, then $f(x)$ will be identical to $f(t)$ except for a time delay $\tau = x - t$. The signal will not be distorted.

Let the phase characteristic be represented by an Nth-order polynomial:

$$\phi(\omega) = \sum_{i=0}^{N} a_i \omega^i$$

When this operates on (1.18), the arguments of the cosine terms of (1.19) become

$$\theta_n + a_0 + a_1(n\omega_0) + a_2(n\omega_0)^2 + \cdots + a_N(n\omega_0)^N \tag{1.21}$$

and the conditions of (1.20) can only be met if $N = 1$, and a_0 assumes one of the values of $2\pi m$. The conditions for no phase distortion then are that the phase characteristic of the transmission system must be linear, and the phase intercept, defined below, must be zero or an integral multiple of 2π.

A normalized phase characteristic of a real, nonlinear, band pass filter is represented by a cubic polynomial $\phi(x)$, shifted by $\phi(0)$ to force a zero phase intercept, in Figure 1.9(a), where

$$x = (\omega - \omega_{lc})/(\omega_{uc} - \omega_{lc})$$

and lc and uc = lower- and upper-cutoff frequencies, respectively. The shift by $\phi(0)$ in the drawing is to facilitate the discussion of phase intercept considerations only.

As indicated, the curve may be approximated by a linear function over a center portion of the usable bandwidth, but in general the linear approximation over that portion will not pass through the origin, or through $2m\pi$ radians as required by (1.20) for no distortion. The radian value of the linear approximation at zero is called the *phase intercept*. The difference between the phase intercept and the nearest value of $m\pi$ is called *phase intercept distortion,* denoted by k in the drawing. The departure of $\phi(x)$ from linearity at either end is the phase distortion and is highlighted in the drawing. These two properties of a real phase characteristic, nonlinearity and phase intercept distortion, cause undesired changes in any complex signal passing through the filter by modifying the phase relations between its component frequencies.

The phase change through a network determines the relative time of arrival of the frequency components of a signal. If one divides the phase characteristic $\phi(\omega)$ by ω ($\phi(x)$ by x for the normalized case), the result is the delay, in seconds, of each component:

$$D = \phi(\omega)/\omega \quad \text{(phase delay in seconds)} \tag{1.22}$$

If this delay is not constant with ω, then different components of the signal undergo different delays, the relative phases change, and the waveform changes shape. It is said to have suffered from delay distortion. *Delay distortion* (DD), the departure from uniform delay, is commonly plotted by first subtracting a constant value from the delay curve, D, taken where the delay curve is minimum. Examples of a nonlinear delay curve and the delay distortion are illustrated in Figure 1.9(b).

In *single sideband suppressed carrier* (SSBSC) transmission, commonly used in communication systems, a carrier signal is created at the receiver to demodulate the incoming signal. No reference modulating carrier phase is available for this demodulating signal, so k is arbitrary and phase intercept distortion is intrinsic to these systems. This is the major reason that baseband digital signals must use a modem before entering a network. Modulation schemes in modems make use of a transmitted carrier to retain the phase information.

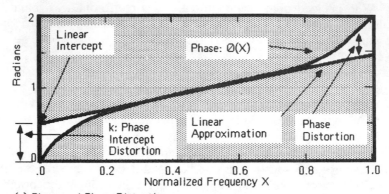

(a) Phase and Phase Distortion

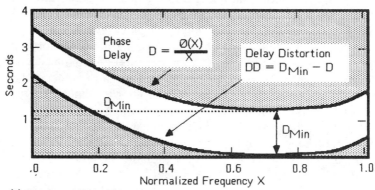

(b) Delay and Delay Distortion

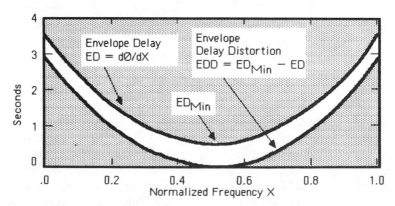

(c) Envelope Delay And Envelope Delay Distortion

Fig. 1.9 Phase-Related Parameters of a Band-Pass Filter

With no reference carrier phase available in SSBSC systems, the incidental k is actually time variable, because the demodulating carrier will wander in phase (and frequency) with respect to the original modulating carrier.[11] This wandering relative phase makes it very difficult to measure a channel phase characteristic on a built-up circuit, but the slope of the phase characteristic can be estimated (see Section 1.4.3). Because it is phase linearity that is important, changes in the slope, from a constant value, are a measure of distortion. This slope, the derivative of phase, is known as *envelope delay* (ED), because it determines the relative phases of sidebands in the envelope about the carrier in *double sideband amplitude modulation* (DSAM) systems. Envelope delay is referred to as group delay outside of North America. The departure from linear phase causes the derivative (ED) to be nonconstant. Just as with delay distortion, the change in ED from some value used as a reference, is called *envelope delay distortion* (EDD). Examples of ED and EDD are illustrated in Figure 1.9(c).

Note that absolute delays of single frequencies transmitted over a channel are not determined from EDD, but from the DD curves. This is a common source of confusion, but note the differences in the shapes of the two, as derived from a cubic phase curve, shown in Figure 1.9(b) and (c), especially at the higher frequencies.

Phase characteristics are odd functions about the center of bandpass elements, or about $\omega = 0$ otherwise. Therefore, envelope delay curves are even and the minimum usually corresponds to the minimum delay value and occurs near the center of the channel. Pure cable facilities are an exception to this generality because they are low pass filters and the center frequency is at zero. If required, the phase characteristic of a channel can be derived, except for a constant term, by integrating the EDD over the frequency band of interest.

As discussed above, data transmission is impaired if the slope of the phase curve is not constant (EDD flat) across the spectrum occupied by the data signal. EDD is encountered when the envelope delay departs from a flat line, so such departures must be controlled and permitted departures specified. All of the problems with the specification of attenuation distortion are present in the specification of EDD. The practical solution is similar. Envelope delay characteristics always have a U-like shape within band (see [9, p. 1335] for example) except for the relatively unimportant case of a system of pure cable. Envelope delay, or group delay in CCITT terminology, has the units of time (seconds), and requirements typically specify one to three maxima EDD limits, in milliseconds or microseconds, each with a specified range of frequencies. (Recall that the envelope delay curve is normalized at the bottom of the U.) An EDD limit means that the curve must not rise above the specified number of microseconds within the range of frequencies stated. Limits are determined in the same way as for attenuation distortion, a little theory and a lot of experimentation. Typical voice band limits for EDD range from approximately 500 ms over a band of approximately 1.0 kHz, to about 3500 ms over a band of approximately 2.4 kHz. Limits of this magnitude would be suitable for an end-to-end connection, as opposed to those that might be specified for a trunk, for example. The specified frequency bands are normally centered at approximately 1.7 or 1.8 kHz, near values frequently used for modem carrier frequencies. An example of an EDD requirement is shown in Figure 1.10. The mask shown there is not an actual requirement but is intended to be representative of magnitudes of limits that might be imposed on a connection intended for use at the faster voice band data bit rates.

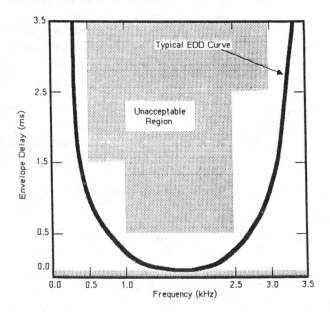

Fig. 1.10 Example of Envelope Delay Distortion Limits

The amount of EDD encountered is determined by the number of filters in a circuit because they are the dominant sources of phase distortion. The number of filters is largely determined by the number of individual carrier facilities encountered or approximately, by the number of trunks in a connection. Requirements for EDD, and some other data parameters, cannot be administered in a generalized switching network because of this dependency on the number of trunks in a particular connection. These requirements are used in dedicated circuits and in some private networks in which the maximum number of trunks in tandem is limited, for various reasons, to a small number.

Filters in transmission system components other than trunks, codecs in digital systems, for example, may be allocated an EDD of perhaps 80 to 150 ms over a band of 2.0 kHz.

The specification of envelope delay has a significant problem that arises in instrumentation for its measurement. It is discussed later in Section 1.5.3.

Tables IX, X, and XI in [4] record many values taken from the distributions of EDD at several frequencies, with respect to that at 1.7 kHz on connections between class-5 offices. Each of the three tables is devoted to data from connections within certain mileage bands, referred to as short (0 to 180 airline miles), medium (181 to 720 airline miles), and long (721 to 2576 airline miles). The tables give values of EDD at three percentage points on the *cumulative distribution function* (CDF) corresponding to the frequency specified. Thus, a table entry of 526 in the 99% column, for the row labeled 1.004 kHz and the short band, means that 99% of all EDD measurements between 1.004 and 1.704 kHz were less than or equal to 526 ms, for connections in the short mileage category. A few values from those tables are reproduced here in Table 1.4:

Table 1.4
EDD in Microseconds. With Respect to 1.704 kHz

Frequency	Band	50%	75%	99%
1.004 kHz	Short	201	239	526
	Medium	240	402	775
	Long	275	470	921
2.404 kHz	Short	212	261	493
	Medium	204	261	585
	Long	214	275	515
2.804 kHz	Short	584	727	1021
	Medium	557	768	1588
	Long	504	807	1573

One might expect that longer connections involve more trunks and, therefore, have more envelope delay distortion. These data seem to support such an hypothesis only in part. There are at least two other factors at work that tend to have the opposite effect. First, network studies have shown that calls between large cities tend to have few trunks regardless of the distance between them. This is not surprising because the relatively high volume of traffic between large cities demands high capacity, and it is economical to use few long trunks rather than many short ones to meet the demand. Second, short distance calls are more likely to encounter carrier facilities with designs that are not as elaborate as those used in long distance service facilities, and so introduce relatively large amounts of phase distortion if compared on a length basis. The length comparison is not equitable from that point of view.

The increasing use of digital facilities and digital switches is rapidly changing the network characteristics for this (and other) parameters. This is so because phase distortion has no meaning for the encoded signal in a digital system. It is only after the signal is decoded and returned to analog form that the presence of nonlinear phase elements becomes important, i.e., from the receiving end codec (see footnote 18) on. Thus, as connections become more nearly all digital, the effective EDD is decreasing.

1.3.3 Frequency Offset

Frequency offset or frequency shift arises in single sideband suppressed carrier transmission systems. It means that all frequencies in the transmitted base band signal are found to be changed by a constant number after detection in the receiver. This happens if the demodulating carrier created at the receiver is not identical in frequency to that of the modulating carrier; if the created carrier is not referenced to the modulating one, differences are inevitable. Modern communication systems may provide separate channels which transmit a reference of the modulating carrier to the receiver to avoid this problem. Networks, such as the North American analog telephone network, for example, may synchronize carrier generators to a master clock ([1, p. 487]). However, the sync network is itself subject to failures, so at times some generators may be free running. The generator oscillators are designed to be very stable by themselves;[12] nevertheless, occasional shifts of a fraction to perhaps 2 Hz do occur (see [4, p. 2102] and [9, p. 1338]). Most data transmission modulation schemes are insensitive to shifts of this magnitude, and this impairment is considered negligible. Speech transmission can tolerate shifts of 10 Hz or more before the effects begin to become noticeable.

1.3.4 Gain, Phase, and Delay Hits

Occasional abrupt changes in the amplitude, phase, or relative time of arrival of a received signal occur. These events are collectively referred to as *hits*. The most commonly cited cause is automatic switching to standby channels in microwave radio systems using frequency or space diversity[13] protection.

A sudden change in amplitude is called a *gain hit*. A *phase hit* is defined as the event in which all received frequencies incur a uniform step change in phase. This can happen if a modulating carrier experiences a sudden shift in phase or frequency.

A delay hit is said to have occurred if there is a step change in relative time. The phase change for each frequency component within the signal is unique in this case. A delay hit can occur in the case of a diversity switch on a microwave link. The primary and standby channel path lengths are not identical. Their propagation time is different, and so the signal suffers a jump in relative time. Digital networks are also susceptible to disturbances known as slips that result in delay hits. These are discussed in Section 1.4.1.

Hits are measured by simply recording the number of occurrences of the event that exceed some threshold during a specified time interval. Five or fifteen minutes is typical. For gain hits, changes in the amplitude of a test tone of 2, 4, 6, 8, or 10 dB may be chosen. Gain hits that exceed 12 dB, and last for more than 4 ms, are classed as dropouts because of their often catastrophic effects on the data-transfer session. For phase hits, values of five to 45 degrees, in five-degree increments, are usually taken as threshold settings for phase-change detectors and counters. No instrumentation has been specified to measure delay hits as such, although they do trigger and are recorded on phase-hit counters.

The effects of various types of hits on data transmission cannot be predicted with any accuracy because the detail of their structure has not been determined and is known to be variable. For example, one does not know the rise time or shape of hits, nor how long a new value of amplitude or phase will be in effect. All one can say is that, given that a hit greater than some magnitude has occurred, the probability that at least one error was generated is significant.

Some data on the percent of five-minute intervals that contain a certain number of hits were reported in [4, p. 2107ff], and some of those are reproduced in Figures 1.11(a), (b), and (c). The figures represent CDFs of the number of events occurring in an observation interval. The ordinates are normal probability scales, i.e., a normal distribution would

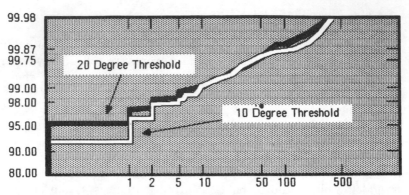

(a) Number of Phase Hit Counts in Five Minutes

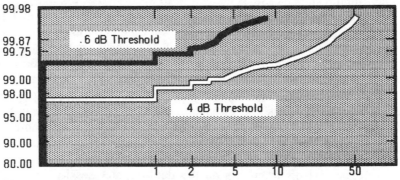

(b) Number of Gain Hit Counts in Five Minutes

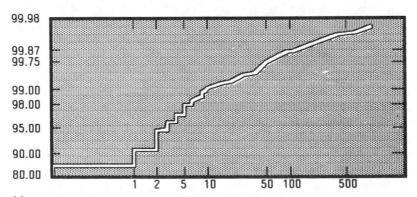

(c) Number of Dropout Counts in Fifteen Minutes

Percent of Intervals With Counts Less Than Abscissa

Fig. 1.11 Distributions of Frequency of Occurrence of Hits
Source: Carey, *et al.*, "1982/83 End Office Connection Survey: Analog Voice and Voiceband Data Transmission Performance of the Public Switched Network," *AT&T Bell Laboratories Tech. J.*, Vol. 63, No. 9, November 1984. Reprinted with permission from the Bell System Technical Journal. Copyright 1984 AT&T

plot as a straight line. From these, the relative frequency of occurrence of hits of various magnitudes can be estimated and used in impairment studies.

A comparison of Figures 1.11(b) and (c) indicates that there are many more dropouts than gain hits even if the different observation periods, five and fifteen minutes, are considered. However, the characteristics of the measurement sets must be considered (see [6]). To register as a gain hit, the measured tone amplitude must remain below the threshold setting for at least 3.6 ms and return to its original value *any time* after 4.4 ms. There is also an automatic gain control (agc) action defined by the statement that a gain hit shall *not* be recorded if the rate of change of amplitude is less than 2 dB in 600 ms. This reflects a time constant for typical agc circuits in modems. The dropout counter has no agc action; the requirement is simply that the change in level must remain below the 12-dB defining threshold for at least 4 ms. In view of these constraints on the kinds of events that cause counters to trigger, it is not clear just what the large difference in activity reflected in those data really mean. Without some measure of the temporal length of dropouts, speculative interpretations would be ill-advised.

Impairment evaluation for these parameters is a matter of establishing thresholds of damage. For example, for a given data rate and transmission scheme (type of modem used), it can be said that errors will occur if the (type of) hit exceeds a certain magnitude. Requirements or objectives are stated as a maximum expected number of hits exceeding some threshold value during a specified time interval. The interval chosen may relate to the length of an expected interactive data communication session or a reasonable measurement time, five or fifteen minutes as mentioned above, and the maximum number of hits stated is small, often less than five. From the CDFs in Figure 1.11, the probability of experiencing a high rate of occurrence is quite small, more than 90% of all intervals are hit free. Even if hits occurred continuously at a maximum rate of five every fifteen minutes, and every hit caused one error, the contribution to error rate would be less than half of a common long-term maximum error rate specification of one in 10^5 bits transmitted, even at the modest speed of 1200 bits per second.

1.3.5 Impulse Noise

The term *impulse noise* refers to distinct sporadic occurrences of large magnitude noise events on a channel. Large, in this context, has come to be defined as excursions of the noise waveform in excess of 12 dB above the rms noise level on the channel (this number holds for a voice bandwidth channel).[14] The task of characterization and impairment evaluation is similar to that of hits. In this case, however, large amounts of data have been collected to aid in characterization, and more elaborate evaluation schemes have been devised and used. (See [10] and [11], for example.) The evaluation techniques relate the peak amplitude of an impulse to its probability of causing one, two, *etcetera,* errors for a given type of data transmission scheme. The data on impulse noise enable estimates of the distributions of peak noise amplitude. Together, these two results can be used to predict error rates in data transmission due to impulse noise. Requirements are stated simply as an expected maximum number of occurrences above a specified voltage threshold in a fixed time interval, five or fifteen minutes. Historically, the contribution to error rate from impulse noise has been much greater than that from hits, but the situation is changing rapidly as explained next.

Most impulse noise in networks originates from switching and other relay activity in central offices. The associated radiating fields induce noise transients into vulnerable analog carrier systems, as well as directly onto cable pairs. With the increase in electronic and digital switches that produce relatively little impulse noise, and digital carrier facilities, that by design are less vulnerable than analog, the relative significance of impulse noise as an impairment has diminished by at least an order of magnitude, as can be seen by comparing data reported in [4] with those from 12 years earlier in [9]. Networks that utilize older electromechanical switching systems may, however, exhibit large amounts of this noise.

The dependency of impulse noise activity upon the type of switching system used in a connection is illustrated by the cumulative distributions of the peak amplitudes of impulses as excerpted from [4] and shown in Figure 1.12. A crossbar type switch, referred to in the figure, is an electromechanical switching machine. These distributions of impulse noise peaks were measured on connections that were established exclusively by the type of switch indicated. Because of the counter used in the test sets, impulse noise occurrences are referred to as counts. Note that 90% of the five-minute intervals observed on connections through electronic switches had five or fewer counts, while those using crossbar switches contained up to about 70 counts.

1.3.6 Nonlinear Distortion

Nonlinear distortion arises when a signal passes through a device with a nonlinear input-output characteristic. No real devices are perfectly linear so some of this distortion is always present. In principle, the nonlinear transfer function can be represented as an Nth-order polynomial with

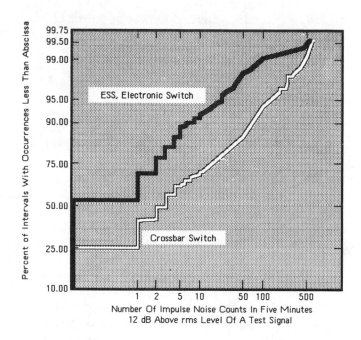

Fig. 1.12 Distributions of Impulse Noise Counts Above a Threshold
Source: Carey, *et al.*, "1982/83 End Office Connection Survey: Analog Voice
and Voiceband Data Transmission Performance of the Public Switched Network," *AT&T Bell Laboratories Tech. J.*, Vol. 63, No. 9, November 1984.
Reprinted with permission from the Bell System Technical Journal. Copyright 1984 AT&T

the input signal as the variable. Each power term in the polynomial will
give rise to one or more harmonics of the input, and N harmonics of the
signal appear in the output.

This suggests a method of nonlinear distortion measurement: use a
selective filter to measure the relative amplitude of the output harmonics.
The result of such a measurement, expressed in power in the harmonics
relative to that in the original signal, is called the *total harmonic distortion*
of the device. Total implies that all harmonics are found and measured; in
practice, the number of harmonics included is not specified, and often
only the second and third are actually included in the measurement. The
assumption is made, but seldom stated, that only the second and third are
important.

Sometimes the maximum power contained individually in the second and third harmonics is specified. The corresponding measurements
are then labeled accordingly as second or third harmonic distortion. Harmonic distortion measurements have a major shortcoming: the input signal must be a single sinusoid and the distortion may not be uniform with
frequency. This means that measurements using many (all?) frequencies,

one at a time, may be needed to cover the entire spectrum of interest. The measurement procedure is quite limited by the long time required to obtain a detailed characterization of the nonlinearity.

A variation of this approach is to use at least two input frequencies and measure sums and differences created by the nonlinearity in the output signal. If one input frequency is labeled A and another B, one speaks of measuring $A + B$, $A - B$, $2A + B$, *etcetera,* output frequencies. The results, expressed in a power ratio as for harmonic measurements, are termed *intermodulation distortion.* An advantage of this method is that a wider range of input frequencies may be chosen and still have the measured products fall within the frequency band of interest. For example, the difference frequency, $A - B$, can be fixed and chosen to be in-band, while the pair is permitted to range over the complete band. (Note that the signal distortion arises from in-band products adding to the original signal, not the generation of harmonics *per se.*) Bell Telephone Laboratories introduced a variation of this technique in the mid-1970s, known as the four-tone method and described in [6, pp. 17–18] and [12, pp. 16–23]. It uses fundamental frequencies selected to avoid certain conditions, described in Section 1.4.5, encountered in telephone channels, and the measurements correlate well with data transmission performance at rates up to 4800 and 9600 bits per second. The power in selected second- and third-order intermodulation products is measured, and the ratio of total fundamental power to the measured power, in decibels, is called R2 for the second-order measurement and R3 for the third. In these measurement terms, it is necessary for R2 and R3 to be as large as 35 to 60 dB for satisfactory data transmission, depending on the type of modulation used and the magnitudes of other parameters on the channel.

An extension of the sinusoidal intermodulation methods uses a low-frequency relatively narrow band of noise as input, and a relatively wide filter is used in the selection and measurement of all the resulting noise power above that band in the output. Yet another uses a broad band of noise with a narrow midsection filtered out. The detector measures output noise power that falls in that narrow band. These last two techniques are also referred to as intermodulation measurements, but may include the word noise in the name. All of these methods measure some aspect of the effects of the nonlinearity and can be related to data transmission degradation empirically or analytically.

Requirements for nonlinear distortion specify the method, the frequencies and test signal level[15] used, and a minimum amount of suppression of generated harmonics or intermodulation products in decibels below the fundamentals, or a maximum percentage of harmonic content in the output. Measuring instruments can be calibrated to read decibels of

suppression or percent (of voltage) harmonic content. Typical requirements for suppression range from 25 to 60 dB. Harmonic content numbers
range from as high as 5% for speech transmission, to about 0.1% for data.
In 1983 in the North American telephone network, it was reported that
about 90% of connections were found to have four-tone intermodulation
products suppressed by more than 40 dB. See [4, p. 2097]. Figure 1.13
shows some of the distributions presented in [4].

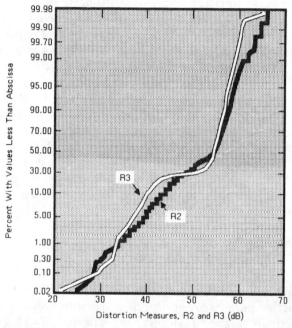

Fig. 1.13 Distributions of Intermodulation Distortion on Long Connections
Source: Carey, et al., "1982/83 End Office Connection Survey: Analog Voice
and Voiceband Data Transmission Performance of the Public Switched Network," *AT&T Bell Laboratories Tech. J.*, Vol. 63, No. 9, November 1984.
Reprinted with permission from the Bell System Technical Journal. Copyright 1984 AT&T

1.3.7 Tracking Error

Before tracking error can be discussed, we must understand what it
is and why devices that give rise to it are used. The threshold of hearing at
1.0 kHz is taken as 10^{-16} watts per square centimeter, measured at the
eardrum. With respect to this level, the human voice, measured at a
distance of three feet, varies in loudness from about 15 dB for a quiet
whisper, to nearly 80 dB for the loudest shout, a dynamic range of 65 dB.
For purposes of telephony, the usable range is taken as 40 dB. Many
transmission facilities are designed to handle such a range, but there are

cases where the cost of devices to reduce it, called *compressors,* are small compared to economies to be gained. Digital and short-haul-analog facilities are two such cases. For digital systems, the decrease in dynamic range means that better signal-to-quantizing-noise ratios can be achieved for a given number of quantizing levels or number of bits in the encoding process (see Section 1.4.3). For analog systems, it eases linearity problems in the design of repeaters and, as shown below, compressors can be implemented in ways to effect better signal-to-noise ratios.

The compression process, followed at the system output by complementary expansion, is called *companding.* Devices that implement it are called *companders.* Signals are compressed at the input to a channel and expanded to original values at the output. The word compander is a contraction of compressor and expander. Companders for analog systems use a variable gain circuit controlled by a short time (approximately 10 ms) average of the input signal. At the lowest frequencies used, the gain per cycle appears to be constant, so, in principle, there is no nonlinear distortion. Companders for digital systems use nonlinear elements with no delay and no averaging. They are known as instantaneous companders. The nonlinearity is designed into the quantizer by using a number of different quantizing intervals or step sizes. This operation is equivalent to multiplying each sample by a constant determined by the input amplitude. The inverse multiplication is completed at the system output before the required digital-to-analog conversion and again, no nonlinear distortion is encountered, if the multipliers are perfect.

Receiving amplifiers in compandered analog systems incorporate a squelch[16] action with attack and release times in the order of 5 to 15 ms. The squelch action permits lower operating signal-to-noise ratios because the noise effectively disappears during pauses in speech and is masked by the speech during the active periods. This relaxation of signal-to-noise requirement allows longer repeater spacing and affects economies in short-haul systems that must serve low-capacity trunking needs.

As illustrated in Figure 1.14, the compression in the analog companders works in a manner to predominantly amplify low-level signals rather than attenuate high-level signals. At the expander, the amplified low-level signals are attenuated to restore the original values. (The popular Dolby circuits in commercial radios and tape decks are companders.) The compressor is located at the transmitting terminal and most noise enters the system beyond that point, in the transmission medium and repeaters. With the low-level input signals amplified in the compressor, noise originating in the transmission medium and repeaters represents an extremely low input signal, too low to deactivate the output squelch circuitry. Thus at the expander, the noise signal is attenuated so much, if no input signal is present, that it is nearly insignificant. Noise suppression of

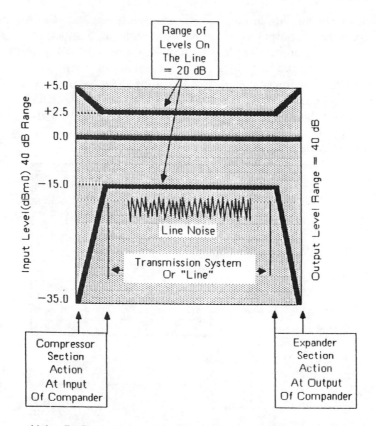

Fig. 1.14 Diagram Illustrating Analog Compander Action

about 27 dB is often achieved when the squelch action is in effect. Because the amplitude of noise at the output is signal dependent, the signal-to-noise ratio resulting from the use of companders and the squelch arrangement is a complicated matter. Limited test results have indicated that the subjective improvement in signal-to-noise ratio is only about dB, instead of 27. A discussion of the measurement of signal-to-n ratio on compandered facilities is given in Section 1.3.9.

One measure of compander performance is how well the signal amplitude follows or tracks changes in input amplitude. simple means of examining how close the output expander ch comes to being an exact inverse of the input compressor char is a measure of amplitude fidelity. Requirements are stated as an allowable departure of output from input amplitude as a function of input amplitude. Such numbers are less than plus or minus 0.5 dB over much of the dynamic range of the compander, changing to plus or minus 1.0 to 3.0 dB at the upper or lower portions of that range. An example is shown in Figure 1.15. As one might expect, compandered systems usually introduce relatively large amounts of nonlinear distortion, through imperfect multiplication in the case of digital systems, and the difficulty of building highly linear variable gain devices in analog systems.

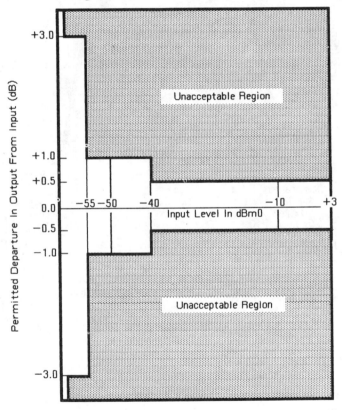

Fig. 1.15 Example of Tracking Tolerance Requirement
Source: Red Book, Volume VI—Fascicle VI.5, CCITT, Geneva, 1985. Reprinted with permission

1.3.8 Phase Jitter

Phase jitter is the name applied to small continuous undesired variations in the normal progression of the instantaneous phase of a single frequency signal or component frequency of a complex signal. Noise on a demodulating carrier is one possible cause of this. The variations are often sinusoidal in nature. In the North American analog network, phase jitter is most commonly found to have a jitter frequency of between 3 and 20 Hz. No reasons for this particular band of frequencies have been identified. Telephone networks in other countries exhibit small amounts of jitter at frequencies down to 1 Hz. Though no actual causes have been identified, phase jitter is usually associated with single sideband systems, and the frequencies indicate that the underlying process is probably complex. The observed frequencies do not appear to be fundamentals or simple products of known sinusoidal sources. Jitter frequencies above 20 Hz exist but they have not been of interest because data transmission using modems has been impervious to such an impairment. Phase-locked loops, for example, common elements in modems, typically are not affected by higher-frequency phase perturbations.

Requirements may be based on analyses of phase-locked loops or on empirical performance evaluations. They are usually stated as a small maximum number of degrees peak-to-peak of phase change within a specified, low and small, frequency band (e.g., a few degrees at 3 to 14 Hz).

The amount of phase jitter present on a connection is a function of its physical length because longer distance calls are subject to more stages of modulation as more multiplexors are encountered. In the North American network, approximately 50% of short-haul calls were found to have jitter of less than four degrees peak to peak, and for long-haul calls, that number was about six degrees peak to peak (see [4, p. 2098]).

1.3.9 Signal-to-Noise Ratio

Control of signal levels and psophometric or C-MESS noise does not guarantee an adequate signal-to-noise ratio in communication networks for two reasons: first, short-haul-analog carrier facilities often use companders, the devices described earlier in Section 1.3.7, that suppress noise in the absence of a signal; second, digital systems contain quantizing, or so-called correlated noise, described in Section 1.4.3. In both cases, relatively little noise is measured if no signal is present, and a standard noise measurement does not reflect the working *signal-to-noise ratio* (S/N).

A different kind of noise measurement is made to assess that ratio. A sinusoidal tone is transmitted from the far end. A standard noise measuring set, with a filter that has a C-MESS shape plus a deep notch at the transmitted tone frequency, is used to measure the noise ([6, p. 7]). The measurement is known as *noise with tone* or *C-NOTCHED* noise and, used with loss data, provides a good estimate of expected signal-to-noise ratio. Because C-NOTCHED noise only exists when a signal is present, speech masks it and listeners tolerate much more of this than they could of C-MESS noise that, by definition, is not suppressed during idle periods. In addition, many nonvoice systems can tolerate much poorer signal-to-noise ratios than one might calculate using C-MESS noise. For example, companders are designed for 27 dB of noise suppression during periods of no transmission. However, when a signal is present the signal-to-noise ratio is less than 27 and may be as small as 12 or 13. Nevertheless, many compandered systems are quite suitable for data transmission from a signal-to-noise point of view. Signal-to-noise requirements are stated in terms of C-NOTCHED noise limits, and these allow much more noise than do C-MESS noise requirements.

Most measurements of signal-to-noise ratio fall in the range of 25 dB to more than 40 dB. Reference [4, p. 2088] presents distributions of S/N on connections as measured in the United States. Some of these are reproduced in Figure 1.16.

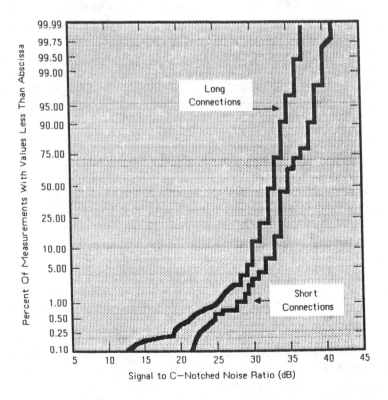

Fig. 1.16 Distributions of Signal-to-Noise Ratio on Connections
Source: Carey, *et al.,* "1982/83 End Office Connection Survey: Analog Voice and Voiceband Data Transmission Performance of the Public Switched Network," *AT&T Bell Laboratories Tech. J.,* Vol. 63, No. 9, November 1984. Reprinted with permission from the Bell System Technical Journal. Copyright 1984 AT&T

1.3.10 Single Frequency Interference

Single frequency interference (SFI) causes errors in data modems if it is of sufficient magnitude. The magnitude required for errors to appear decreases as the frequency of the interference approaches that of the data carrier where a carrier-to-tone ratio of 4 or 5 dB may cause problems. Near the band edges, tones with amplitudes that are larger than the carrier, negative carrier-to-tone ratios, may be unimportant. In sharp contrast, for voice transmission, single tones are extremely annoying. The human ear can detect a tone *buried* in up to 30 dB of noise, i.e., $20 \log[(\text{tone level})/(\text{noise level})] = -30$ dB. As a result, the prevention of SFI has been of great importance historically, so it is almost never found. It is mentioned here for completeness only. It docs occur, but the cause is almost always crosstalk from an adjacent circuit in a trouble condition; it is singing. Such a trouble is not only uncommon, it is detected and repaired quickly.

Even though SFI is seldom of concern, there are requirements, and they specify that single frequency tones be held to levels more than 55 dB below the signal. This seems to be more a voice rather than data transmission parameter, but the consequences of too much SFI are greater for the data user because no useful throughput is possible, all the data are destroyed, and a new connection is mandatory. The consequences of such a catastrophic event are usually much greater than simply redialing a voice connection or tolerating the annoying tone for a time.

1.3.11 Eye Patterns

Consider a digital data stream converted to a simple continuous analog signal that can assume the values plus or minus one volt. Let $+1$ represent a binary one, and -1 a binary zero. Pass the signal through a filter with a high-frequency cutoff, f_0, adequate to follow the bit rate but low enough to cause a significant transition time when the signal jumps between the high and low values, on the order of one third of a bit interval. Let the bit rate, Br, have the nyquist rate $2f_0$. Such a signal is illustrated in the upper portion of Figure 1.17.

Figures 1.17(a), (b), and (c) illustrate a digital sequence converted to dc pulses that, when sent through a low-pass filter, yield the analog signal shown in Figure 1.17(c). Consider an oscilloscope with a long persistence phosphor, triggered by a clock recovered from the bit stream, used to continuously view successive three-bit intervals of such a signal. If the bit stream is derived from a random data generator, the pattern on

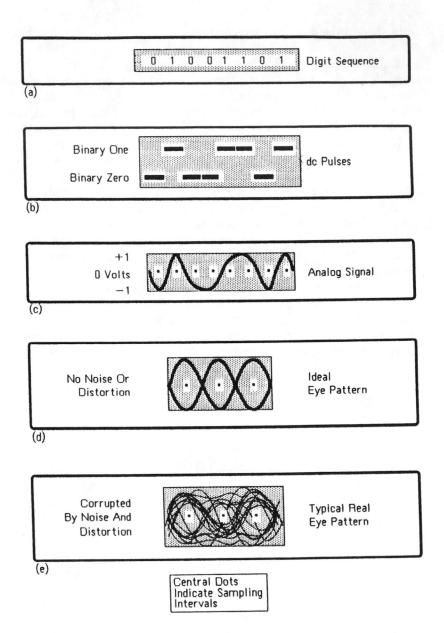

Fig. 1.17 Formation of Eye Patterns; Ideal and Distorted

the screen should appear approximately as shown in Figure 1.17(d). This is an idealized picture, representative of noise- and distortion-free transmission. Because of their shape and the importance of the center of each loop shown, as described below, they are called eye patterns.

Noise will make individual traces forming eye patterns on the screen of the oscilloscope appear to thicken. Attenuation, envelope delay, and nonlinear distortions produce characteristic departures from the ideal pattern. These effects cause the open center portion of the eye to shrink, shift left or right, or move up or down from its unperturbed position, as illustrated in the patterns in Figure 1.17(e). Effects of some impairments taken one at a time are illustrated in idealized form in Figure 1.18.

Eye patterns can be observed in real systems near the detector output. The eye opening, or the dimensions of the clear area near the center of each eye, is an indication of the total noise and distortion in the system. Increased impairments result in a smaller or shifted eye. A detection system, designed for a signal of the type described above, will typically be turned on, or gated, for a short time interval near the expected time of arrival of the center of the eye, and make a decision as to whether the observed signal is positive (a one) or negative (a zero) to construct the digital output. These decision locations are called sampling intervals and are indicated by the central black dots in the drawings. The vertical position of the dot is indicative of a threshold level defining a positive (above) or negative (below) output voltage. As the eye decreases in size, the clear area becomes less extensive vertically, and, with a noise corrupted trace, the chances for decision errors obviously increase. The eye pattern is a dynamic picture showing the conditions under which the detector is operating; it, therefore, has great intuitive appeal, and various attempts to quantify and measure the *eye opening* for use as a transmission-rating tool have been made. One of these is described in the next section.

1.3.12 Peak-to-Average Ratio

As shown in the last section, the eye pattern reflects the environment in which a sampling detector is working. A trained observer can estimate the general quality of a transmission system and note the dominant impairments by watching such a display for a few moments. A comparison of the eye resulting from an undistorted analog signal with a distorted eye, as in Figures 1.17(d) and (e), may lead the reader to conclude that, in the distorted case, many of the peaks of the analog signal

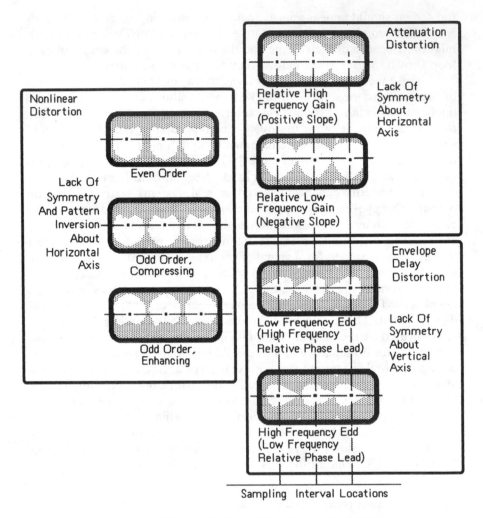

Fig. 1.18 Effects of Particular Distortions on Eye Patterns

have been depressed, or perhaps shifted so far as to be mixed with near-zero crossing values of prior portions of the signal. Most often, such compression and shifting is due to intersymbol interference resulting from attenuation and envelope-delay distortion.

The peak-to-average ratio (P/AR) measurement is very sensitive to these types of distortion. If one compensates for total loss on a channel, then a measure of the average peak value of a signal, similar to the one illustrated in Figure 1.17, compared to its full wave average, is an instrumentable measure of one observed characteristic of eye patterns, the average opening of the eye.

The P/AR system uses a train of pulses as a test signal. The transmitted P/AR signal is shaped to provide a peak of the time waveform that is near the theoretical maximum possible in an ideal voice band channel. The peak, therefore (almost always), decreases as distortion increases. In a case of no distortion, but added noise, the average value of the signal appears to increase while the peak remains nearly constant, and, again, the peak-to-average ratio will decrease. Such a measuring instrument has been built and given extensive field trials as a rapid method to obtain a single number overall quality rating of a channel for data transmission. The instrument transmitter sends a continuous bipolar type pulse train with spectral components selected to provide a high initial peak-to-average ratio ([6, p. 19]). The receiver is designed and calibrated so that a perfect channel, delivering a noise-free undistorted signal will be measured as having a P/AR rating of 100. A distortion or noise condition, resulting in an effective decrease of the peak of 50%, will produce a P/AR rating of 50.

The P/AR signal is responsive to numerous steady state impairments, as intended. It does not respond to transient phenomena such as impulse noise and hits because it works on average values, and it provides a single number estimate of channel quality.

Acceptance of the P/AR system was poor for a number of reasons: First, nonlinear distortion can enhance the peaks and produce ratings in excess of 100. Second, P/AR ratings are not additive, and, in fact, two channels in tandem may produce a higher rating than either channel separately even though transmission over the tandem connection may be poorer than over either. Such occurrences, though not common, are disconcerting. Third, because many impairments affect the meter reading, it is not possible to decode a P/AR rating to determine what impairment(s) gave rise to a particular reading. This may be considered a deficiency in trouble analysis.

In addition, the answer to the question, "What is a good (or bad) P/AR rating?", is not simple. It depends upon the type of modulation and detection methods used for the data transmission, which is generally a function of bit rate. Very slow rate systems, up to about 300 bps, may use amplitude modulation (AM); and as the bit rate increases, one encounters frequency modulation (FM), phase modulation (PM), quadrature amplitude modulation (QAM), and elaborate encoding techniques.[17] Slow speed systems can typically tolerate channels with P/AR ratings as low as 50; moderate rate, 1200 to 2400, bps systems, require ratings from about 65 to 75, and those operating at 9600 bps or more demand ratings of about 85 or greater. Some typical values of P/AR readings on channels on various types of facilities are given in Table 1.5.

Table 1.5
Typical P/AR Ratings

Type of Facility	P/AR Range
Cable, nonloaded	70–100
Cable, loaded	85–99
Short analog	75–97
Long analog	92–98
Digital	91–98

In spite of these problems, correlations of P/AR ratings with data error rates over real channels, for fixed bit rates, are very good, typically above 0.95. P/AR is considered by some maintenance personnel to be a reliable quick check on the health of a channel if bench mark readings are made and recorded when a system is first turned up for service. References [13] and [14] contain detailed information on the instrumentation and use of a P/AR measuring system.

1.4 DIGITAL PARAMETERS

Most of the parameters already discussed for analog transmission are also of interest in digital (pulse) transmission, except those specifically related to analog modulation and *frequency division multiplexing* (FDM) such as frequency offset. Parameters unique to digital systems are listed in Table 1.6 and discussed in the following sections.

Table 1.6
Digital Parameters

Parameter	Section
Slips	1.4.1
Aliasing	1.4.2
Quantizing noise	1.4.3
Total distortion	1.4.4
Jitter	1.4.5

1.4.1 Slips

Networks using switched digital facilities must be synchronized to enable efficient *time division multiplexing* (TDM). The precise instant of rise or fall of an observed pulse relative to the synchronization clock or timing source depends, however, on the electrical distance or transmission time of that pulse from the synchronization source. Temperature changes over large geographic distances change that electrical distance with time. When the electrical distance or transmission delay changes, the number of pulses in transit (in the pipe) changes. Even though buffers operating as elastic storage devices are used to compensate for some drift, occasional loss of synchronization is unavoidable. Systems are designed to recognize these events and recovery is automatic and nearly instantaneous. The recovery process results in what are known as *slips*. A slip is the loss or repetition of one complete digital frame of information and occurs when a buffer becomes empty or overflows and calls for a reset. Whether a slip is a loss or repetition is determined by what type of correction is needed to alleviate the buffer condition.

Figure 1.19 illustrates how slips may occur. In Figure 1.19(a), two sequences of frames (fixed-length block of bits) are shown entering a switch location from each of two routes, A and B. The time sequence

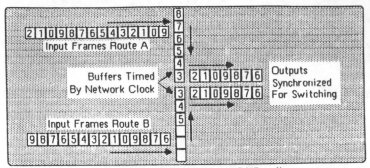

(A) Route A Length Shrinking, Buffer On Verge Of Overflow

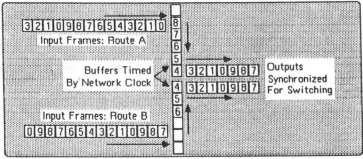

(B) Frame 9 Discarded For Buffer Relief

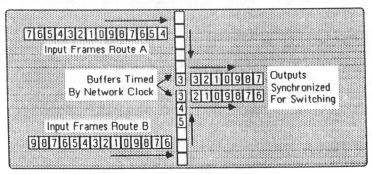

(C) Route A Expanded, Frame 3 Repeated To Await 4

Fig. 1.19 Illustration of Slips Due to Electrical Length Changes of Two Geographically Diverse Digital Routes Entering a Switch

of frames is indicated by the numbering sequence: 0 · · · 9, 0 · · · 9, · · · .The frames are shown leaving the two transmission systems, routes *A* and *B*, and entering separate buffers, the vertical segments. The two sequences of entering frames are not synchronized as shown by the right-most frame numbers, 9 on route A and 6 on route B.

Normal operation of the two buffers is shown compensating for the timing difference between the two streams. The buffer outputs are timed by a network clock and provide synchronous outputs for switching; however, the buffer serving route A is about to overflow (it is full) because route A has become electrically shorter. A buffer reset is imminent. Figure 1.19(b) illustrates the situation immediately after the reset. The frame numbered 9 has been discarded to relieve the overflow condition. The two outputs remain synchronized, but at a cost of the loss of one frame in one channel (route). A slip that appears as a forward jump in time has occurred. Figure 1.19(c) illustrates the opposite extreme. In this case, route A is assumed to have increased in electrical length or acquired greater delay, and the buffer is void of fresh input. The output frame stream must be continuous so the buffer has repeated frame number 3 to allow the input from route A to catch up. A slip that appears as a time repetition has occurred.

Figure 1.19 illustrates slips on one route due to length changes of that route. A significant change of length of the synchronization link from a master clock to a switching machine could force a slip on every route being switched at that location. In this case, every buffer would appear to overflow or underflow as the clock driving it appeared to slow down or speed up. Depending on their relative occupancy at the start of the apparent time change, the slips need not occur on every buffer at the same instant of course.

Objectives for allowable slips on global networks are on the order of five slips in a 24-hour period ([7, Rec. G.822]) in normal operation. Under severe conditions, less than 0.1% of the time, more than 30 slips are allowed in one hour. Smaller networks should have more stringent requirements. Depending upon a number of factors, including the sophistication of the users' software, a slip may have a number of consequences, up to and including a disconnect. For transmission impairment purposes, a slip is simply classed as a delay hit.

1.4.2 Aliasing

Aliasing or frequency fold-over distortion is intrinsic in the sampling process that is the basis of digital transmission of analog signals. The sampling pulses represent a continuous pulse train with an infinite periodic spectrum. This pulse train is multiplied by the analog signal to produce a *pulse amplitude modulated* (PAM) pulse train. This multiplication results in the spectrum of the sampled signal being replicated at multiples of the sampling frequency. In other words, the input signal spectrum is replicated in the frequency domain an infinite number of times.

Formally, the sample pulses are represented by a train of unit impulses:

$$s(t) = \sum_{n=-\infty}^{+\infty} \delta(t - nT), \qquad n \text{ an integer}$$

The pulses are spaced T seconds apart. The spectrum of this sequence is itself a sequence of impulses in the frequency domain, spaced n/T Hz apart, given by

$$S(x) = \frac{1}{T} \sum_{n=-\infty}^{+\infty} \delta\left(x - \frac{n}{T}\right)$$

When $s(t)$ is multiplied by a signal, $g(t)$, with spectrum $G(y)$, bandwidth-limited from y_L to y_u, the spectra are convolved and the result in the frequency domain is

$$F_{sg}(f) = \sum_{n=-\infty}^{+\infty} G\left(f - \frac{n}{T}\right) \quad \text{(sampled spectrum)} \tag{1.23}$$

The spectra, $S(x)$, $G(y)$, and that of (1.23), $F_{sg}(f)$, are illustrated in Figures 1.20(a), (b), and (c).

Only one of the replications illustrated in Figure 1.20(c) is desired. Separation of the desired one from all the others can only be accomplished if they do not overlap; that is, the input, $g(t)$, must be band limited to a width smaller than the spacing of the sampling pulse spectral elements, $(y_u - y_L) < n/T$. Since the input signal cannot be strictly band limited, unwanted frequency components of the sample interfere with the wanted component. This interference, called aliasing, is a function of the quality of the input bandlimiting (antialiasing) filters. Some amount of aliasing, though negligible, is always present. The measurement of the phenomenon is discussed in Section 1.4.4.

1.4.3 Quantizing Noise

In a real sampling process, the sampling pulse is narrow and considered to be of unit height; after sampling, its height is the magnitude of the waveform at the sampling instant. If the magnitudes of the samples are restricted to a finite set of values, y_i, separated by a uniform amount, s, then all values of the signal falling between $(y_i - s/2)$ and $(y_i + s/2)$ will

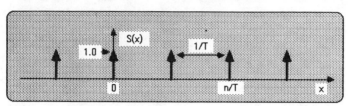

(a) Sampling Pulse Spectrum

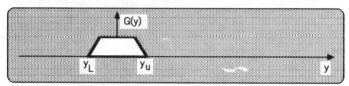

(b) Bandlimited Signal Spectrum

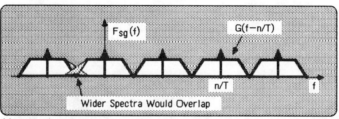

(c) Sampled Spectrum

Fig. 1.20 Spectra for the Sampling Process

give rise to the same sample value, y_i, and the signal is said to be linearly quantized. Each sample then has an error associated with it of a magnitude between zero and $s/2$. The quantity s is called the quantizer step size. The process is illustrated in Figure 1.21(a). In 1.21(b), a sinusoid with a peak-to-peak amplitude just matching the full range of the set of available quantized values is shown. Any signal exceeding the range will be clipped to the peak value, and the quantizer is said to be overloaded. Figure 1.21(c) illustrates a variation of the process in which the step sizes are not uniform. Such an arrangement serves to decrease the effective dynamic range of signals applied to it and is referred to as a nonlinear quantizer. Devices of this type are popular in digital transmission systems and are discussed in detail below. First, however, the error signal, called *quantizing noise,* is addressed further.

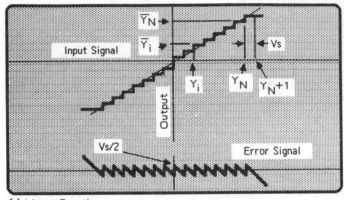

(a) Linear Quantizer

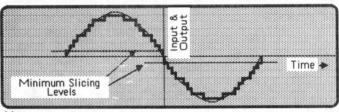

(b) Quantized Full Load Sine Wave

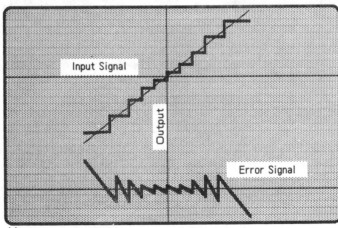

(c) Nonlinear Quantizer

Fig. 1.21 Linear and Nonlinear Quantizer
Source: Telecommunications Transmission Engineering, Volume 2, Facilities, Second Edition, American Telegraph and Telephone Company, 1980. Reprinted with permission of AT&T—All Rights Reserved

There are two types of noise arising from quantizers and the quantizing process. First, in a back-to-back pair, analog-signal-to-analog-signal, of real codecs,[18] if there is no input signal then the output is usually not zero, even if the input noise is neglected. Note in Figure 1.21(b) that the smallest quantizing levels are not zero, but plus and minus a small value. Sensible design practice dictates that this smallest quantizer value be set just above internal system noise. In operation, with the system input terminated, the codec output typically jitters back and forth between the opposite polarity minimum levels due to internal noise fluctuations. This effective steady state noise is held to values on the order of 15 dBrnC or less and may be considered negligible.

Of more interest is true quantizing noise from the error signals illustrated in Figure 1.21. The small variable errors occurring about each slicing level can be treated as additive random noise added to a nonquantized signal. The statistics of the noise are determined by the dynamics of the signal being sampled and the values are restricted to those encompassed by the saw-toothed error signals illustrated in the figure. In a linear quantizer with the voltage between slicing levels equal to Vs, the rms quantizing noise, n_q, will have a value of

$$n_q = Vs/12^{1/2} \quad \text{(rms quantizing noise)} \tag{1.24}$$

for an input signal that uniformly covers one or more quantizing intervals.[19]

To limit the 40-dB dynamic range of speech, a nonlinear (logarithmic) quantizer is used to implement range compression. This was referred to as instantaneous companding in the discussion of tracking in Section 1.3.7. Two major types of compression are in use. They are named mu-Law and A-Law from the mathematical functions describing their actions. The mu-law companding characteristic is illustrated in Figure 1.22. Let y_n be the value of the output of a quantizer slicing level, n, and let $x = v/V$, where v is the instantaneous input voltage, V, be the maximum (peak) input voltage for which clipping will not occur. The number of possible y values is determined by the number of quantizing steps, N, on either side of zero, $2N$ total. In a linear quantizer, the output y_n occurs for any x in the range y_n to $y_{n+1} - \varepsilon$, where ε approaches zero. (Or if $|V| = 1.0$, then $y_n = n/N$ is the output voltage for any signal level, v, encompassed by the slicing levels designated n to $n + 1$.)

The mu-law compression characteristic is used in the countries of North America and several others; its discrete quantizer outputs are related to the input by the expression:

$$y_n = [\ln(1 + \mu x_n)]/[\ln(1 + \mu)] \quad \text{(mu-Law companding)} \tag{1.25}$$

The notation ln is used to indicate natural logarithm, base e. In older

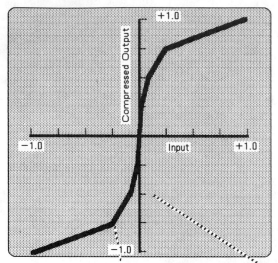

One Approximation To A Mu—law Compressor Characteristic

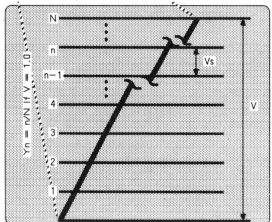

Expansion Of One Segment To Define Terms
V Shown Here Is For Illustration Of The Linear Case Only
See Text For The Actual Interval Definitions

Fig. 1.22 Illustration Of A Mu-Law Compressor
From: Members of Technical Staff, Bell Telephone Laboratories, *Transmission Systems for Communications*, 4th Edition, Western Electric Company, 1970

systems $\mu = 100$ and 255 in later ones. The nonlinear characteristic is approximated with a piecewise linear fit as illustrated in the drawing. Elsewhere, the A-law compression characteristic is used. In this case:

$$y_n = Ax/(1 + \ln A) \quad \text{for } 0 \le Ax \le 1 \tag{1.26a}$$

and

$$y_n = (1 + \ln Ax)/(1 + \ln A), \qquad 1 \le Ax \le A \quad (A\text{-Law companding})$$
$$(1.26b)$$

with $A = 87.6$.

The mu-law and A-law compressors yield identical outputs for $v = V_{\max}$, and the mu-law output is about 3 dB larger than the A-law for the minimum (about 40 dB below the maximum) output. The quantizing noise calculation is complicated by the compression and by the amplitude distributions of the signals applied. Note that the rms value of the noise will be different for each segment of the linear approximation to the nonlinear characteristic. Larger applied signals generate more noise because of the larger quantizing interval. The noise is truly signal dependent. For mu-255 quantizers with eight-level PCM encoding, the effective noise improvement for quiet talkers[20] is about 35 dB, compared to an equal-level linear codec designed for the same dynamic range. The statistics of quantizing noise differ significantly from those of thermal noise because it is peak limited to one half of the maximum quantizing interval. For data transmission, any difference in their effects has not been quantified. For speech, careful tests indicate different subjective effects, but it is not clear that this is a property of the noise alone or includes effects of the encoding/decoding process as well. The subjective effects of quantizing noise are treated differently than those of steady (psophometric or C-MESS) noise through the use of weighting factors that convert quantizing noise values to equivalent steady noise values. A number of such weighting factors exist, one for each of several types of encoding, and are discussed in Chapter 4.

1.4.4 Total Distortion

Interference due to aliasing and quantizing noise, as well as miscellaneous other small noise sources in digital systems, are lumped together in a measurement referred to as total distortion. It is made in exactly the same manner as C-NOTCHED noise (Section 1.3.9) and cannot be distinguished from other signal dependent noise in a casual measurement. In digital systems it is used as an overall quality check. The amount of noise varies with the magnitude of the exciting signal because of compandor action, so the required signal-to-distortion ratio is stated for several values of input. In requirements, a plot of signal-to-distortion as a function of level is usually provided. See Figure 1.23 for an example. Older seven-level encoding systems (seven-binary-digit encoding) maintain an average working-signal-to-total-distortion ratio on the order of 28 to 30 dB. Full eight-level encoding improves the numbers to 34 to 36 dB. A six-decibel improvement for each level added is expected because each increase multiplies the number of voltage slicing levels by two and the step size, Vs, in the relation for the noise magnitude, (1.24), is divided by two.

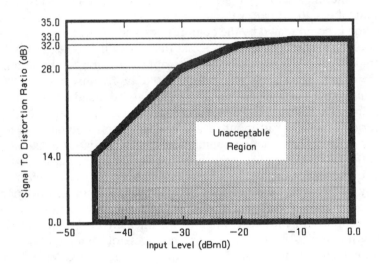

Fig. 1.23 Example Of Total Distortion Requirement
Source: Red Book, Volume VI—Fascicle VI.5, CCITT, Geneva, 1985. Reprinted with permission

1.4.5 Jitter

So-called digital transmission systems actually transmit an analog signal in the form of a pulse stream or pulses modulated onto a carrier. The signal on the medium suffers from all the problems of any analog signal. The pulses gradually disperse into one another, primarily because of delay distortion, and the signal-to-noise ratio deteriorates. The pulse stream must be reconstructed periodically, lest the eye pattern become perilously closed. This is accomplished by means of a type of amplifier known as a pulse regenerator. Regenerators sample the pulse stream to recapture the bits and generate a new pulse waveform. The timing for the sampling is recovered from the arriving signal in order to avoid the problems of synchronizing large numbers of the devices to a synchronization network. The timing recovery is not perfect and minor random differences in relative sampling times exist within regenerators. These differences can accrue along a digital system and give rise to small variations in the instants between zero crossings of the digital signal delivered at the end. These variations are known as jitter. There are other causes of jitter, for example, any real master timing source has a small amount of jitter in its output, but that due to regenerator action is usually dominant. Jitter accumulation can give rise to excessive error rates in the reconstructed bit stream and slips in some types of equipment.

Total jitter is subdivided into two classes. Short-term variations of the zero crossings are referred to simply as *jitter,* and long-term variations are known as *wander.* Jitter detectors measure the effective peak-to-peak amplitude of variations in the time between successive zero crossings. This time period is referred to as a *unit interval* (UI). The UI is, of course, dependent upon the bit rate at which the system is operating, and this in turn depends upon the capacity of the system or level of (digital) multiplexing.

Wander, or relatively long-term variations in the UI, can be accommodated by detectors more easily than rapid variations or jitter. Requirements on tolerable variations are, therefore, given as a function of rapidity of variation. This translates into the bandpass of filters employed in the measuring apparatus. Filters with very low, low-frequency cutoffs pass the long-term variations, and those with higher low-frequency cutoffs separate out the rapid changes. Permitted variations in the UI are, therefore, presented as a mask showing peak-to-peak amplitude *versus* bandpass filter low-frequency cutoff. Such a mask, from [7, Rec. G.824], applicable to a digital system operating at a bit rate of 1.544 megabits per second is shown in Figure 1.24. Frequency f_4 is the high-frequency cutoff of the filter. It remains fixed, and the low-frequency cutoff is changed, as indicated, to check the prescribed maximum jitter for each bandwidth.

1.5 MEASUREMENT TECHNIQUES

Transmission measurements are affected by the object being measured and the measurement environment as well as the characteristics of the test set employed. In addition, the quantity displayed by the test set may not be exactly what the name of the measurement implies. This section presents various details and concepts associated with a number of measurements. The purpose is to point out exactly what is being measured and what factors influence the results. A knowledge of these issues is important in the establishment of requirements or in designing to meet

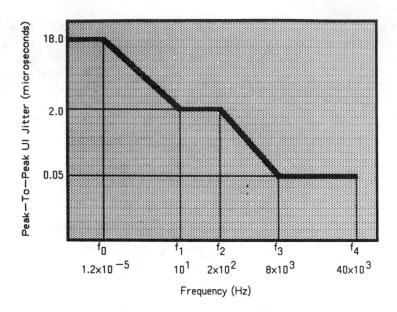

Fig. 1.24 Example of a Jitter Requirement for Digital Systems

requirements, and provides an improved understanding of how and why the parameters being measured affect transmission quality. The measurements considered are listed in Table 1.7.

1.5.1 Balance and Common Mode Rejection

The measurement of balance can be difficult, and variations in the test arrangement can yield significantly different results. A strongly recommended standard procedure is detailed in ANSI/IEEE Std 455-1978.

Balance is crucial to noise and crosstalk control in communications systems. The term is used freely, but what constitutes good balance and how difficult it is to achieve is not widely appreciated. Figures 1.25, 1.26, and 1.27 are used to explain this. The first two of these figures show a level meter connected across the legs of a balanced transmission line to make a bridging[21] measurement.

Figure 1.25 is used to illustrate bridging loss and introduce the discussion of balance. The perfectly balanced source, with output impedance Z_S, is delivering a signal to a perfectly balanced load of the same impedance. The transmission line is short and may be considered to have zero loss. Recall that a level measurement is a voltage measurement with the meter calibrated in dBm for the impedance Z_S. The meter responds to the current that flows through it, I_2, that is caused by the voltage across

Table 1.7
Measurement Techniques

Measurement	*Section*
Balance and common mode rejection	1.5.1
Damping	1.5.2
Envelope delay approximations	1.5.3
Impulse noise	1.5.4
Nonlinearity; four-tone method	1.5.5
Phase jitter	1.5.6
Amplitude jitter	1.5.7
Phase and gain hits	1.5.8
Dropouts	1.5.9

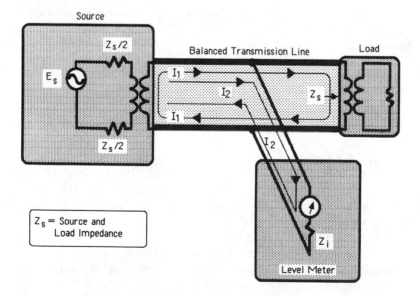

Fig. 1.25 Bridging Measurement

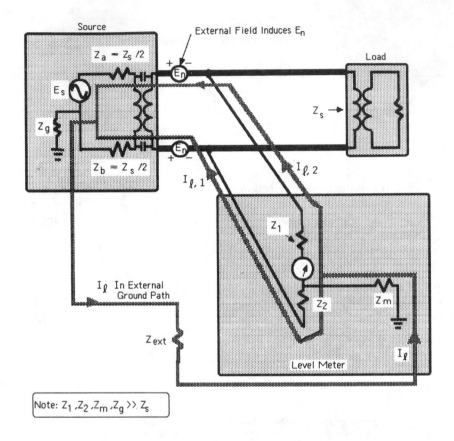

Fig. 1.26 Balance Factors in a Bridging Measurement

the line. The level measurement will be a good one if the effect of connecting the meter is negligible.

What is negligible must be determined by the person using the data obtained and will involve at least two considerations: first, the value displayed by the instrument must be a representation of what the user believes is being measured (not due to some extraneous factor introduced by the connection); second, the value displayed should be accurate to within a tolerance acceptable to the user.

The current, I_1, is the normal current through the load due to the source voltage, E_S, that divides between the source and load impedances. The connection places the meter impedance, Z_i, across the line and I_2 flows through the meter. The current I_2 is a disturbance to the circuit,

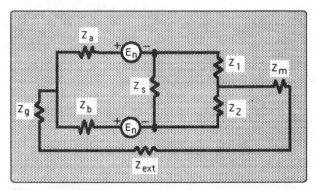

(a) Circuit Diagram

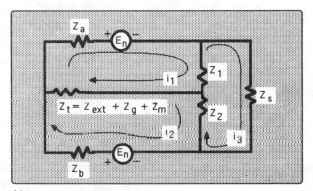

(b) Diagram Simplified And Rearranged

Fig. 1.27 Determination of Load Current, I_3, Due to Longitudinal emf, E_n

causing an additional voltage drop across the source impedance, so the accuracy of the reading depends on the ratio of I_2 and I_1; ideally I_2 should be zero. If Z_i is very large compared to Z_S, then approximately

$$I_1 = E_S/2Z_S \quad \text{and} \quad I_2 = E_S/2Z_i$$

For an accuracy of P percent, the ratio of these two must be less than or equal to $P/100$, and so

$$Z_i \geq 100Z_S/P$$

Or the accuracy, P percent, is given by

$$P = 100Z_S/Z_i \text{ percent   (bridging accuracy)} \qquad (1.27)$$

With source and load impedances of 600 Ω, the input meter impedance should be greater than 0.6 MΩ for an accuracy of 0.1%.

The second figure, Figure 1.26, is used to illustrate the balance problem from a more practical view than that presented in Section 1.2.7. Several elements, ignored in Figure 1.25, have been added. For example, shunt impedances, dominantly capacitive at frequencies of interest, exist between the windings of any transformer. These are indicated by the two capacitors connecting the primary to the secondary in the source. In general, they will not be equal in size. The impedances Z_g and Z_m are net impedances to ground as seen from *inside* the source and the level meter. Italics are used because it is difficult to define what that means. Within the source and the meter, these impedances are distributed among stray capacitances and various insulation resistances. A measurement taken with one probe on the source chassis, or on the third (grounded) conductor of a three-wire power cord, depends upon precisely where the other measurement probe is placed. For example, with the source illustrated, if one wanted to know the resistance to ground from the balanced output terminals, different values would obtain for each terminal because the leakage across the source transformer cannot be made perfectly symmetrical with respect to ground. Note also that the output impedance of the source has the two components, Z_a and Z_b, that are only approximately equal to $Z_S/2$ in this drawing.

The impedances, Z_1 and Z_2, represent the input impedance, Z_i, of the level meter of Figure 1.25, with some path leading to ground through Z_m. All impedances except Z_a, Z_b, and Z_{ext} are very much larger than Z_S.

The impedance, Z_{ext}, in the shaded line, is that of the ground path impedance between the two ground points shown internal to the equipment. It may be nearly zero if the meter and the generator are powered from a common alternating current source, or thousands of megohms or more if both are battery operated. So Z_{ext} can range from essentially zero to practically infinity. For this example, the load is still considered to be perfectly balanced and totally isolated from ground, as in Figure 1.25.

External electrical fields are always present, with power line radiation being the most prevalent. The electromotive force (EMF) induced on the transmitting line from such fields is shown as two identical longitudinal noise sources, E_n. The E_n are actually distributed along each conductor of the transmitting line, and cause a longitudinal current, I_1, to flow as shown, part in each side of the transmission line, and then as a single current through the ground path.

Because of the relative sizes of Z_S and the other impedances, I_1 is determined almost completely by E_n and the sum of Z_m, Z_{ext}, and Z_g. This can be shown by an analysis of the circuits in Figure 1.27. The upper figure, (a), is the circuit for the diagram of Figure 1.26, and in (b), Z_g, Z_{ext}, and Z_m are combined into Z_t and the circuit redrawn to show the three loops. The loop equations are

$$-E_n = i_1(Z_a + Z_1 + Z_t) - i_2 Z_t - i_3 Z_1 \tag{1.28a}$$

$$E_n = -i_1 Z_t + i_2(Z_b + Z_2 + Z_t) - i_3 Z_2 \tag{1.28b}$$

$$0 = -i_1 Z_1 - i_2 Z_2 + i_3(Z_1 + Z_2 + Z_S) \tag{1.28c}$$

Solving these for i_3, we have

$$i_3 = \frac{E_n[Z_1 Z_b - Z_2 Z_a]}{(Z_a + Z_b)[Z_t(Z_1 + Z_2) + Z_x] + Z_S Z_y + Z_a Z_b(Z_1 + Z_2)}$$

$$Z_x = Z_1 Z_2 + Z_S Z_t; \qquad Z_y = Z_1 Z_b + Z_a Z_b + Z_2 Z_a \tag{1.29}$$

Equation (1.29) shows that there will be no current in the load only if the difference of cross products $Z_1 Z_b$ and $Z_2 Z_a$ is zero.

Some simplifications result if $Z_1 = Z_2$ and use is made of the usual condition that Z_t and $Z_1 \gg Z_a$ and Z_b. Thus,

$$i_3 = E_n \frac{Z_1(Z_b - Z_a)}{(Z_a + Z_b)(2Z_t Z_1 + Z_1)} \tag{1.30}$$

showing the need for balance of the source.

If, in addition, $Z_t \gg Z_1$, then,

$$i_3 = \left(\frac{E_n}{2}\right) \frac{(Z_b - Z_a)}{(Z_b + Z_a)Z_1 Z_t} \tag{1.31}$$

which explicitly shows the benefit of isolation from ground.

Finally, if Z_t is zero, (1.29) reduces to

$$i_3 = E_n \frac{(Z_1 Z_b - Z_2 Z_a)}{(Z_a + Z_b)Z_x + Z_S Z_y + Z_a Z_b(Z_1 + Z_2)} \tag{1.32}$$

and then, if $Z_1 = Z_2$, (1.32) becomes with $Z_1 \gg Z_a$,

$$i_3 = E_n \frac{(Z_b - Z_a)}{Z_1(Z_b + Z_a)} \tag{1.33}$$

Note that in these expressions for i_3, it is the absolute value of the unbalance, $Z_a - Z_b$, that determines the current, not a ratio dependent upon the magnitude of the source impedance. Thus, with low impedance circuits it is generally easier to achieve good balance than it is with high impedance ones because the absolute difference is smaller for a given percent tolerance on components.

The voltage produced by i_3 across the load is called E_m, m for metallic, and appears as interference at the load. It is used as a measure of balance for the system. Balance, B, is defined as

$$B = 20 \log(|E_n|/|E_m|) \text{ dB}   \text{(balance)} \tag{1.34}$$

Equations (1.29)–(1.33) were derived with reference to Figure 1.26, showing a level meter in use, to emphasize that the measurement device can cause erroneous readings if precautions are not taken. The load was assumed to be perfectly balanced and isolated. Any real load will have an input circuit similar to that of the level meter, or the output circuitry of the source. If the meter were removed, the paths to ground of the load would be represented by Z_1 and Z_2 and the analysis and results are identical, so the equations above apply in this case also.

In Figure 1.26, the E_n were shown induced on the transmission line. They can also appear, of course, in ground wires and as contact potential in an improperly prepared grounding system.

Signal sources and loads are not the only portions of a system that are critical in maintaining balance. The transmission line, especially if a physical pair of wires, has nonuniformly distributed capacitance, and series and leakage resistances, that contribute to unbalance. Distributed inductance will usually be negligible at voice band frequencies. A high degree of electrical symmetry must be maintained in every component of the transmission system in order to preserve good balance and minimize interference from external sources. Great care is given to the geometry of the wire pairs in the manufacture of cables and real cable balances can approach 100 dB if pair integrity is maintained in use. An infamous economic measure is to use one good wire from each of two bad pairs!

In the discussion above, the undesired voltage was measured at the input to the load (or the level meter as in Figure 1.26). It is possible to have identical voltages at the input and still have a voltage at the output of the load device due to E_n. This can occur when the load is some device more complex than a simple resistor. A *conversion* takes place within the load. This condition is said to arise from poor common mode rejection.

Common mode rejection is measured by applying identical voltage to both sides of the balanced input to a device and measuring the output voltage, E_0, across Z_S. Its magnitude, M, is then given by

$$M = 20 \log|E_n/E_0| \text{ dB} \quad \text{(common mode rejection)} \qquad (1.35)$$

Obviously, one must insure that no unbalance exists in the external circuit.[22]

1.5.2 Damping

When one reads an analog meter, the ballistics of the meter movement, along with any electrical or mechanical damping, can greatly affect the value indicated and the ease with which it is read. These are important design considerations in many noise measuring devices, but there are also other constraints. Sounds are not heard on an instantaneous basis; the ear integrates the arriving sound pressure variations over a period of about 200 ms. In keeping with the desire to make a noise measurement correlate with subjective annoyance, noise meters are designed to have an integration interval of that length. It is fortunate that, for most kinds of noise present on telephone channels, the instruments happen to be fairly easy to read, in terms of mentally averaging the movement of the meter. If crosstalk is an important part of the noise being measured, severe jumping of the needle occurs and reading may be difficult. A *damping* switch is often provided to increase the integration time[23] so more consistent readings can be made when this problem is encountered. Noise meters are typically specified as accurate to plus or minus 1.0 dB ([5, p. 7]). If the damping switch is in the *ON* position when no crosstalk or similar disturbances are present, the readings may be affected by this amount or more. A reading taken with the damping switch on, should be so noted.

Automated network surveillance systems measure noise with equipment using the 200-ms integration interval. There is no provision for noting that the noise was sporadic or crosstalk-like in character. Noise of such a nature is much more annoying than the usual steady type, so automated C-MESS or psophometric measurements may hide a very poor transmission condition. This is simply one aspect of the broader and unresolved problem of objectively measuring subjective annoyance of discontinuous disturbances.

Designers of instruments that use digital readout devices instead of analog must cope with another problem. It is very difficult to design a digital display noise meter because of design conflicts between the desire for a *smooth motion* display, the 200-ms integration period, and the display refresh interval. A relatively long refresh period aids in presenting the smooth motion but defeats the purpose of the integration time. A rapid refresh, more nearly following what an analog display would show, appears so unstable that a decision on the average value is difficult. Subjective tests in which users read both standard analog and *best* digital displays, show average values in reasonable agreement, but the variances of the differences are large (> 2 or 3 dB). Such tests also indicate that a geometric average seemed more natural to obtain than an arithmetic one. That is, the values reported from observations of digital displays tended to be closer to geometric averages. Multielement display devices in which each digit is discreet, such as a Nixie tube, were found to be superior to discreet element displays such as liquid crystal display (LCD) devices.

1.5.3 Envelope Delay Approximations

Approximations is used in the title of this section because in envelope delay, probably more than in any other transmission measurement; it is true that what is recorded is what the instrument sees and not what the user thinks the instrument sees.

In communication channels, the change of phase of component frequencies between input and output is not uniform. It is determined by the time delay for each frequency, and that is primarily determined by the numerous filters encountered in transmission systems. Data transmission suffers because of this nonuniform displacement of component frequencies long before speech transmission. Referring to Section 1.3.11 on eye patterns, we might allow about one fourth of a pulsewidth for the phase distortion to maintain some margin for other impairments. This is 125 ms at a data rate of 2000 bps, so we are interested in relative delays or delay distortion of this order between the fundamental and higher harmonics of the data waveform. The effects of the phase displacement on speech do not become noticeable until the displacements exceed about 10,000 μs.

In a laboratory environment, the measurement of delay of a sinusoid is straightforward. With delay changes significant (about 100 μs) compared to the periods of frequencies of interest (300 to 3000 μs), it would appear that measuring the delay characteristic of a channel with sufficient accuracy would be straightforward. In the laboratory, however, one has the source of the sinusoids to compare phases, and hence delays, of output sinusoids to make the delay measurement at each frequency; this is not the case in a network measurement. The source is not handy at the far end of a channel, so it is unknown where the zero reference is for each frequency. In addition, a frequency shift on the channel will preclude any phase comparison with a reference.

The alternative has been to measure envelope delay or envelope delay distortion. Envelope delay is defined as the derivative of phase as a function of frequency, $d\phi/d\omega$, $\omega = 2\pi f$. The method used to do this requires that the measurement be made at the location of the source signal and that phase information for the channel measured be returned to the source over a path separate from the one being measured. The reasons for this will become clear in the following discussion.

The measurement is accomplished by transmitting a variable in-band carrier frequency, f_c, from the near end. The carrier is amplitude modulated. At the far end of the circuit, the modulation is transferred to a constant frequency carrier and returned to the transmitting end. The modulation sidebands have thus undergone a phase shift in both directions of transmission. Their relative shift in the forward direction will change when the transmitting carrier is changed due to their new position on the phase characteristic of the forward channel. In the return direction, the relative shift is fixed because the return carrier frequency is not changed. The phase difference between the upper and lower sidebands, $\phi_u - \phi_l$, as detected at the transmitter location, can be used as an estimate of the slope of the forward plus return channel phase characteristic. When divided by the sideband separation, upper sideband frequency minus the lower sideband frequency, $f_u - f_l$, an estimate of the derivative is obtained. The carrier frequency, f_c, is, of course, midway between f_u and f_l. As the carrier is swept across the band, the quantities:

$$(\phi_u - \phi_l)/(f_u - f_l) \quad \text{or} \quad \Delta(\phi)/\Delta(f)$$

are plotted or stored. Envelope delay is then the difference in these quantities for any two frequencies. Thus

$$\text{ED} = \Delta(\phi)_2/\Delta(f) - \Delta(\phi)_1/\Delta(f) \tag{1.36}$$

where the subscripts refer to the Δ's at any two carrier frequencies, f_1 and f_2 ($f_1 < f_2$). EDD is generally calculated by subtracting the minimum observed value from the measured ED and so would be a positive quantity. In practice, the frequency at which the minimum value occurs is assumed at the start of a measurement. This may give rise to raw measurement data with negative values. Strictly, these data should be renormalized.

At the start of a measurement, the variable carrier is adjusted to the frequency of assumed minimum ED, and the phases of the modulation sidebands are compared with each other to obtain a reference phase slope; and with the phase of the modulating oscillator to establish a total delay reference. The controls are also physically adjusted to the zero positions. The delay reference is accurate only to within a multiple of 2π radians of the modulating frequency; an absolute delay reference is not possible but is not necessary. These reference initial values are recorded. As the transmitting carrier is varied, the reference differences are subtracted from each new difference, yielding the relative phase shift of the sidebands near the new carrier frequency with the constant unknown phase shift of the return channel removed. As the relative shift passes through multiples of $2n\pi$, these must be accrued and included in the delay calculation. A diagram of this measurement system and the pertinent mathematical relations at each key point is shown in Figure 1.28.

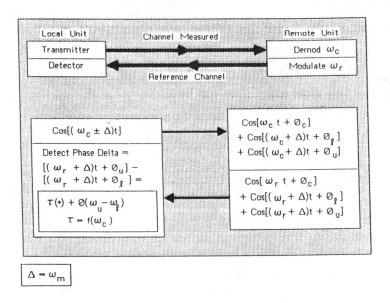

Fig. 1.28 Principle of Envelope Delay Measurement

The record of the initial total delay is also essential because channel electrical length changes, described in Sections 1.3.4 and 1.3.1, may cause the total delay to change during the course of a measurement. Instruments have been made which establish a phase reference at the start of a measurement and assume it to be constant during the measurement interval. These instruments are often found to be very difficult to use because of the wandering reference.

The method has few problems until the phase characteristic begins to show relatively rapid changes with frequency. The Δ-value is now critical because small changes in the modulating frequency can produce large changes in the observed $\Delta(\phi)$ anywhere that the phase characteristic is steep. In other words, the ED or EDD measured is a function of the sideband separation or modulating frequency. This separation is called the aperture, or window of the test set. Lower modulating frequencies effect narrower apertures.

This might still be all right except that there are two standard modulating frequencies. In the United States it is $166\frac{2}{3}$ Hz, elsewhere (CCITT) it is $83\frac{1}{3}$ Hz. For the same channel, envelope delay plots made with two sets using the different apertures may be strikingly different in the representations of the bends in the typical U-shaped envelope delay curve such as was illustrated in Figure 1.10. The CCITT standard may be said to have greater resolution than the United States standard because of the narrower window, but that is not the only criterion to use. Which representation is better is a matter of opinion and application, as explained in the following material.

As shown in the following, echoes due to reflections at impedance mismatches show up as ripples in the phase characteristic of a channel. A narrow aperture shows these ripples more readily than a wide aperture, but it may magnify some that are of no real consequence because of a relation between the aperture, the period of the ripple, and the measurement.

Refer to Figure 1.29 which shows a sinusoidal ripple in a phase characteristic with the measurement signal, consisting of the carrier ω_c and its two sidebands from the modulating frequency ω_m, superimposed. The ripple in the phase has a frequency on the ω-axis of

$$f_r = \Omega/2\pi, \qquad 1/f_r = 2\pi/\Omega \tag{1.37}$$

where Ω is the period of the ripple in radians per second.

The modulating frequency can be expressed as

$$f_m = \omega_m/2\pi = \Delta\Omega/2\pi, \qquad \Delta\omega = 2\pi f_m \tag{1.38}$$

The phase angles of the lower and upper sidebands are

$$\phi_1 = A \sin[(2\pi/\Omega)(\omega_c - \Delta\omega)] \tag{1.39a}$$

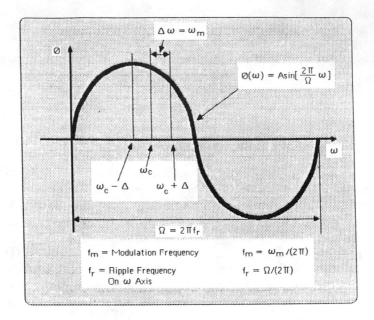

Fig. 1.29 Sketch for Envelope Delay Error Derivation

and

$$\phi_2 = A \sin[(2\pi/\Omega)(\omega_c + \Delta\omega)] \tag{1.39b}$$

with the phase of the carrier:

$$\phi(\omega_c) = A \sin[(2\pi/\Omega)\omega_c] \tag{1.40}$$

Recall that the envelope delay is the derivative of phase with respect to ω:

$$\text{ED} = (d\phi/d\omega) = A(2\pi/\Omega) \cos[(2\pi/\Omega)\omega_c] \tag{1.41}$$

which is going to be approximated by

$$(\phi_2 - \phi_1)/2\Delta\omega \tag{1.42}$$

The ratio of the approximation to the true value will be

$$\frac{(\phi_2 - \phi_1)/2\Delta\omega}{\{(2\pi/\Omega)A \cos[(2\pi/\Omega)\omega_c]\}} \tag{1.43}$$

but

$$\phi_2 - \phi_1 = A\{\sin[(2\pi/\Omega)(\omega_c + \Delta\omega)] - \sin[(2\pi/\Omega)(\omega_c - \Delta\omega)]\}$$
$$= 2A \cos[(2\pi/\Omega)\omega_c] \sin[(2\pi/\Omega)\Delta\omega] \qquad (1.44)$$

and so the ratio of the measurement to the true derivative is

$$\frac{\sin[(2\pi/\Omega)\Delta\omega]}{(2\pi/\Omega)\Delta\omega} = \frac{\sin(x)}{x} \text{ or } \mathrm{sinc}(x) \qquad (1.45)$$

with

$$x = (2\pi/\Omega)\Delta\omega = 2\pi(f_m/f_r) \qquad (1.46)$$

Thus, intrinsic to the method, is a multiplication of the result by a function that is dependent upon the modulating frequency and the ripple frequency. The function approaches unity as the ripple frequency increases, and has a number of zero crossings at low-ripple frequencies. Whenever the modulating frequency, f_m, is a multiple of the ripple frequency, f_r, then

$$\frac{\sin[2\pi(nf_r)/f_r]}{x} = \frac{\sin(2n\pi)}{2n\pi} = 0 \qquad (1.47)$$

and there is a zero output from the test set. Thus ripples of specific periodicities are missed entirely and plot as a flat line. These ripples are referred to as blind spots (in the ripple-frequency domain), and correspond to the modulating sidebands traversing the sinusoidal ripple in the phase characteristic with a spacing that is equal to the period, or a multiple of the period, of the ripple. Even though they ride up and down on the ripple, their difference in phase is always zero and no envelope delay distortion is registered. This is illustrated in Figure 1.30.

The period or frequency of echo ripples is determined by the round trip delay from the detector in the remote unit (Figure 1.28) to the impedance mismatch and back; a listener echo. Reflected echoes add to the original signal on a vector basis with a phase determined by a delay equal to the round trip transmission time. The net phase seen by the detector is, therefore, that of the vector sum. The round trip delay is constant so the shift incurred by the echo, and so its phase with respect to the original, is dependent upon frequency. As the carrier frequency is swept across the band, the relative phase of reflected-to-incident signal sweeps through a number of complete revolutions. The two signals will be exactly in (or out) of phase whenever the delay is a multiple of $2n\pi$ ($n = 0, 1, 2, \ldots$) at

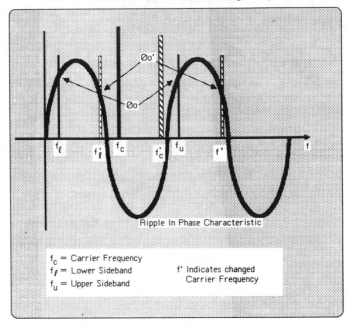

Fig. 1.30 Occurrence of a Zero in an Envelope Delay Measurement

the frequency being transmitted. The out-of-phase point represents the maximum shift of the net phase compared to that of the incident wave. If the delay is T_d, then this or any other given phase relation will repeat for

$$2\pi f T_d = 2n\pi \quad \text{or} \quad f = n/T_d, \qquad n = 0, 1, 2, \ldots \tag{1.48}$$

The difference between two received carrier frequencies giving rise to echoes with this same phase relation to the incident signal is then

$$f_2 - f_1 = 1/T_d \tag{1.49}$$

which is the observed period (in hertz) of ripple in the measurement.

Note that the period of the frequency ripple in hertz is simply $1/T_d$.[24] Thus, long delays produce high-frequency ripples. Echo suppressors or cancelers are used in circuits when the round trip delay reaches about 45 ms. So, when echo suppressors or cancelers are used, assuming that they effectively eliminate the echo, they place a lower limit of about 22 Hz on the period of the ripples of interest. These would represent a high-frequency ripple bound on a plot of EDD of 0.045 cycles per Hz. A rough lower bound might be determined by the fact that voiceband data modems use carriers in the center of the band, about 1700 Hz. So, from the Nyquist criterion, the minimum baud length is on the order of just under 1 ms. In view of the eye patterns and sampling interval concept of Figure 1.18, echoes returning in less than about $\frac{1}{3}$ ms would not by themselves cause errors. This suggests a lower-ripple frequency of interest of 0.0003. The period of this ripple would be about 3300 Hz on the frequency scale, or about one cycle over the voiceband. Figure 1.31 shows the response of EDD sets, using the United States and the CCITT apertures, to ripple periods in this range as a result of the $[\sin(x)]/x$ multiplier.

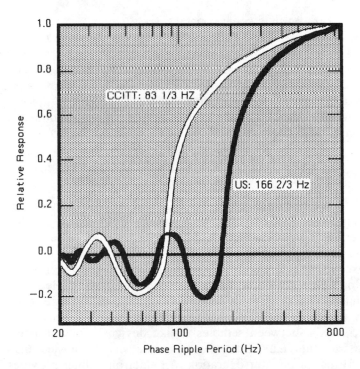

Fig. 1.31 Relative Response of EDD Measuring Sets with Two Different Modulating Frequencies
Adapted from: Bell System Technical Reference, AT&T Pub. 41008, American Telegraph and Telephone Company, 1974

Practical instrument calibration results in an effective multiplier of unity for midrange ripple periods. Those periods above this value have their magnitude amplified. For those below, it is attenuated. Narrow apertures miss fewer ripples (have fewer blind spots), but they accentuate sharp curvature in an EDD plot. The aperture choice is a design trade-off with no solid guidelines. Details of an instrument and an analysis of some of the peculiarities of this measurement technique are given in [5, pp. 10ff] and [11, pp. 2ff].

Approximating the derivative with the delta functions as in the expression for ED (see (1.36)), appears cumbersome at best and leads to distortions in the data. Not being able to measure the true derivative is a mixed blessing, however. The frequency of the ripples in the ED measurement caused by echoes is determined by the delay as shown above. Designate the delay as T_d. Then the ripple can be expressed as a constant times a sinusoid:

$$\phi(\omega) = A \cos(T_d\omega + m)$$

where $\omega = 2\pi f$ and m is some constant.

The derivative of this function with respect to ω has T_d as a multiplier, and so its amplitude becomes proportional to the delay even if the magnitude of the echo is constant. The true interference, the echo signal, is then measured with an error that increases with the delay, a situation worse than the error in the approximate method described. A plot of EDD could be totally obscured by the large magnitude of high-frequency ripples even though their interference value might be insignificant.

1.5.4 Impulse Noise

As mentioned in Section 1.3.5, impulse noise requirements are stated in terms of a maximum expected number of impulses in a specified time interval. The instrument used to measure the noise is a voltage threshold detector. Whenever the noise waveform on a circuit being measured exceeds the threshold, a trigger is generated which advances a counter one digit. The first instruments were built using electromechanical counters. These devices require some amount of time to operate, typically about 100 ms, and because of normal tolerances, noticeable differences exist between samples of like devices. For consistency between instruments, a maximum counting rate was established, 6.5/s, that allows 154 ms for counter operation. An electronic timer provides input blocking for an interval of this length, allowing the counter to reset.

The limited counting rate means that some impulses will not be counted because more may arrive within the fixed interval or so-called dead time of the instruments. The interarrival times of impulses, which may be as small as a few milliseconds, are important in the statistics used to estimate resulting data transmission error rates. The dead time causes two or more impulses that occur within the 154 ms to be counted as one, so a percentage of them are missed. The percent varies from essentially zero to as high as ten in a measurement period of 15 minutes and is roughly proportional to the total number recorded in a given measurement interval. On average, for a 15-minute period, about 10% are missed if the total number counted is about 20. The statistics relevant to missed impulses may be included in calculations to establish requirements for this impairment. The resulting adjustments to requirements are small however, usually less than 10% of the maximum number stated.

For impulse noise characterization purposes, instrumentation that is not subject to the limitations described in the last paragraph may be used. In these cases, a definition is needed to distinguish between two impulses which are very close together and one impulse with a long oscillating trailing edge (see [11, p. 3253]). The need for a definition is illustrated in Figure 1.32. The lower three drawings, Figures 1.32(b), (c), and (d), show the impulses in the top one, Figure 1.32(a), rectified and subject to detection whenever the waveform exceeds a voltage slicing level. Two possible slicing levels are shown. In the first of the rectified waveforms, Figure 1.32(b), there are eight separate occurrences of the waveform exceeding level 1, so eight impulses would be counted. Raising the slicing level to 2, Figure 1.32(c), there would be a count of six. Use of a guard interval as shown in the lower drawing, Figure 1.32(d), alleviates this potential uncertainty in counting. Once the threshold is exceeded, the counter input is blocked until the input waveform has decreased below the active threshold for a period of time indicated by the black regions. As shown, when the counter opens after the first two peaks of each separate impulse, the input is *still* above threshold so the blanking cycle is reactivated. Only after the third peak has the waveform remained below threshold for a time in excess of the guard interval. This then yields the correct count of two. Studies of the autocorrelation function of recorded impulses showed that 4 ms was an optimum choice of guard interval time for voiceband impulse noise, so that is the current recommendation in [6].

As with continuous noise measurements, impulse noise measuring sets are equipped with weighting filters that initially were patterned after the bandwidth of data modems. This gave rise to a proliferation of available filters and confusion concerning requirements. One such filter, called

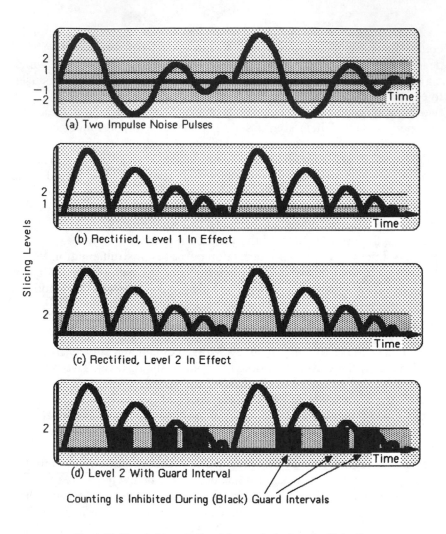

Fig. 1.32 Thresholds and Guard Intervals for Impulse Noise Detectors

voice band (VB) weighting, was used most often and requirements were written as N counts in X minutes with the threshold set at so many dBrnVB(0). The continuing desire to reduce the number of filters required for general noise measurements, coupled with more experience with impulse noise as an impairment, led to replacement of the VB filter with the standard C-MESS or C-NOTCHED filters. The latter types have complex phase characteristics and significantly modify the peak amplitudes of impulses passing through them. As a result, correlations between measurements with VB and C-MESS weightings are not strong. This means that error rate estimates made from two sets of data, one taken with each of them, will be different. Impulse noise tests have come to be used more as a go-no-go gauge rather than an error rate estimation tool, so this poor correlation is now of little interest. Additional background material for impulse noise measurement and evaluation criteria can be found in [10] and [11].

1.5.5 Nonlinearity; Four-Tone Method

1.5.5.1 Basic Considerations

Of the many ways to measure nonlinearities, the four-tone intermodulation distortion test method is chosen for description because the considerations involved in its design bring out many factors of importance in telecommunication transmission over nonlinear channels. Several of these are discussed in the following.

The four-tone measuring system arose from a need to provide a test set that was easy to use and yielded measurements that related accurately to the effects of nonlinearities on data transmission. Studies relating to the technique, and reported in [12], provide some graphic examples of the interplay between linear and nonlinear elements and insight for the interpretation of nonlinearity measurements. Before describing these, some basic phenomena relevant to nonlinearities will be illustrated.

One model of a nonlinearity is a third-order polynomial:

$$y = a_1x + a_2x^2 + a_3x^3 \tag{1.50}$$

If this model were accurate, intermodulation distortion and harmonic distortion measurements, described in Section 1.3.6, would be equivalent because the coefficients of the polynomial could be computed from either. The polynomial is only an approximation, however, and the two types of measures do not always agree if the comparison is made. In practice there are three major departures from the model: frequency dependency, time variability, and the existence of higher order terms. The first two are the most important and warrant some discussion.

Consider the channel illustrated in Figure 1.33(a). It has two nonlinear elements separated by a linear one. The two polynomials producing distortion are identical, but the distortion power out of the channel will depend upon the linear segment. Assume that there is no direct current (dc) coupling between elements, the usual situation for a channel of any appreciable length. The manner in which the distortion products from

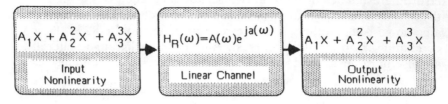

(a) Model Illustrating Addition Of 3rd Order Distortion

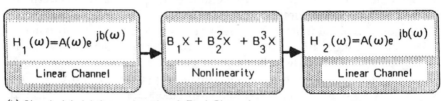

(b) Simple Model Approximating A Real Channel

$$H_1(\omega)\, H_2(\omega) = H_R(\omega) = \text{Typical Channel}$$

Fig. 1.33 Two Models of Nonlinearities on Channels
Adapted from: Bell System Technical Reference, AT&T Pub. 41008, American Telegraph and Telephone Company, 1974

each nonlinearity add at the output is determined by the change in magnitude and phase angle of the products from the first while traversing the path to the second. Or simply stated, the addition is dependent upon the linear element. As a simple example let the input to the first nonlinearity, $s_i(t)$, be

$$s_i(t) = A\,\cos(\omega_1 t) + B\,\cos(\omega_2 t) \tag{1.51}$$

and let the nonlinearities be devices with unit gain and a perfect squaring device, $f(x) = 1 + x^2$, rather than the cubics shown in Figure 1.33. If ω_1 and ω_2 are close together, then

$$\omega_1 - \omega_2 \ll \omega_1 \text{ and } \omega_2$$

and

$$\omega_1 + \omega_2 \text{ is near } 2\omega_1 \text{ and } 2\omega_2$$

which permits us to define the second-order products of interest to be those with frequencies equal to twice the fundamentals and the sum of the fundamentals.

Now, with the dc terms blocked, the input to the linear element, $s_1(t)$, is

$$
\begin{aligned}
s_1(t) = \ & A \cos(\omega_1 t) + B \cos(\omega_2 t) \\
& + (A^2/2)\cos(2\omega_1 t) + (B^2/2)\cos(2\omega_2 t) \\
& + AB \cos[(\omega_1 + \omega_2)t] + AB \cos[(\omega_1 - \omega_2)t]
\end{aligned}
\tag{1.52}
$$

If the linear element has unit gain and a nonlinear phase characteristic, $k + \omega^2$, the signal at the input to the second nonlinearity, $s_2(t)$, is

$$
\begin{aligned}
s_2(t) = \ & A \cos(\omega_1 t + k + \omega_1^2) + B \cos(\omega_2 + k + \omega_2^2) \\
& + (A^2/2)\cos(2\omega_1 t + k + \omega_1^2) \\
& + (B^2/2)\cos(2\omega_2 t + k + \omega_2^2) \\
& + AB \cos[(\omega_1 + \omega_2)t + k + (\omega_1 + \omega_2)^2] \\
& + AB \cos[(\omega_1 - \omega_2)t + k + (\omega_1 - \omega_2)^2]
\end{aligned}
\tag{1.53}
$$

The second-order products out of the second nonlinearity will comprise all the squares and cross products of the terms in (1.53) with frequencies as defined above: Twice the fundamentals and the sum of the fundamentals. These are found to be

$$
\begin{aligned}
(A^2/2)&\cos(2\omega_1 t + 2k + 2\omega_1^2) + (B^2/2)\cos(2\omega_2 t + 2k + 2\omega_2^2) \\
& + AB \cos[(\omega_1 + \omega_2)t + 2k + \omega_1 + \omega_2^2] \\
& + [(A^3 B)/2]\cos[(\omega_1 + \omega_2)t + k + \omega_1^2 - (\omega_1 - \omega_2)^2] \\
& + [(AB^3)/2]\cos[(\omega_1 + \omega_2)t + k + \omega_2^2 - (\omega_1 - \omega_2)^2] \\
& + A^2 B^2 \cos[2\omega_1 t + k + (\omega_1 + \omega_2)^2 + (\omega_1 - \omega_2)^2] \\
& + A^2 B^2 \cos[2\omega_2 t + k + (\omega_1 + \omega_2)^2 - (\omega_1 - \omega_2)^2]
\end{aligned}
\tag{1.54}
$$

Equation (1.54) has seven terms. The first and sixth are at a frequency of twice one fundamental, ω_1. The phases of these two terms are dependent upon $2k$ and k and upon ω_1 squared and the sum and difference frequencies squared. How they are added is thus dependent not only on the choice of ω_1 and ω_2, but on the fixed phase shift, k, of the linear element. The same addition dependency exists for the $2\omega_2$ product, the second and seventh terms, and for the sum frequency, the third, fourth, and fifth terms.

Some of the cross products resulting from the second nonlinearity produce third-order components of the type $2\omega_i - \omega_j$, $i, j = 1, 2$, that will fall near the fundamentals and must be rejected in the measurement of the ratio of distortion power to fundamental power. This is another consideration in the selection of ω_1 and ω_2.

If this exercise is repeated using a cubic nonlinearity, the amplitude of the fundamental is found to be affected because cubing a sinusoid produces a fundamental component equal to three times that of the third harmonic produced. Also, depending upon the resultant phase, the sign of the cubic may be positive and add to the fundamental, thus producing *enhancing* distortion, or negative, yielding *compressive* distortion. This is also discussed in more detail later.

The preceding has illustrated the role of the phase of *linear* elements in the addition of nonlinear distortion. Now recall from the discussion in Section 1.3.2 that the phase characteristic for a channel includes the parameter, k, the phase intercept. (This is the same k that was used in the second-order addition example just presented.) In general, k is indeterminate and is time variable when frequency offset is present. An offset of f Hz causes k to range through 2π radians, at a rate of f times per second.[25] To see this, a sinusoid of frequency f and phase ϕ, that is shifted by f_s can be written

$$\cos(2\pi ft + 2\pi f_s t + \phi) \tag{1.55}$$

to show that f_s can be viewed as a continual phase shift or as a continual increase in k. This is an appropriate manner of thinking of frequency offset when $f_s \ll f$, and as shown in Section 1.3.3, f_s is nearly always less than two. Thus, whenever k is not constant, the nonlinear distortion is also time variable because the entire phase characteristic is moving through 2π radians f times per second.

Some of the results of the study reported in [12] have to do with the effects of this variable parameter k on nonlinear distortion measurements. The study involved analysis of a model comprising one nonlinear and two linear elements as illustrated in Figure 1.33(b). Three phase characteristics, representative of commonly encountered carrier facilities, were used in the linear element. The results from all three are similar and are not distinguished in the following. In all cases, the total phase characteristics were varied by control of the parameter k.

Comparisons between distortion outputs for two types of input signals are reported. One with a Gaussian amplitude distribution and raised cosine spectrum, with the first zero at 2600 or 2800 Hz, about one and one-half times nominal data modem carrier frequencies. The spectra were rolled off at frequencies below 500 or 700 Hz. These signals are said to be representative of data modem signals but not desirable for use in a test set because second- and third-order products would not be separable. The second type of signal, permitting such a separation, is combinations of small numbers of sinusoids. The comparisons were directed at selecting a set of sinusoids yielding measurement results similar to those using the Gaussian signals. An imposed constraint was that the total input power be the same for both signal types because nonlinearities are level sensitive. Results are presented as graphic displays of distortion products out of the system as k ranges over 360°.

The behavior of second-order distortion from the system illustrated in Figure 1.33(b) is reported as shown in Figure 1.34(a). The curves in this figure show the ratio of second-order distortion power to fundamental power for three input signals, as a function of k. Two of the three are the Gaussian signals, the third is a four-tone combination with frequencies selected to force a near match of results with one of the Gaussian. Though there are steep drops in output even for the complex Gaussian signals for k in the vicinity of 160°, they do not go to zero. Combinations of sinusoids tend to have much deeper nulls than do complex signals. For the case illustrated, the depth of the null for sinusoids is reduced by several decibels by power adding two of the products resulting from two separate, closely spaced pairs, of tones. The A and B in the figure refer to these two pairs. Differences in the power in second-order terms are apparent even for the two similar Gaussian signals, and they increase as the null is approached. This suggests that accurate nonlinearity measurements can only be made with the actual signal to be transmitted. However, since the large differences occur near the null, where the total distortion power is small, the errors implicit in a measurement are acceptable.

Third-order products of Gaussian signals were found to be relatively insensitive to k, and added approximately on a voltage basis as illustrated in Figure 1.34(b). Third-order distortion of sinusoidal signals does exhibit nulls, however, as illustrated for a single sinusoid by the black curve. The solution to this is discussed below after a further description of the measurement technique.

1.5.5.2 Choice of Frequencies

As noted above, there are a number of considerations that restrict the choice of frequencies that may be used in a test set and in the selection of products to measure. This section discusses these considerations. First, however, we list the frequencies chosen for the four-tone method and describe precisely what is measured.

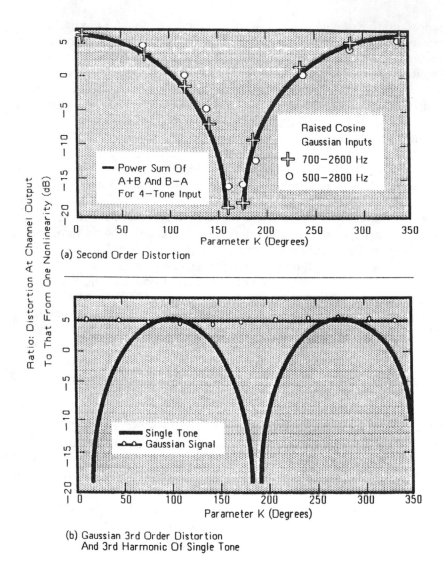

(a) Second Order Distortion

(b) Gaussian 3rd Order Distortion
And 3rd Harmonic Of Single Tone

Fig. 1.34 Phase Intercept Effects on Nonlinear Distortion Addition
Adapted from: Bell System Technical Reference, AT&T Pub. 41008, American Telegraph and Telephone Company, 1974

The implementation uses two pairs of tones, referred to as the low group and high group, or simply as A and B (see [12, pp. 16ff] and Figure 1.35(a)). The low-group frequency pair, A, has frequencies of 856 and 863 Hz, a separation of 7 Hz. The high-group frequency pair, B, has a separation of 11 Hz with frequencies of 1374 and 1385 Hz. The separations are

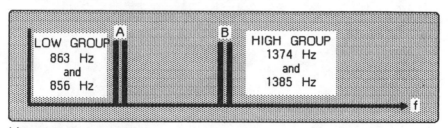

(a) Transmitted Spectrum

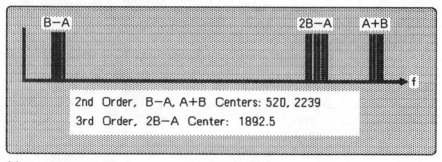

(b) Receive Spectrum

Fig. 1.35 Four-Tone Signal Transmit and Receive Spectra

not critical but must be small and different for each pair to prevent an unwanted condition noted below. The four tones are then: A_1, A_2, B_1, and B_2. The intermodulation products chosen for the measurement are shown in Figure 1.35(b). There are two groups of selected second-order products of the $A + B$ and $B - A$ type. They are at

$$A_1 + B_1 \qquad B_1 - A_1$$
$$A_1 + B_2 \qquad B_1 - A_2$$
$$A_2 + B_1 \qquad B_2 - A_1$$
$$A_2 + B_2 \qquad B_2 - A_2$$

If the pair spacings were the same, then $A_1 + B_2 = A_2 + B_1$ and these components of the same frequency and nearly identical phase would always add on a voltage rather than a power basis and yield an incorrect result.

One group of third-order products, of the $2B - A$ type, is used:

$$2B_1 - A_1$$
$$2B_1 - A_2$$
$$2B_2 - A_1$$
$$2B_2 - A_2$$
$$B_1 + B_2 - A_1$$
$$B_1 + B_2 - A_2$$

Other third-order terms, such as $2A + B$, fall out of band or too close to band edges to use. A little arithmetic will show that all of the selected products within each of the three groups lie within 29 Hz of each other.

The receiver uses filters with 50-Hz bandwidths to select the $A + B$, $B - A$, and $2B - A$ products. The second-order distortion result is called $R2$ and is the ratio, in decibels, of the received fundamental power to the average power of the two groups of second-order terms. Besides reducing the depth of the null illustrated in Figure 1.34(a), averaging aids in reducing time variability if frequency offset is present.

The ratio of fundamental power to the power of the selected third-order products is referred to as $R3$. In the study it was found that for such pairs of tones, if the frequencies A and B are chosen such that the $2B - A$ products are fairly close together and in a portion of the frequency characteristic that has minimal envelope delay distortion, the cancellation of third-order products shown in Figure 1.34(b) can be avoided, and the addition of the components remains approximately on a voltage basis. Note that the third-order products selected all lie near 1900 Hz which is most often in the flat portion of the U-shaped envelope delay curves, the portion of the frequency band which has nearly linear phase. This choice minimizes the chances for cancellation of third-order products and the resultant measurement nulls as k varies.

Another constraint on the choice of frequencies is due to phase jitter. As described in the next section, phase jitter frequency components, as found on real channels, may fall within 300 Hz of a single transmitted frequency. Thus, the distortion components chosen for measurement must be at least that far removed from any fundamental so as to avoid confusing components of phase jitter about the fundamentals with the nonlinear distortion products.

The measurement must also be able to discriminate between distortion components and noise. This is accomplished with the narrow (50 Hz) filters and an implementation that allows a measurement of the noise in the filter pass bands to be used in calculating a correction factor applied to the measurement. The noise power in the narrow band is subtracted from the measured distortion power before the ratio is computed. Noise in the filter pass bands is measured by reading the received power with one pair of tones disabled at the transmitter and the second pair increased in power by three decibels to maintain a constant signal level.[26] The resulting reading is due to noise alone because now there are no intermodulation products falling within the filter pass bands.

The need for noise discrimination is greatest for third-order distortion measurements. For example, suppose a connection of four identical links in tandem were measured and found to have a third-order distortion-to-noise ratio of 6 dB. On each link, the third-order products would be down 12 dB from their total (as stated earlier, voltage addition applies for third-order distortion), and the noise would be only 6 dB below its total (power addition applies). The third-order distortion-to-noise ratio would be zero for each link and measurements on them would be meaningless without some correction capability.

Digital channels pose an additional problem. Quantizing noise has discrete spectral components that can beat with the distortion tones being measured and cause errors or added time variability. More about that problem in the discussion of phase jitter in the next section.

1.5.6 Phase Jitter

An adequate description of jitter phenomena is obtained by considering a generalized narrow band signal.[27] If a tone of amplitude A_0, and frequency f_0, is transmitted over a linear channel, the received signal, $s(t)$, may be described by the expression:

$$s(t) = A_0 G(f_0)[1 + m(t)]\cos[2\pi f_0 t + \phi(f_0) + b(t)] + n(t) \qquad (1.56)$$

where $G(f_0)$ and $\phi(f_0)$ are the channel amplitude and phase characteristics at f_0, and $m(t)$ and $b(t)$ represent incidental amplitude and phase modulation. (The derivative of $b(t)$ with respect to t is incidental frequency modulation.) The last term is additive, uncorrelated, noise and interference.

The phase term, $b(t)$, can be represented in the frequency domain by

$$b(t) = k_0 + b_0 2\pi f_{0s}(t) + \Sigma_i B_i \cos(2\pi f_i t + b_i), \qquad 1 < i < N, N > 0$$
$$\qquad (1.57)$$

In this expression, k_0 is the phase intercept illustrated earlier in Figure 1.9(a), f_{0s} is frequency offset, and the first two terms combined describe the quantity k referred to as phase-intercept distortion. Note that the time variability of k is described explicitly in this representation as a constant times the frequency offset. The f_i are modulation sidebands that appear about the carrier frequency, f_0, with amplitude B_i and phase b_i. The N terms are independent and not necessarily harmonically related, so the expression is not a Fourier expansion but simply a sum of sinusoids. The modulation of the phase of the carrier by each of the N terms may be represented as a harmonic sum with Bessel coefficients but that is irrelevant to the present discussion. In addition, the B_i terms need not be symmetrical about the carrier. The source of this interference is often found in single sideband carrier supplies where, due to inadequate filtering, the ac power source is found to be changing the instantaneous frequency of the carrier. The f_i are, therefore, often related to the commercial power frequency of 60 Hz and its first five harmonics (up to 300 Hz), or other low frequency power sources such as ringing generators (20 Hz). From the low frequency components observed, it appears that the mechanisms are more complex than this simple description indicates. The B_i are small, with the total usually less than 0.2 radians (see [4, p. 2098]). The net signal, $b(t)$, thus constitutes low-index phase modulation with significant sideband energy within 300 Hz of the carrier, the desired signal. Note that $G(f)$ is considered flat, and $\phi(f)$ linear within 300 Hz of the carrier. Test sets for measuring phase jitter may use filters with passbands of 600 Hz or less [6, p. 15]. Phase jitter is measured and specified in terms of the total effects of all the B_i included in the passband, as net (average) degrees peak to peak. As noted in Section 1.3.8, the phase jitter frequencies of interest are usually very small. Common bandwidths used in practical measurements include $3 < f_i < 20$ Hz.

Phase jitter measurements may be obscured by still other factors. Figure 1.36 shows representative measurements of phase jitter on a PCM channel as the test tone is swept across the voice band. The jitter bandwidth used is about 100 Hz. The sharp peaks in the measurement on the PCM channel are caused by the sampling. It is a property of the sampling process that the energy in a tone of a frequency at or near a submultiple of the sampling frequency is spread into a very large number of harmonics. For a test tone f_0, whenever the expression:

$$Nf_0 = f_{\text{sampling}}, \qquad N \text{ an integer} \tag{1.58}$$

is satisfied, these harmonics appear. In Figure 1.36, the results are apparent for $N = 8, 5, 4,$ and 3. This measured jitter is called spurious because it apparently has no effect on data modems.

Noise can also be a problem in phase jitter measurements. To date, no practical means has been found to permit test sets to distinguish between $b(t)$ of (1.56) and noise on the channel. This is taken into account in

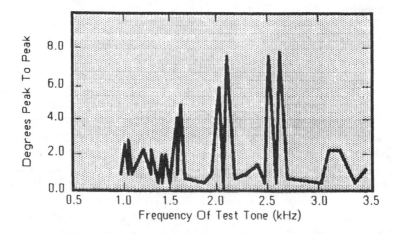

Fig. 1.36 Apparent Jitter Measured on a PCM Channel
Adapted from: Bell System Technical Reference, AT&T Pub. 41008, American Telegraph and Telephone Company, 1974

the design requirements for test sets, with a typical one being that noise 30 dB down from the carrier or test tone, must produce a peak-to-peak phase jitter indication of no more than 4° when the full bandwidth, 600 Hz, is used. The effects of white Gaussian noise on a phase jitter set are illustrated in Figure 1.37.

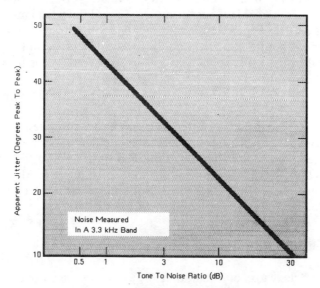

Fig. 1.37 Response of a Phase Jitter Meter To Noise
Adapted from: Bell System Technical Reference, AT&T Pub. 41008, American Telegraph and Telephone Company, 1974

1.5.7 Amplitude Jitter

The $m(t)$ term in the expression for $s(t)$, from (1.56), represents incidental amplitude modulation, and can be represented by a summation identical to the one for incidental phase modulation. In this case, double sideband AM products can be identified. No dc terms are used in current instrumentation because dc is not transmitted in a bandpass channel. (In the special case of baseband data, or some other types of transmission, over pure cable, such terms might be of interest.) Incidental AM, at frequencies less than 300 Hz, is found on channels, as reported for example in [4, p. 2100], but is usually less than 10% of the test tone amplitude in magnitude. It is of little interest currently because there have been no reported cases of problems as a result of it. This is most likely due to the fact that AM is not used in modems operating at bit rates in excess of about 600/s. Specifications for incidental AM test sets are provided, however (see [6, p. 26], for example).

1.5.8 Phase and Gain Hits

Phase and amplitude transients, or hits, and methods for evaluating their effects on systems, were described briefly in Section 1.3.4 and some of the unknowns, such as rise times, mentioned. It should be apparent that a complete description of one of these transients is difficult and its effect on transmission requires a detailed analysis of the implementation of the receiver in use. Hit measuring sets record the peak amplitude of the transients as they appear at the internal detector outputs. Time constants associated with automatic gain control (AGC) circuits and phase locked loops used in the sets are chosen to be representative of those found in data modems. The maximum counting rate (of transient occurrences) is also specified. This is chosen to be the same as for impulse noise counters, about 6.5/s, for reasons of consistency only. On real channels, transients of magnitudes of interest are even less frequent than impulse noise occurrences. Test sets record the number of occurrences greater than a selectable peak value, in a specified time interval. For phase hits, the levels are often 10- to 45° peak, in steps of 5°. Gain hit thresholds may be 2, 3, and 6, or 4, 8, and 10 dB with respect to the amplitude of a transmitted test tone.

1.5.9 Dropouts

Momentary loss of signal, or a large decrease in amplitude, has been culled out as a special type of gain hit because of the potentially severe effect on transmission. These events are called dropouts and, for discussion and measurement purposes, are defined as a decrease in amplitude in excess of 12 dB and lasting for at least four milliseconds. As with the other transients, the definition is based on observations of data modem performance in the presence of such events. A confounding factor in the design of dropout detectors is the fact that on microwave systems, rapid fades can cause momentary loss of signal power, which is replaced with noise power because of AGC action during the absence of signal. The noise power, during the signal dropout, can rise to values equal to or even greater than what the signal was. A very narrow band selection for the tone is required to detect the dropout in these cases.

NOTES

1. Although it can lead to confusion with other quantities, it is common practice to use the word "level" when referring to transmission power.
2. As mentioned in the Preface, terms that are not common to all branches of electrical engineering are defined in the Glossary. Some of these are, however, believed to be relatively well known. Occasionally such a term is used without an accompanying definition, or the reader may desire more detail. In these cases, please refer to that section.
3. The reasons for 600 and 900 Ω are not known to the author but may relate to convenient dimensions for the construction of open wire transmission lines in the early days of telegraphy.
4. See Chapter 6.
5. See [1, p. 48].
6. For some systems engineering purposes, the loss is defined as from the center of one switch to the center of the next, but measurements at the *center* of a switch are not physically possible.
7. Abbreviation for the French name, Comité Consultatif International Télégraphique et Téléphonique, known in English as the International Telegraph and Telephone Consultative Committee, the international standards organization for telecommunications.
8. Hybrids are devices used as converters between four-wire and two-wire systems. They are discussed in detail in Chapter 5.
9. The general conditions for stability involve phase considerations and allow the gain to be positive, but are not applicable in treating echo problems because phase cannot be controlled.
10. The noise source shown in Figure 1.5 is normally replaced by an oscillator for the test.

11. The time variability is discussed in some detail later in this chapter.
12. At least as good as one part in a million per day.
13. Arrangements that switch to a different RF carrier frequency or a different antenna when a fade occurs on the working channel.
14. In a voice bandwidth channel, the 12 dB provides just enough separation from the background noise to assure practically no false registrations by impulse noise detectors due to the background noise.
15. Level specification is important because nonlinear distortion is level sensitive (see Section 1.5.5).
16. A squelch circuit acts as a signal controlled switch that places a large loss in the output if no signal is present.
17. The notation ASK, FSK, *etcetera,* for shift keying instead of modulation is commonly used, but the principles are the same.
18. Codec is the name applied to a device comprising a quantizer and a means for encoding the derived sample for transmission as a digital signal, or one that accomplishes the reverse process, producing an analog signal at the output.
19. $Vs/12^{1/2}$ is the standard deviation of a uniform distribution of height Vs and equals the rms value because the distribution is symmetric about zero.
20. A quiet talker is one whose average speech is in the range of 40 to 50 dB.
21. A measurement made with a high-impedance well-balanced instrument is called a *bridging measurement* and is assumed to disturb the circuit being measured in an insignificant manner.
22. This may be done by strapping the input terminals together and then applying the voltage to the strap, providing the strap does not harm the device or its operation.
23. A standard damped integration period is not defined but is often on the order of several seconds.
24. Frequency on a frequency axis has been referred to as *quefrency*.
25. This effect can occasionally be observed during the measurement of EDD as a slow wander in the zero setting.
26. One pair of tones is transmitted to keep companders, echo suppressors, and quantizers active and so avoid hiding noise that is present during the distortion measurement.
27. Narrow band means that the full bandwidth channel amplitude and phase can be represented by the characteristics of a single sinusoid at its center.

REFERENCES

1. AT&T, *Telecommunications Transmission Engineering, Volume 1, Principles,* Second Edition, Bell System Center for Technical Education, 1977.
2. Cochran, W.T., and Lewinski, D.A., "A New Measuring Set for Message Circuit Noise," *Bell System Technical Journal,* Vol. 39, July 1960.
3. Ingle, J.F., Park, K.I., and Redman, R.C., "1983 Exchange Access Study: Analog Voice-Frequency Transmission Performance Characterization of the Exchange Access Plant," *IEEE Transactions on Communications,* Vol. COM-35, No. 1, January 1987, pp. 104–112.
4. Carey, M.B., *et al.,* "1982/83 End Office Connection Study: Analog Voice and Voiceband Data Transmission Performance Characterization of the Public Switched Network," *AT&T Bell Laboratories Technical Journal,* Vol. 63, November 1984.
5. AT&T, *Telecommunications Transmission Engineering, Volume 3, Networks and Services,* Second Edition, Bell System Center for Technical Education, 1977.
6. *IEEE Std 743-1984,* "IEEE Standard Methods and Equipment for Measuring the Transmission Characteristics of Analog Voice Frequency Circuits," Institute of Electrical and Electronic Engineers, 1984.
7. CCITT, *Red Book, Volume III—Fasicles III.1 to III.4,* Rec. G.101– G.956, ITU, Geneva, 1985.
8. Sunde, E.D., "Theoretical Considerations of Pulse Transmission," *Bell System Technical Journal,* Vol. 33, May–June 1954.
9. AT&T, "1969–70 Switched Telecommunication Network Connection Survey," *Bell System Technical Reference,* AT&T Pub., April 1971.
10. Fennick, J.H. "A Method for the Evaluation of Data Systems Subject to Large Noise Pulses," *IEEE International Convention Record,* March 1965.
11. Fennick, J.H. "Amplitude Distributions of Telephone Channel Noise and a Model for Impulse Noise," *Bell System Technical Journal,* Vol. 48, December 1969.
12. AT&T, "Transmission Parameters Affecting Voiceband Data Transmission—Description of Parameters," *Bell System Technical Reference,* AT&T Pub. 41008, July 1974.

13. Campbell, L.W., "The P/AR Meter-Characteristics of a New Voiceband Rating System," *IEEE Transactions on Communications Technology,* Vol. COM-18, No. 4 (April 1970), pp. 147–153.
14. Fennick, J.H., "The P/AR Meter-Applications in Telecommunications Systems," *IEEE Transactions on Communications Technology,* Vol. COM-18, No. 2 (February 1970), pp. 68–73.

Chapter 2
Measures of Loudness

For most people, loudness is an everyday experience and hardly warrants a definition. It is obvious when something is loud, but it is not so obvious what loud really means or how to determine how loud it actually is. A great deal of research has been devoted to the tasks of defining, quantifying, and measuring this everyday psycho-acoustic phenomenon. Much of that research was directed at understanding the principles of the concept of loudness and provided the foundation for the development of measures useful in communication engineering. Some of that former work is summarized in this chapter and two measurement techniques based on it are presented. Much of the material in the following sections is based upon the comprehensive paper by John L. Sullivan, published in 1971 ([1]).

Before delving into the details of the development of measures of loudness, a review of the elements required in any measurement scheme may be helpful. First, one needs a detection system, a mechanism that is sensitive to changes in the quantity to be measured. For loudness, a subjective phenomenon, the basic detector is a listener. A person, or groups of people are used to judge changes in one sound or equality between two different sounds. Subjective tests and facilities for conducting them are discussed in Chapter 3. Second, one needs a reference standard, *i.e.,* there are physical pieces of material that are deemed to be exactly one meter in length. For loudness of speech, standard reference acoustic spectra have been defined which, when presented at specified pressure levels, are deemed to exhibit a defined amount of loudness. The development of one such standard is described in the following sections. Third, a scale and a method for its calibration is needed. A loudness scale is defined in the next section.

These three elements of a measurement system are relatively easy to come by for the subjective phenomenon of loudness. The challenge in the work to be discussed comes about because the detectors for loudness, people, are not readily available for everyday use as meters. An objective system, meaning a means of translating acoustic pressure into loudness, is needed. This chapter describes some of the fundamental work that led to the formulation of a general equation for this translation, and two adaptations of it for engineering applications. The basic Loudness Equation, as it is called, was developed primarily by Harvey Fletcher and W. A. Munson at Bell Telephone Laboratories in the early 1930s.

The loudness calculation procedures involve the establishment of a reference system. The system is assumed to have a variable feature that permits description of a rating scale in terms of adjustments of the feature. Use of the reference system permits absolute measures of loudness; however, the computational procedures are readily extended to many engineering measures of relative loudness that are independent of any such reference.

Loudness studies make use of a number of physical measures and derived metrics that are not often encountered in other disciplines. Many that will be used in this and later chapters are defined in the next section.

2.1 DEFINITIONS

Pascal. A measure of acoustic pressure. One Pascal is a pressure of one newton per square meter.

dBt. dBt = dB relative to $2.04 \cdot 10^{-5}$ newtons per square meter or -93.81 dB (re: 1 Pascal), often taken as -94. dBt = $20 \log[p/(2.04 \cdot 10^{-5})]$; p in newtons per square meter.

The reference pressure is casually related to the threshold of hearing. It is claimed that the t in dBt refers to the first letter of the *two* in $2.04 \cdot 10^{-5}$ newtons per square meter, and so identifies the reference value. The use of dBt is becoming obsolete, being replaced by dBSPL for *decibel sound pressure level*. (Strictly, dBSPL is a general measure and requires a descriptive adjective.)

Sone. The unit sone is the loudness of a 1-kHz tone 40 dB *above the listener's threshold of hearing.* The sone is thus the bridge between the subjective phenomenon of loudness and an objective measure. It is used as a measure of loudness (see loudness below).

Phon. The loudness level of a sound in phons is numerically equal to the median level, in dBt, of a 1000-Hz tone judged, in many trials, to be equally as loud as the sound in question.

Loudness (n sones). A number n, of sones, equal to the number of times that a sound is louder than one sone (n may be a fraction). The loudness of a sound, n, is determined by subjective testing. Loudness has been described as the intensive attribute of an auditory sensation. It can be expressed, as stated, on a scale of sones, thus making it agree with the common experience of observers. However, it is ordinarily expressed as loudness level (phons) instead of loudness (sones) (see loudness level below).

Loudness level (phons). The loudness level of a sound, in phons, is numerically equal to the median value of the sound pressure, in dBt, of a 1000-Hz free progressive wave[1] judged to be equally loud by a number of listeners in many trials. The listeners are facing the sound source. In tests for standards work, it is common practice to use young people with unimpaired hearing as test subjects. Loudness (*n*), and loudness level (dBt) do not have a linear relation.

Loudness unit. The term loudness unit, apparently proposed by Sullivan, is defined in the same manner as loudness level above, but is reserved for the case in which the *sound* compared to the 1000-Hz tone is speech. This definition and usage eliminates any ambiguities or concerns over equivalent loudness of noises versus speech. Speech of *n* LUs, of course, has a loudness of *n* sones.

Loudness rating. The basic definition for loudness rating is: a ratio, expressed in dB, of the loudness of speech at the entrance to a listener's ear to the loudness of the speech as it leaves the talker's mouth. The definition may be extended, and preceded by an adjective, to include the ratio (dB) of any two signals that represent a speech waveform. The ratio need not be of acoustic pressures only. Transmit loudness rating, a ratio of acoustic pressure to a voltage in a system, is an example. Common practice sets up the ratio in such a manner that a ratio representing a decrease in loudness is a positive quantity (Convention used with loss, but loudness rating is not necessarily a loss. It may be a transducer conversion for example).

The quantities defined above are summarized in the Appendix to this chapter with a pictorial representation of their definitions and use. A reading of the Appendix now, with these definitions in mind, should help to clarify the relations among the various quantities.

Orthotelephonic. Orthotelephonic is an adjective describing the acoustical relationship between two people of equal height facing one another at a distance of one meter in a quiet nonreverbatory room. The distance is measured as that between the parallel lines drawn through the centers of the participant's ears. (Normally, a listener in this situation would be using both ears, not one as when a telephone receiver is being used. One ear of the listener may be acoustically shielded to simulate the telephone situation. The two cases are distinguished by referring to binaural and monaural listening.) This orthotelephonic situation (environment) is used as the reference condition for the terms defined next.

Orthotelephonic system. An orthotelephonic system is a communication system that simulates the orthotelephonic environment, defined explicitly through the orthotelephonic measurement and (2.1) below.

Orthotelephonic measurement. An orthotelephonic measurement is a measure of the acoustic loss between a talker's lips and a listener's ear(s). It is expressed as a ratio, in decibels, of the loss in a system to that in the orthotelephonic environment. For this purpose, the loss in the orthotelephonic environment is measured from two inches in front of the talker's lips to the center position between the listener's ears, with the listener not present. This pressure difference, or loss, is taken to be 25 dB, constant with frequency. A transmission system, including a microphone and a receiver is ordinarily involved in the system measured. In this case, the measurement must include, or take into account, the acoustical coupling between the speaker's lips and the transmitter and that between the receiver and the listener's ear.

An orthotelephonic system is defined as any system in which the subjectively measured loss is 25 dB. For example, if the net loss of a transmitting element, including the acoustic path to the element, were k_1, and that of a receiver, including the acoustic path, were k_2, and $k_1 = -k_2$ (they are perfect complements), and that of an intervening medium were 25 (dB), then the sum of these would define the loss of a perfect orthotelephonic system.

Overall orthotelephonic system (OOS).

$$OOS = k_1 - k_2 + 25 = +25 \quad \text{overall orthotelephonic system loss}$$
$$(2.1)$$

The *orthotelephonic system is the model* for the reference system mentioned at the beginning of this section. The model is, therefore, an ideal or perfect system because it simulates an acoustical environment deemed most suited to easy conversation.

Orthotelephonic response. For a general telephone system with orthotelephonic transmitter response, T_{ot}; orthotelephonic receiving response, R_{ot}; and line response, L, the overall response, OTS, is

$$OTS = T_{ot} + R_{ot} + L \quad \text{orthotelephonic response} \quad (2.2)$$

and the departure, D, of the general system from an orthotelephonic system is

$$D = OTS - OOS$$
$$= T_{ot} + R_{ot} + L - 25 \quad (2.3)$$

Masking. Let b equal the level of a pure tone that is just audible in the presence of a specified noise, and let b_0 be the just-audible level of the same tone when the noise is absent. Masking, M, is defined as

$$M = b - b_0 \quad \text{masking} \tag{2.4}$$

Masking is dependent upon frequency and noise level.

Masking Spectrum. Masking spectrum is the function defined by a plot of masking *versus* frequency of the tone.

2.2 LOUDNESS COMPUTATION CONCEPT

Author's Note: The rest of this chapter details research of a mathematical relation between subjective loudness and objective measures. The readers who do not wish to follow a complex chain of reason and experiment may skip to Chapter 3. However, those who do so will not appreciate the ingenuity of the relation or the credibility of its results.

The absence of a complete and verified theory of auditory perception, particularly for the psycho-acoustic phenomenon known as loudness, leaves room for variety in the subject of this section. A number of methods have been proposed for the computation of loudness; [2] cites eight of them, including one of the two discussed in the following sections. The methods are perhaps equally credible. Most of them are formulated to answer different specific questions, including such things as annoyance value and the loudness of telephone conversations, the subject in this case.

The concept of the method of computation of speech loudness to be discussed is diagrammed in Figure 2.1. A speech[2] source with a defined reference spectrum L_s, measured in dBt, is transmitted over a system with a frequency response R, measured in dB. A positive value for R indicates a loss or attenuation. The received spectrum is $L_s - R$ dBt (per hertz).

A junction is shown at the output of the block labeled received speech spectrum. Details of the events at this junction are illustrated in Figure 2.2(a). Curve 1 represents the speech source L_s. Curve 2 represents the threshold of hearing for continuous spectra waveforms, X. Curve 3, labeled R, is the loss-frequency characteristic of a system in decibels. The loss attenuates the speech, L_s, which has the same effect as raising the threshold to curve 4, $R + X$ dBt. The effect of the attenuation is drawn in this fashion to more clearly illustrate the derived effective spectrum as explained next.

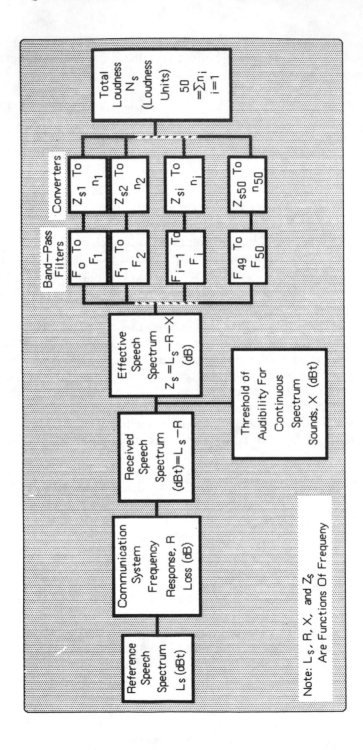

Fig. 2.1 Computation of Speech Loudness

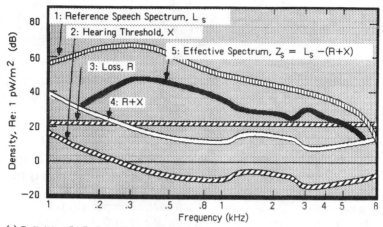

(a) Definition Of Z, Acoustic Spectral Density
Effective In The Perception Of Loudness

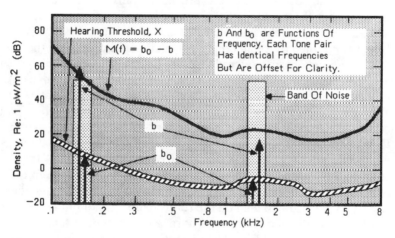

(b) Masking, Difference In Tone Thresholds,
b_0 Without Noise And b, With Noise

Fig. 2.2 Definitions of the Z Spectrum and Masking

Not all of the received (attenuated) spectrum contributes to the sensation of loudness. Sound pressure below the threshold is not effective in producing auditory nerve pulses. The portion that is effective, the difference between the received spectrum and the threshold, is called the effective speech spectrum, Z_s. It is that portion of the original spectrum that exceeds the threshold, $L_s - R - X$, or

$$Z_s = L_s - (R + X) \tag{2.5}$$

This is shown as curve 5 on Figure 2.2(a). By using the attenuation to raise the threshold instead of attenuating the original spectrum, the intersection of Z_s with X is made explicit. The total effective acoustic pressure is the area bounded by the black and the white curves, 4 and 5.

(Figure 2.2(b) has to do with the masking discussion in the next section but is a drawn here with 2.2(a) so direct comparisons may be made between the two.)

To understand the approach taken to the computation of loudness, it is helpful to have a basic understanding of the structure and function of the human ear. The part of the ear that is sensitive to sound is the basilar membrane, a relatively long narrow tissue that supports nerve endings that translate the received sounds into nerve pulses. The membrane is contained within the cochlea of the ear and divides it into an upper and lower chamber. The cochlea is filled with a fluid. Sound waves striking the ear drum are coupled by small bones, the hammer and anvil, to the stirrup which vibrates against a covered opening in the upper chamber of the cochlea. An acoustic wave then travels the length (about 30 mm) of the cochlea, passing over the upper surface of the basilar membrane. At the tip of the cochlea it passes through a small hole and returns through the lower chamber, passing along the lower surface of the basilar membrane and then exiting through a second covered window. The nerve endings of the membrane are ordered such that the frequency response of the structure is monotonic with the distance along the membrane. In the frequency domain, the wave in the cochleal fluid travels from high frequencies to low, and then from low to high on the return path. A plot of the frequency sensitivity *versus* distance on the membrane (in percent of total) is shown in Figure 2.3, reproduced from [3]. (The distance is measured in a direction from low to high frequency.) The curve should be thought of as a plot of frequency response maxima rather than indicating ideal frequency selectivity. A wave of low frequency causes some excitation over the entire length. In general, however, a sound with a large range of frequencies excites more of the membrane than does a sound containing a smaller range. It is, therefore, assumed that loudness may be dependent upon frequency as well as amplitude, and that a spectrum can be parceled such that each segment (of a uniform spectrum) would contribute equally to the sensation of loudness. These segments are referred to as *unit loudness elements*. For example, the total loudness of a sound with a flat spectrum might be subdivided into 50 frequency bands, each of which was deter-

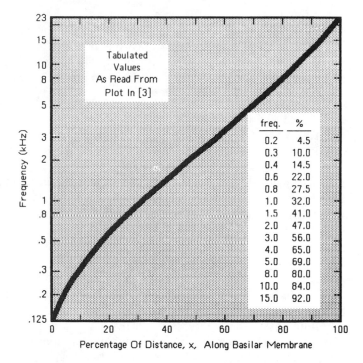

Fig. 2.3 Relation Between Frequency and Position along the Basilar Membrane
Source: F. Fletcher and W.A. Munson, "Relation Between Loudness and Masking," *J. Acoust. Soc. Am.*, Vol. 9, No. 1, July 1937. Reprinted with permission

mined to contribute 2% of the total loudness. This is the rationale behind the set of filters shown toward the right side of Figure 2.1. Note that we have said 2% of loudness. This does not necessarily equate to 2% of distance. The length of these postulated equal loudness intervals is to be determined.

The boxes labeled converters in Figure 2.1 represent mechanisms that translate the properly selected segments of the effective spectrum into elements of loudness, n_i. These are summed, as indicated at the far right of the figure, to yield the total loudness N_s.

To implement the concept, the following tasks must be completed:

1. determine a reference speech spectrum;
2. determine the threshold function, $X(f)$;
3. determine the sizes of the unit loudness spectral elements;
4. determine the converter mechanisms;
5. validate the derived procedure.

Note that the validation process, listed as the last item, is a proper part of each of the four preceding steps. Also, at each step, the question of reasonableness of the proposed task must be satisfied. For example, is there such a thing as a (meaningful) reference spectrum? After all, almost no two people sound identically alike. The question of reasonableness is not addressed directly, however, the data are presented or summarized, and the reader is invited to decide.

2.3 LOUDNESS EQUATION AND FLETCHER-MUNSON METHOD

The following development of a method for the computation of loudness is taken from that due to Harvey Fletcher and W.ʾA. Munson. It is based on research conducted by them and others in the 1930s and reported in the Journal of the Acoustical Society of America in 1937, ([3]).

The five steps needed to implement the computation were listed in a sequence according to the block diagram of Figure 2.1. The ordering of the steps is logical, once a model for the procedure is in hand. The actual development came about in a different manner, beginning with a broad statement of the question implied in step four: How are acoustic waves converted to loudness? The elements of all five steps are interwoven in this one, however, and become explicit as the development proceeds.

Recall from the description of the basilar membrane given above, how an acoustic wave travels from one end to the other and back again. It seems reasonable that the total excitation of the membrane by the wave must be a function of the distance traveled. Thus, if F is the excitation (from both sides) per unit length at a position, x, along the membrane, then the total excitation, N, is given by

$$N = \int F \, dx \qquad (2.6)$$

where the integration is along the entire length of the membrane. If the excitation, F, can be identified with loudness, (2.6) forms the basis of an expression for loudness. The desired relation between F and loudness was determined through the masking function, (2.4), $M = b - b_0$, where the b and b_0 are the just audible levels of a tone with and without noise present.

Figure 2.2(b) is an idealized representation of the masking phenomenon. The threshold for pure tones in a quiet room is illustrated by the wide dashed black and white curve of height b_0 (b_0 a function of frequency), the same curve as in Figure 2.2(a). If narrow bands of noise of constant amplitude, centered about the frequency of any tone, cover up or mask the tone, its level must be increased to a value b to be audible. This increase in level, $b - b_0$, is the masking, M. The masking function, $M(f)$, is indicated by the solid curve. Note that b_0 is negative over much of the frequency range shown so $b - b_0 > b$. It is assumed that membrane excitation by the noise is greater where masking is greater, and that whenever masking levels are equal, then the excitations must be equal. This is equivalent to assuming that equal elements of loudness are sent to the brain from all elements, dx, on the membrane, where masking is the same. If this is true, then excitation F is a function only of masking, and F of (2.6) can be replaced with $F(M)$.

Now, if the relation between distance along the membrane and frequency, shown in Figure 2.3, is called $\tilde{S}(f)$ (distance $x = \tilde{S}(f)$), this can be used in (2.6) to change the integration variable from x to f, and it can be written as

$$N = \int F(M)\tilde{S}'(f) \, dF \quad \text{loudness equation} \tag{2.7}$$

where the symbol $'$ indicates the derivative of $\tilde{S}(f)$ with respect to f. The function $\tilde{S}(f)$ need not be known. At this point it serves merely as a means of keeping track of the variable of integration. Frequency can be measured easily; measuring x is difficult.

The task is now to determine the function $F(M)$ that will map excitation into loudness at any frequency. Excitation can be measured, as acoustic pressure in dBt, for example. If the masking, M, is constant over an interval of f from f_1 to f_2, and zero elsewhere, (2.7) becomes

$$F(M) = N/[\tilde{S}(f_2) - \tilde{S}(f_1)]$$
$$= N/(x_2 - x_1) \tag{2.8}$$

and $F(M)$ can be determined by measuring the loudness of the noise causing this masking. Such a noise is not possible because, even though noise can be sharply band limited, the masking caused by it trails off, particularly at the frequencies above the band of noise. That is, the fine definition in the construction of $M(f)$ indicated in Figure 2.2(b) is not realizable, the measured $F(M)$ would be fuzzy or thick. In other words, the right-hand side of (2.8) cannot be determined by direct experimentation.

If, however, a wide band of noise, effective over practically the entire length of the basilar membrane is used, and if the resultant masking is nearly constant with the frequency of the tone, then errors due to relatively small departures of M from a constant will not be significant. Such a band of noise, yielding sufficiently constant masking over the range 100 to 10,000 Hz, was created with appropriate filters. Then masking spectra, corresponding to seven levels of the wide band of noise were

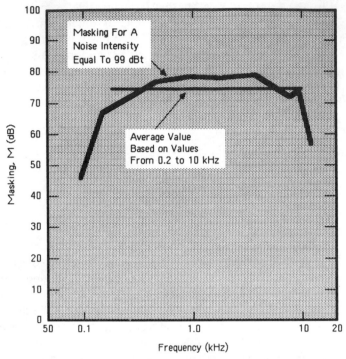

Fig. 2.4 Sample of the Masking Spectra Obtained with Wide-Band Noise
Source: F. Fletcher and W.A. Munson, "Relation Between Loudness and Masking," *J. Acoust. Soc. Am.*, Vol. 9, No. 1, July 1937. Reprinted with permission

generated. One of the masking spectra, obtained by varying the frequency of the tone, when the total noise intensity was 99 dBt is shown in Figure 2.4.[3] An average, or equivalent constant masking value for each level of the noise was obtained by averaging over the observed masking values from 200 to 10,000 Hz. The result for the 99-dBt noise intensity is indicated by the line on the figure. The results for all seven noise levels, plotted against intensity, are shown here as Figure 2.5 with the data approximated by the curve drawn in.

Loudness levels tests were made using the same noise as for the masking tests. The results are shown as the data points in Figure 2.6(a). Recall that the loudness level is the level of a 1-kHz tone that is judged to be equally as loud as the sound with which it is compared. So, while Figure 2.6(a) related intensity to loudness level, it is still required to relate loudness level, in dBt (or phons) to loudness (sones).

At the time of this work, considerable data were available relating the two measures, loudness level and loudness. From these, the curve shown in Figure 2.6(b) was constructed. This was used to convert the

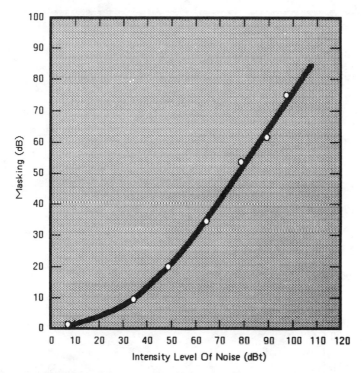

Fig. 2.5 Masking *versus* Noise Intensity
Source: F. Fletcher and W.A. Munson, "Relation Between Loudness and
Masking," *J. Acoust. Soc. Am.,* Vol. 9, No. 1, July 1937. Reprinted with
permission

loudness levels of the data points in Figure 2.6(a) to loudness numbers
(sones). The loudness numbers provided seven values of N_i, $1 \leq i \leq 7$,
one for each of the seven masking values, M_i, to use in (2.8). By successive trials, including adjusting the effective bandwidth $f_2 - f_1$, various
forms for $F(M)$ were found until a best fit was obtained to the measured
loudness values. The final form of the function $F(M)$ is shown in Figure
2.7.

2.4 MASKING CONTOURS AND THE Z FUNCTION

In repeated experiments of masking using relatively narrow bands of
noise instead of the wide band used to derive $F(M)$, Fletcher and Munson
found that, when the noise bandwidth was about 200 Hz, and the band
was centered around the frequency of the tone being masked, essentially
no increase in masking could be obtained by making it wider, provided the
same spectral density was maintained. The conclusion was that if the

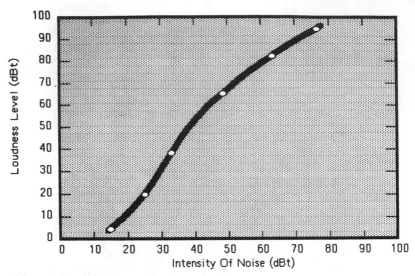

(a) Loudness Level of Noise Masking Spectra

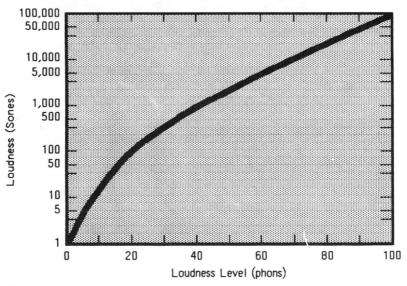

(b) Data Relating Loudness Level to Loudness

Fig. 2.6 Results of Loudness Tests with Wide-Band Thermal Noise
Source: F. Fletcher and W.A. Munson, "Relation Between Loudness and Masking," *J. Acoust. Soc. Am.*, Vol. 9, No. 1, July 1937. Reprinted with permission

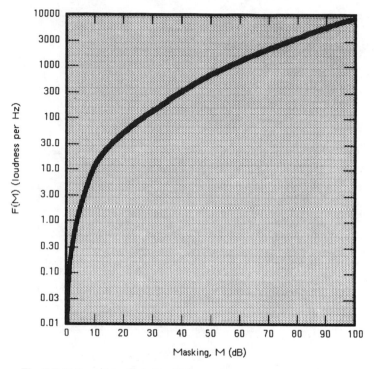

Fig. 2.7 Values of the Function $F(M)$
Source: F. Fletcher and W.A. Munson, "Relation Between Loudness and Masking," *J. Acoust. Soc. Am.*, Vol. 9, No. 1, July 1937. Reprinted with permission

spectral density of noise is known for a bandwidth of 200 Hz, its masking for a tone within the band could be predicted.[4] These results are summarized in Figure 2.8 illustrating some of the masking contours they constructed from their data. The parameter for these contours is masking, M. For example, the curve labeled 40 shows the center frequency, f, and amplitude, B, of a 200-Hz bandwidth of noise that will produce 40 dB of masking for a tone at its center frequency. If the density spectrum for a sound is known, then its masking spectrum can be constructed from these curves.

The investigators found that a quantity called Z, given by the expression:

$$Z = B - b_0 + k \tag{2.9}$$

could be used as an alternate parameter for the masking contours. In this expression, B is an ordinate value from the masking contours, a narrow

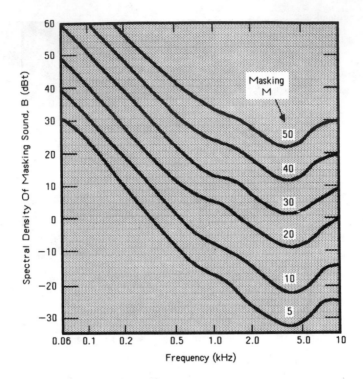

Fig. 2.8 Masking Contours
Source: F. Fletcher and W.A. Munson, "Relation Between Loudness and Masking," *J. Acoust. Soc. Am.*, Vol. 9, No. 1, July 1937. Reprinted with permission

band noise amplitude, b_0 is the hearing threshold for a pure tone in the absence of masking, and k is an empirically derived function of frequency. The quantities, Z and M, are related through the ordinate value of B. This relation between M and Z was found to be as shown in Figure 2.9.

It was further found that the empirically evaluated parameter k was approximated very closely by the slope of the function relating frequency and basilar membrane distance, shown in Figure 2.3. That is, if the function in Figure 2.3 is designated $\tilde{S}(x)$, $\tilde{S}$ = frequency, then, in decibel terms:

$$k = 10 \log[d\tilde{S}(x)/dx] \qquad (2.10)$$

$d\tilde{S}(x)/dx$ is $\tilde{S}'$ of (2.7), of course, but this detailed notation is needed below. Figure 2.10 is a plot of k as a function of frequency for the empirically derived values and as calculated by (2.10) using measured values of the slope of $\tilde{S}(x)$. With the relation of (2.10) used for k in (2.9), the expression for Z becomes

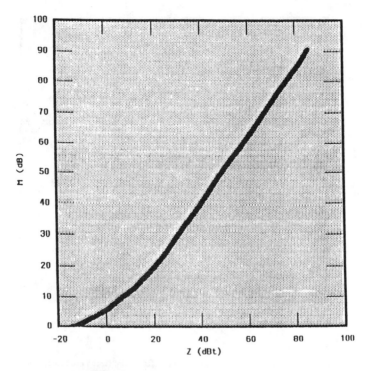

Fig. 2.9 Relation Between *M* and *Z*
Source: F. Fletcher and W.A. Munson, "Relation Between Loudness and Masking," *J. Acoust. Soc. Am.*, Vol. 9, No. 1, July 1937. Reprinted with permission

$$Z = B - b_0 + 10 \log[d\tilde{S}(x)]/dx \tag{2.11}$$

Using this relation, the parameter, Z, can be interpreted as follows: Consider a sound density spectrum I, where I is the density per hertz. The power in an elementary frequency interval is then $I \cdot df$. Note that B in (2.11) is the amplitude of a density spectrum, and that b_0 defines a threshold (spectrum). Now assume that the interaction of the nerves on the basilar membrane with the sound wave traveling along it is such that essentially all of the power within a frequency interval df at f is absorbed in the increment dx at x. With f and x related as in Figure 2.3, $df = d\tilde{S}(x)/dx$. The ratio of the power absorbed in this increment, to that necessary to achieve threshold is (B and b_0 are decibel terms)

$$10^{B/10}/[10^{b_0/10} \cdot [d\tilde{S}(x)/dx] \cdot dx] = 10^{(B-b_0)/10} \cdot [d\tilde{S}(x)/dx] \cdot dx \tag{2.12}$$

Let J_0 be the power necessary to achieve threshold, then the incremental power, dP, delivered to the small segment is

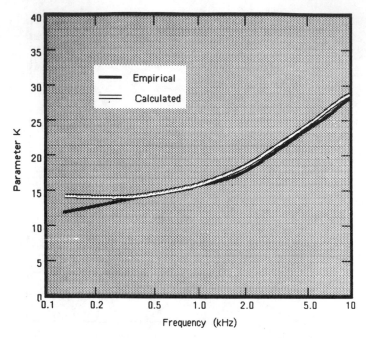

Fig. 2.10 Values of the Parameter K
Source: F. Fletcher and W.A. Munson, "Relation Between Loudness and Masking," *J. Acoust. Soc. Am.*, Vol. 9, No. 1, July 1937. Reprinted with permission

$$dP = J_0 10^{(B-b_0)/10} \cdot [d\tilde{S}(x)/dx] \cdot dx$$

or

$$10 \log(dP/dx) = B - b_0 + 10 \log[d\tilde{S}(x)/dx] + 10 \log J_0 \qquad (2.13)$$

Comparing the right-hand sides of (2.13) and (2.11) shows that they are identical if J_0 is taken as unity. It is seen that Z is the power per unit length, delivered to the basilar membrane, expressed in decibels above that necessary to achieve threshold. In other words, Z is the *effective* spectrum of a sound.

The loudness relation given above as (2.7) can, therefore, be rewritten as

$$N = \int Q(Z)\tilde{S}'(f) \, df \quad \text{loudness equation in } Z \qquad (2.14)$$

The range of integration would generally be the entire audible spectrum, but other frequency bands might be selected, *i.e.*, that of a communication channel.

Values of $Q(Z)$ can be obtained from the curves of Figures 2.7 and 2.9 by choosing values for Z and finding the loudness through the parameter M. A plot of $Q(Z)$ *versus* Z is shown in Figure 2.11. $Q(Z)$ has the same units as $F(M)$, loudness per hertz, but is based on the effective spectrum of the sound, *i.e.*, that portion that lies above the threshold instead of the total. The function, $Q(Z)$, is fundamental to the two most common methods in use today for calculating the loudness of speech for telecommunication systems. The development of these equations is roughly parallel to that of Fletcher and Munson as described in this section.

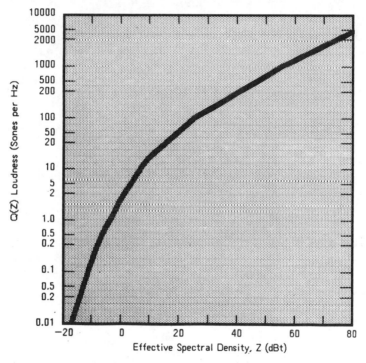

Fig. 2.11 The $Q(Z)$ Function
Source: F. Fletcher and W.A. Munson, "Relation Between Loudness and Masking," *J. Acoust. Soc. Am.*, Vol. 9, No. 1, July 1937. Reprinted with permission

2.5 THE EARS OR IEEE LOUDNESS EQUATION

The first engineering adaptation of the loudness equation to be described is that known as the *IEEE Loudness Equation* because it is published by that organization as a standard ([5]). Examples of its application appear in another IEEE standard ([6]). The equation was developed at

Bell Telephone Laboratories in 1941 as a refinement of the Fletcher-Munson results discussed in the preceding sections. It was not reported, at least in detail, until the appearance of the paper by John L. Sullivan in 1970 ([1]). The purpose of the paper was to present the development of an equation for application in an automatic loudness measuring facility called the *electro acoustical rating system* (EARS). The equation, known as the EARS equation, was adopted, with minor changes by the IEEE. The following sections trace the development from the loudness equation (2.14) to the computational form of the IEEE equation.

In the expression for Z in (2.9), $Z = B - b_0 + k$, B is the noise density, b_0 the threshold of hearing for pure tones, and Z, as shown in the previous section, is the portion of the noise density that is effective in exciting the basilar membrane. (Remember that all these quantities are functions of frequency.) If (2.9) is rewritten as

$$Z = B - (b_0 - k) \tag{2.15}$$

then the quantity, $b_0 - k$, may be considered as the threshold of hearing for continuous spectrum sounds. That is, the threshold for continuous spectrum sounds is k dB less than b_0 (for pure tones). Let $b_0 - k = X$; then the effective portion of a continuous spectrum sound is given by

$$Z = B - X \quad \text{effective level} \tag{2.16}$$

where B is the density of the sound and X is the effective threshold for hearing. B and Z are both expressed in units such as dBt and may be referred to as pressure spectra or levels (as functions of frequency) as in the name given to (2.16). The function X may be computed by subtracting k (Figure 2.10) from the threshold for pure tones and is shown in Figure 2.12, reproduced from [1].[5]

If B in (2.16) is uniform with frequency, (2.14) can be written as

$$N = \int Q(B - X)\tilde{S}'(f)\, df \tag{2.17}$$

which explicitly shows that loudness is a function of the area bounded below by the curve of Figure 2.12 and above by a straight line, B, but weighted by $S'(f)$.

Now consider a spectral density, X', such that $X' - X$ is a constant, C. X' could be called a *flat effective spectrum* because Z would then be constant and equal to C. If X' were substituted for B in (2.17) and N were plotted as a function of frequency, the result would be the cumulative contribution to loudness for a flat effective spectrum. Note that in such a case, loudness as given by (2.17), is simply a constant, $Q(C)$, times the

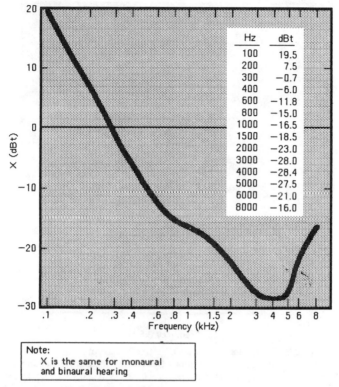

Hz	dBt
100	19.5
200	7.5
300	−0.7
400	−6.0
600	−11.8
800	−15.0
1000	−16.5
1500	−18.5
2000	−23.0
3000	−28.0
4000	−28.4
5000	−27.5
6000	−21.0
8000	−16.0

Note:
X is the same for monaural
and binaural hearing

Fig. 2.12 X, The Threshold of Audibility for Continuous Spectrum Sounds
Source: J.L. Sullivan, "A Laboratory System for Measuring Loudness of Telephone Connections," *Bell System Tech. J.*, Vol. 50, No. 8, October 1971. Reprinted with permission from the Bell System Technical Journal. Copyright 1971 AT&T

integral of $\tilde{S}'(f)$. An example of such a distribution of loudness contribution, normalized by division by total loudness, N, and expressed in percent, is shown in Figure 2.13. The curve shown was actually derived in a rather different fashion, to be discussed. The significance of this curve is that the slope at any point represents the incremental contribution to loudness due to an element of the spectrum, df, at f. There is nothing new here except that the function $\tilde{S}'(f)$ takes on more meaning. $\tilde{S}'(f)$ is the relative contribution to loudness due to any frequency component, f, contained in a sound spectrum. It is the frequency weighting function for loudness. In other words, loudness is proportional to the derivative, with respect to distance, x, of the function relating frequency to distance along the basilar membrane; the slope of the curve in Figure 2.3. From the relative steepness, it is observed immediately, that the frequencies below about 1 kHz are most important. This is also apparent in Figure 2.13,

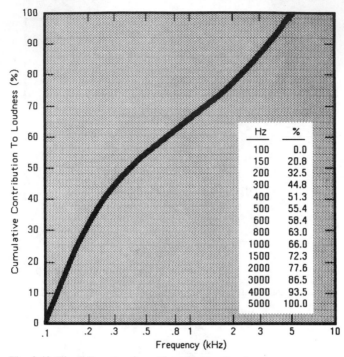

Hz	%
100	0.0
150	20.8
200	32.5
300	44.8
400	51.3
500	55.4
600	58.4
800	63.0
1000	66.0
1500	72.3
2000	77.6
3000	86.5
4000	93.5
5000	100.0

Fig. 2.13 The S Function for a Flat Effective Spectrum
Source: J.L. Sullivan, "A Laboratory System for Measuring Loudness of Telephone Connections," *Bell System Tech. J.*, Vol. 50, No. 8, October 1971. Reprinted with permission from the Bell System Technical Journal. Copyright 1971 AT&T

particularly for the range of frequencies used in most communication systems, below about 3 kHz.

We can now consider task 3 of Section 2.2, determining unit loudness elements. For a discrete computation, $\tilde{S}(f)$ might be broken into increments of equal weight to determine frequency bands of equal or unit contribution. If this is done, the results could be accumulated by these elements of unit weight and plotted as a function of frequency. It would be a step function approximation to the curve in Figure 2.13.

For computational purposes then, the loudness equation (2.14), can be rewritten as

$$N = \sum q_s \Delta \tilde{S} \quad \text{computational loudness equation} \tag{2.18}$$

where the $\Delta \tilde{S}$ are the unit loudness elements of bandwidth just mentioned, and the q_s are the portions of a sound spectral density that fall within those limits. The summation is taken over the bandwidth of the sound. Note that each q_s is equal to the integral of $Q(Z)$, as in (2.14), over its corresponding unit loudness bandwidth.

Equation (2.18) is general with regard to the spectrum used to find the q_s, and, because it was derived from tests using thermal noise, its validity for speech may be suspect. This question was part of the impetus for the study culminating in 1941. The availability of significantly more data accumulated in the late 1930s made the study feasible.

For continuity, the reader is referred back to Figure 2.1, the loudness computation concept presented in Section 2.2, and the five steps needed for its realization:

1. determine a reference speech spectrum;
2. determine the threshold function, $X(f)$;
3. determine the sizes of the unit loudness spectral elements;
4. determine the converter mechanisms;
5. validate the derived procedure.

One method of determining the threshold function, $X(f)$, was described above. The unit spectral elements were found to be the unit intervals associated with the $\Delta \tilde{S}$ of (2.18), and the converter mechanisms, the $\Delta \tilde{S}$ themselves. Item 5, validation, was as mentioned, a motivation for the work completed in 1941 and to be described below. So far nothing has been said about the reference speech spectrum, item 1, and that will be addressed next.

2.5.1 Reference Speech Spectrum

To derive a shape for a reference speech spectrum, L_s of Figure 2.1, sample spectra were gathered by long-term-average power measurements of continuous speech. (The averaging included pauses between words.) Samples were obtained from 13 males and 12 females. Some differences in the spectra were noticed for individual voices but the data were sufficiently similar to justify smoothing and averaging.

For the level of the reference spectrum, data from five sources reporting average speech pressure level were available. The levels reported in four of these range from 88.0 to 92.6 dBt when converted to equivalent values at a distance of two inches from a talker's mouth. The fifth source reported an average pressure of 81.7 dBt for males and 79.7 for females. Reference [1] summarizes the data and discusses the relatively large difference between the first set of four and the fifth.[6] The reference pressure is taken as 90 dBt, a rough average of values reported within the group of four. The resulting reference spectrum is illustrated in Figure 2.14. It is called the B_{90} *spectrum* because the integral over the band is equal to the desired 90 dBt total pressure.

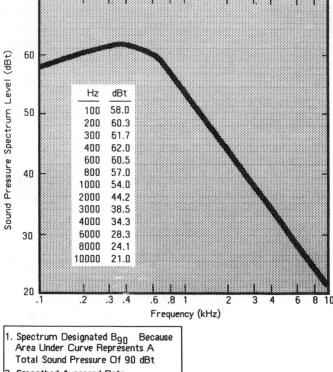

Hz	dBt
100	58.0
200	60.3
300	61.7
400	62.0
600	60.5
800	57.0
1000	54.0
2000	44.2
3000	38.5
4000	34.3
6000	28.3
8000	24.1
10000	21.0

1. Spectrum Designated B_{90} Because
   Area Under Curve Represents A
   Total Sound Pressure Of 90 dBt
2. Smoothed Averaged Data
   For 13 Males & 12 Females

Fig. 2.14 Sound Pressure of Continuous Speech Two Inches from the Lips of
a Talker
Source: **J.L. Sullivan, "A Laboratory System for Measuring Loudness of
Telephone Connections,"** *Bell System Tech. J.,* **Vol. 50, No. 8, October
1971. Reprinted with permission from the Bell System Technical Journal.
Copyright 1971 AT&T**

2.5.2 The S Function

The $\tilde{S}$ function as described in Section 2.2 and shown in Figure 2.3 is
one reported by J. C. Steinberg in 1937 ([7]). It is based upon an analysis
with pure tones. The desired S function, not related to pure tones, was
derived as follows.

Data from experiments by two researchers, Steinberg in 1924, and
Van Wynen in 1940, showed loudness effects due to filtering real speech.
In these experiments, undistorted speech was loudness balanced against
speech filtered by low-pass, high-pass, and bandpass filters. That is, the
level of the nonfiltered speech was varied until it was judged to be equally
loud to the filtered speech. Tests were made for many values of the cut-off

frequencies. The amount by which the undistorted speech had to be reduced in level, to be judged equally loud, is called *the loudness loss of the filter*. Some of the results for low-pass and high-pass filters are shown in Figure 2.15. Each point on the graph is the intersection of the loudness loss (ordinate) with the cut-off frequency (abscissa). For example, the lowest white square is a data point from the Steinberg high-pass filter data. When the filter cut-off was 1.5 kHz, 16.6 dB of loss had to be introduced in the undistorted speech path for equal loudness. Therefore, a high-pass filter with a cut-off of 1.5 kHz is said to have a loudness loss of 16.6 dB.

If the filters are good enough to be considered ideal, then the intersections of the curves for a high-pass filter and a low-pass filter for the same speech yields two important quantities:

- the frequency above and below which the contributions to loudness are equal, f_{50};
- the number of decibels by which undistorted speech must be reduced for a 50% reduction in loudness, LL_{50}.

These frequencies and levels are noted on the figure for each set of tests.

If $Z_{s2} - Z_{s1}$ is the difference in decibels before (1) and after (2) a flat change in undistorted speech, and one assumes a logarithmic relation between loudness and effective level, then

$$Z_{s2} - Z_{s1} = k \cdot \log(N_2/N_1) \tag{2.19}$$

where N_2 and N_1 are the corresponding loudness numerics and k is a constant. For a 50% reduction, $N_2/N_1 = 2$, and using the LL_{50} values from the figure, we find that $k = 41.92$ for the Van Wynen tests and 30.2 for the Steinberg tests. Substituting these values for k in (2.19) and applying it to the observed loudness loss values at various filter cut-off frequencies for each set of data, the contribution to loudness for the nonfiltered portions of the spectrum can be computed.

For example, the Van Wynen tests show that for a high-pass filter with cut-off of 550 Hz, $Z_{s2} - Z_{s1} = -10.3$ (loudness loss = +10.3 dB). With $k = 41.2$, N_2/N_1 is found to be 0.562. The filter has passed 56.2% of the loudness and suppressed 43.8%.

When this calculation is done for each of the data points shown in Figure 2.15, the results can be plotted as shown in Figure 2.16.[7] It is interesting to note that the points from the two sets of tests group rather closely around the curve drawn in as an approximate fit, in spite of the relatively large difference between the two k values, 30.2 and 41.2. This curve, showing the cumulative contribution to loudness as a function of

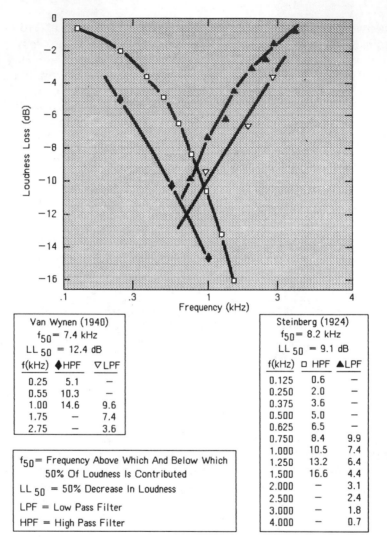

Van Wynen (1940)		
f_{50}= 7.4 kHz		
LL $_{50}$ = 12.4 dB		
f(kHz)	◆HPF	▽LPF
0.25	5.1	—
0.55	10.3	—
1.00	14.6	9.6
1.75	—	7.4
2.75	—	3.6

Steinberg (1924)		
f_{50}= 8.2 kHz		
LL $_{50}$ = 9.1 dB		
f(kHz)	□ HPF	▲LPF
0.125	0.6	—
0.250	2.0	—
0.375	3.6	—
0.500	5.0	—
0.625	6.5	—
0.750	8.4	9.9
1.000	10.5	7.4
1.250	13.2	6.4
1.500	16.6	4.4
2.000	—	3.1
2.500	—	2.4
3.000	—	1.8
4.000	—	0.7

f_{50}= Frequency Above Which And Below Which
50% Of Loudness Is Contributed

LL $_{50}$ = 50% Decrease In Loudness

LPF = Low Pass Filter

HPF = High Pass Filter

Fig. 2.15 Filter Test Results
Source: J.L. Sullivan, "A Laboratory System for Measuring Loudness of Telephone Connections," *Bell System Tech. J.*, Vol. 50, No. 8, October 1971. Reprinted with permission from the Bell System Technical Journal. Copyright 1971 AT&T

frequency, for undistorted speech, may be thought of as an *S* function for speech.

The cumulative loudness curve for real speech shown in Figure 2.16, must now be translated to an equivalent one for a *flat* speech spectrum in order to find the desired *S* function.

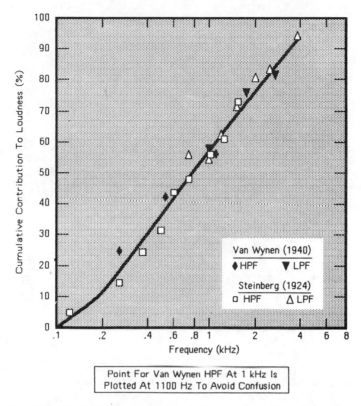

Fig. 2.16 Cumulative Contribution to Loudness of Speech for the Steinberg and Van Wynen Filter Tests
Source: J.L. Sullivan, "A Laboratory System for Measuring Loudness of Telephone Connections," *Bell System Tech. J.*, Vol. 50, No. 8, October 1971. Reprinted with permission from the Bell System Technical Journal. Copyright 1971 AT&T

The method used to derive the S function involves deriving the q_s function of (2.18) simultaneously. The process is iterative. First, estimates for the functions in (2.18) are obtained as described below; then a total loudness is computed and compared to actual data. The functions are modified accordingly, and the process repeated until all the data and the derived functions are sufficiently compatible.

2.5.3 The Weighting Function for Speech, q_s [8]

Data from loudness balance tests of speech against a 1000-Hz tone obtained by Munson and reported in [1] are shown in Figure 2.17. It gives

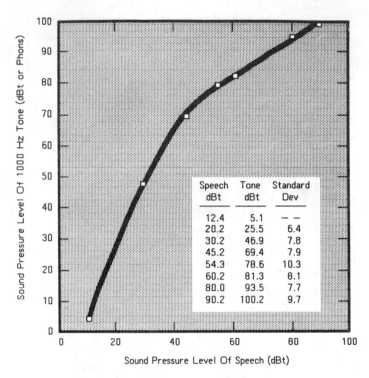

Speech dBt	Tone dBt	Standard Dev
12.4	5.1	– –
20.2	25.5	6.4
30.2	46.9	7.8
45.2	69.4	7.9
54.3	78.6	10.3
60.2	81.3	8.1
80.0	93.5	7.7
90.2	100.2	9.7

Fig. 2.17 Loudness Balance Test Results for a One-Kilohertz Tone and Speech (from Munson Tests, 1936)
Source: J.L. Sullivan, "A Laboratory System for Measuring Loudness of Telephone Connections," *Bell System Tech. J.*, Vol. 50, No. 8, October 1971. Reprinted with permission from the Bell System Technical Journal. Copyright 1971 AT&T

the loudness level of the 1000-Hz tone in dBt (phons) plotted against the measured speech pressure level in dBt. The points shown are averages of results from eleven subjects with about five trials for each. These data could be used as a check on the results of the loudness computation as the iterations proceeded. First, however, it was necessary to convert the loudness level data to loudness.

The relation between loudness level in phons and loudness in sones is shown in Figure 2.18 (Figure 6 of [1]). Two standards exist today for this conversion. The figure shows both, with the explanation that the line was derived to fulfill a need for a simple form. The dark curve is considered to be more realistic and is the one used here. As noted on the figure, this curve was derived by Munson in 1933 and adopted as a standard in 1942. The results of the conversion are shown in Figure 2.19. This is the curve to be matched by the use of (2.18) for speech:

$$N_s = \sum q_s \Delta S$$

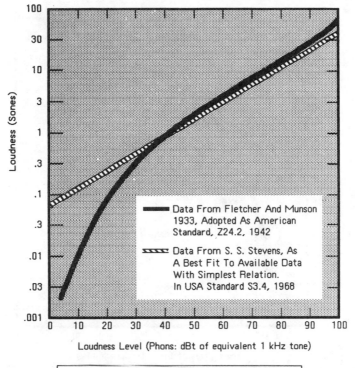

Fig. 2.18 Relation Between Loudness and Loudness Level
Source: J.L. Sullivan, "A Laboratory System for Measuring Loudness of Telephone Connections," *Bell System Tech. J.*, Vol. 50, No. 8, October 1971. Reprinted with permission from the Bell System Technical Journal. Copyright 1971 AT&T

with weighting functions q_s to be determined.

From the work with the flat spectra and masking described in the earlier sections, it was known that effective spectra, Z_s, were required. Recall that

$$Z_s = B_s - X$$

where the X is for continuous spectra sounds.

As a part of the Munson balance tests, Figure 2.17, the spectra of the subjects voices were measured and averaged. Similar data were available from the Van Wynen and the Steinberg filter test. These were used as source spectra, B_s.

The X function for continuous spectra sounds, needed for the conversion of the average measured spectra to effective spectra was shown in

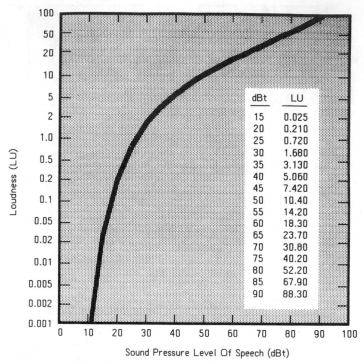

dBt	LU
15	0.025
20	0.210
25	0.720
30	1.680
35	3.130
40	5.060
45	7.420
50	10.40
55	14.20
60	18.30
65	23.70
70	30.80
75	40.20
80	52.20
85	67.90
90	88.30

Sound Pressure Level Of Speech (dBt)

Fig. 2.19 Effective Spectra
Source: J.L. Sullivan, "A Laboratory System for Measuring Loudness of Telephone Connections," *Bell System Tech. J.,* Vol. 50, No. 8, October 1971. Reprinted with permission from the Bell System Technical Journal. Copyright 1971 AT&T

Figure 2.12. It is redrawn here as Figure 2.20. For perspective, a few points from the threshold for single tones have been added.

Three average effective spectra, Z_s, were derived by applying the curve of Figure 2.20 to the measured spectra according to the relation above. Two of these are shown in Figure 2.21. Data from the Van Wynen tests, which would roughly fall between those for the Munson and Steinberg tests, are not shown. The solid black curve, effective *reference* spectrum, was constructed by subtracting the threshold function from the B_{90} reference spectrum of Figure 2.14. It is included to show the difference between the effective spectral density for a system providing orthotelephonic response (Figure 2.1 with $R = 0$) and those that were found in the various subjective tests.

The determination of q_s and ΔS begins by estimating M equal percentile loudness intervals from the curve of Figure 2.16, $\Delta S = f_i - f_{i-1}$. These intervals are marked off on the spectra and the values of Z_s at the midpoints assigned as constants over the intervals, thus building step function approximations to the spectra. After some number of trials, it is

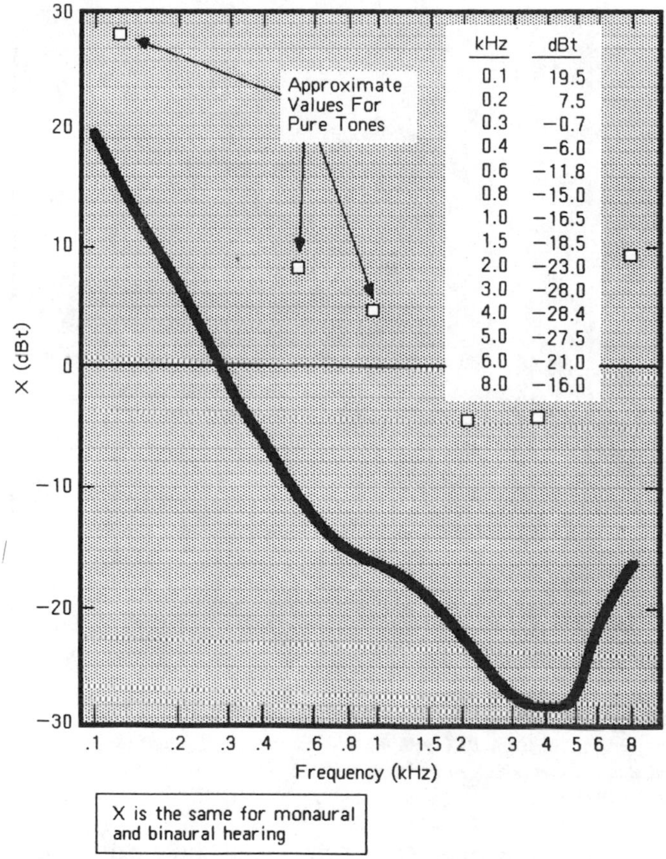
30
20
10
0
−10
−20
−30
X (dBt)
Approximate
Values For
Pure Tones
kHz dBt
0.1 19.5
0.2 7.5
0.3 −0.7
0.4 −6.0
0.6 −11.8
0.8 −15.0
1.0 −16.5
1.5 −18.5
2.0 −23.0
3.0 −28.0
4.0 −28.4
5.0 −27.5
6.0 −21.0
8.0 −16.0
.1 .2 .3 .4 .6 .8 1 1.5 2 3 4 5 6 8
Frequency (kHz)
X is the same for monaural
and binaural hearing

Fig. 2.20 The Threshold of Audibility for Continuous Spectrum Sounds, $X(f)$
Source: **J.L. Sullivan, "A Laboratory System for Measuring Loudness of Telephone Connections,"** ***Bell System Tech. J.*****, Vol. 50, No. 8, October 1971. Reprinted with permission from the Bell System Technical Journal. Copyright 1971 AT&T**

decided that an M of 50, providing 50 intervals each of width two percent of loudness, provides reasonable step representations of the spectra. These two-percent intervals are used throughout the remaining calculations. The S function derived is actually the one presented earlier in Figure 2.13.

Having determined the ΔS, the loudness level weights, q_s, must be found. To do this, the effective speech spectra, Z_s, are adjusted in level to correspond to each value of total pressure in Figure 2.17. (The area under each corresponding B_s would equal to the total pressure.) The areas under the effective spectra are segmented into the 50 intervals. For the first iteration, estimates of q_s are made. The sum of the resulting rectangles,

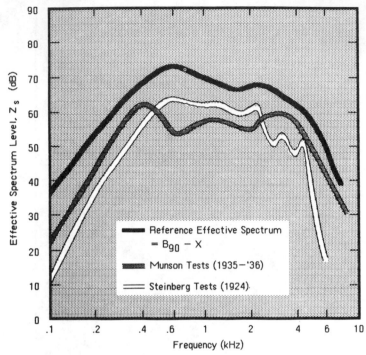

Fig. 2.21 Examples of Effective Spectra
Source: Adapted from J.L. Sullivan, "A Laboratory System for Measuring Loudness of Telephone Connections," *Bell System Tech. J.*, Vol. 50, No. 8, October 1971

each multiplied by the estimate for each level, should equal the measured loudness. The calculations, compared with the curve of Figure 2.19, lead to revised estimates. The iteration process continues until a good fit to the curve is obtained.

The derived function relating loudness per two-percent interval to a flat equivalent density, Z_s = a constant within the interval, is shown in Figure 2.22. Notice that there is some loudness indicated even when the effective spectrum takes on small negative values. French and Steinberg, in [4], attribute this to power addition of the acoustic spectrum with internal ear noise bringing the total into the audible range.

A measure of the accuracy of (2.18) with these ΔS and the q_s is reported in [1]. Six sets of data providing subjective loudness estimates of speech samples, and their average spectral compositions, taken over a period of many years were available. Some of these tests were balance tests against a 1000-Hz tone, others were comparisons of speech common to two channels with different loss-frequency characteristics. Calculated loudness values were compared with the reference reported values. The

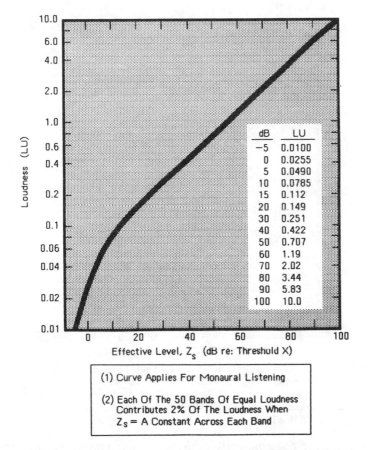

The data table within the figure:

dB	LU
−5	0.0100
0	0.0255
5	0.0490
10	0.0785
15	0.112
20	0.149
30	0.251
40	0.422
50	0.707
60	1.19
70	2.02
80	3.44
90	5.83
100	10.0

Loudness (LU)

Effective Level, Z_s (dB re: Threshold X)

(1) Curve Applies For Monaural Listening

(2) Each Of The 50 Bands Of Equal Loudness Contributes 2% Of The Loudness When Z_s = A Constant Across Each Band

Fig. 2.22 $Q(Z)$, Two-Percent Loudness Bands (q_s in LU) *versus* Effective Level
Source: J.L. Sullivan, "A Laboratory System for Measuring Loudness of Telephone Connections," *Bell System Tech. J.*, Vol. 50, No. 8, October 1971. Reprinted with permission from the Bell System Technical Journal. Copyright 1971 AT&T

results of the comparison are shown in Figure 2.23. The reported loudness values are plotted along the abscissa, labeled observed. The values calculated from (2.18) are plotted along the ordinate, labeled computed. The average difference, computed value minus observed value, and the standard deviation of the differences for each set of tests is also shown.

In [1], Sullivan provides details of all the calculations, and briefly describes the test situations for each of the data bases. The first two sets of data, Steinberg 1924, and Van Wynen 1940, include those that were used to construct the filter effect curves of Figure 2.15, for the derivation of the S function. However, the Van Wynen tests included seven filter

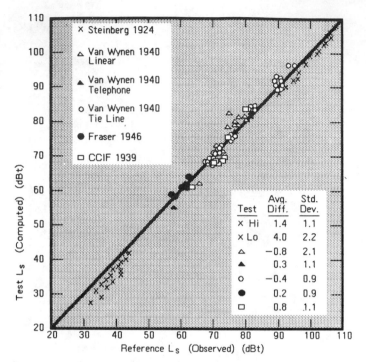

Fig. 2.23 Comparison of Observed and Computed Speech Loudness
Source: J.L. Sullivan, "A Laboratory System for Measuring Loudness of Telephone Connections," *Bell System Tech. J.,* Vol. 50, No. 8, October 1971. Reprinted with permission from the Bell System Technical Journal. Copyright 1971 AT&T

shapes that were not appropriate for the earlier purposes, i.e., resonant circuits, but are included in the comparison. In these tests, the subjects listened to the undistorted speech on one channel and the filtered speech on another. They adjusted an attenuator in the undistorted channel until the two were judged to be equally loud. The attenuator setting is taken as the loudness loss.

Received levels in the Steinberg tests were such that the sensation levels, at reference, were 107.1 dBt, upper cluster of points marked with an x in Figure 2.23, and 42.6 dBt, lower cluster of x points. The agreement between observed and computed values is better at the high sensation levels. In general, the plotted points lying furthest from the curve correspond to higher distortion values or greater filtering of the speech. These represent responses so distorted that they should not occur in practice. Also, these extremes of sensation level are not apt to be found in a well-designed communication system.

The Van Wynen linear system tests, identified by the open triangles in Figure 2.23, were performed with levels closer to those normally encountered in real systems. While the computation method appears more accurate in this case (the average error is -0.8 dB), the precision is about the same as with the Steinberg low-level data, the standard deviations are equal to 2.1 and 2.2 dB. These data were also taken on heavily distorted (filtered) speech. The differences between the measured and calculated values for the other four sets of data are quite small.

The use of (2.18) with the derived $S(f)$ and q_s functions to compute the loudness of speech is probably evident by this point. The speech spectrum is modified by subtracting the threshold of hearing curve, Figure 2.20, from it to obtain the effective pressure, Z_s. It is then divided into the 50 intervals of equal loudness derived from Figure 2.13. The midinterval values of Z_s are converted to q_s values according to the curve of Figure 2.22. These are summed to yield the total loudness, N_s. Reference [1] includes some special forms and other aids that simplify the mechanics of the calculation, as well as details of instrumentation to measure quantities such as the loudness loss of devices (R of Figure 2.1) directly.

Loudness levels for the reference B_{90} spectrum, with effective spectrum illustrated by the solid black curve of Figure 2.21 was computed for several values of total pressure. The result is shown in Figure 2.24. (The value 6.06 in the data table, marked with an asterisk, does not appear to fit the curve and is likely a typographical error because 6.60 does fall in place.) This curve is similar in shape to that of Figure 2.19, but note the difference in the range of loudness values on the abscissae: five orders of magnitude in 2.19, and only three in 2.24. The difference lies in the measure of level used as the ordinate. For Figure 2.19 is it absolute pressure in dBt, while for 2.24 it is loudness level in phons. So, even though phons are numerically equal to the dBt of a 1000-Hz tone, this comparison emphasizes that because the phon value is assigned on the basis of equivalent loudness, a phon is a very different measure than is dBt.

Finally, the desired S function, cumulative loudness as a function of frequency for the B_{90} spectrum, is found as follows: If the summation, (2.18), using the q_s derived from the reference (B_{90}) spectrum, is plotted as a function of frequency and normalized by dividing by total loudness, N_s, the result appears as the circles in Figure 2.25. They show the cumulative contribution to loudness as a function of frequency, in percent. This plot differs in shape from that in Figure 2.13 because the latter was derived for a flat effective spectrum, Z_s = a constant, whereas for the present one, $Z_s = B_{90} - X$. The significant difference is that the plotted points can be approximated by the line from 100 to 5000 Hz on the logarithmic frequency scale. The curve in Figure 2.13 is very nonlinear by comparison.

With the function represented by the circles on Figure 2.25 approximated by the line, the equation for the relation is

$$S(f) = C \cdot \log(f), \quad C \text{ a constant} \tag{2.20}$$

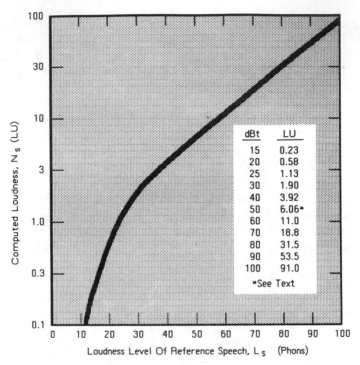

dBt	LU
15	0.23
20	0.58
25	1.13
30	1.90
40	3.92
50	6.06*
60	11.0
70	18.8
80	31.5
90	53.5
100	91.0

*See Text

Fig. 2.24 Loudness *versus* Loudness Level for Reference Speech
Source: J.L. Sullivan, "A Laboratory System for Measuring Loudness of Telephone Connections," *Bell System Tech. J.*, Vol. 50, No. 8, October 1971. Reprinted with permission from the Bell System Technical Journal. Copyright 1971 AT&T

The frequency loudness weighting is simply the logarithm of frequency. For such a line, the 50 equal loudness intervals (each of two percent) will be of equal linear length on the ordinate. For example, if $LU(f)$ is the loudness contribution up to frequency f, and LU_t the total loudness, the equation for the line, percent at f, is $P(f) = 100 \cdot LU(f)/LU_t = a \cdot \log(f) + b$ and for any two frequencies, f_1 and f_2, $f_2 > f_1$:

$$P(f_2) - P(f_1) = a \cdot \log(f_2) - a \cdot \log(f_1)$$
$$= a \cdot \log(f_2/f_1)$$

Thus, for any constant ratio, f_2/f_1, the contribution to loudness will be the same.

To reiterate, the linear curve resulting from the Z_s spectrum due to the B_{90}-reference speech, $Z_s = B_{90} - X$, shows that, in terms of this spectrum, equal log(frequency) intervals are equally important to total loudness. That is, constant increments of q_s, on the ordinate, determine

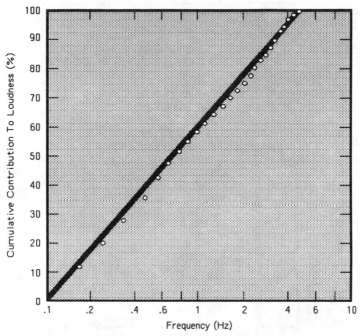

Fig. 2.25 Cumulative Contribution to Loudness for the Reference (B_{90})
Spectrum, $S(f)$

Source: J.L. Sullivan, "A Laboratory System for Measuring Loudness of
Telephone Connections," *Bell System Tech. J.*, Vol. 50, No. 8, October
1971. Reprinted with permission from the Bell System Technical Journal.
Copyright 1971 AT&T

frequency bands of varying sizes in frequency, but constant in log(frequency), that are equally important. Conversely, equal log(frequency)
increments can be assigned equal weights. For such intervals, q_s of (2.18)
will be a constant, independent of frequency, and determined only by the
spectrum amplitude.

2.5.4 The EARS Equation

Derivation of the EARS equation for loudness follows the principles
of the procedure for calculating loudness by (2.18), but begins with a
result of that function and works backwards. In accordance with the
closing paragraphs of the last section, Figure 2.1 may now be redrawn as
in Figure 2.26. In this figure, V_0, a voltage in units of dBV, has a flat
spectrum whether the fundamental source is electrical or acoustical. The
signal passes through an arbitrary linear system, R, and the voltage is
weighted in the frequency domain by $\log(f)$. If the amplitude of the

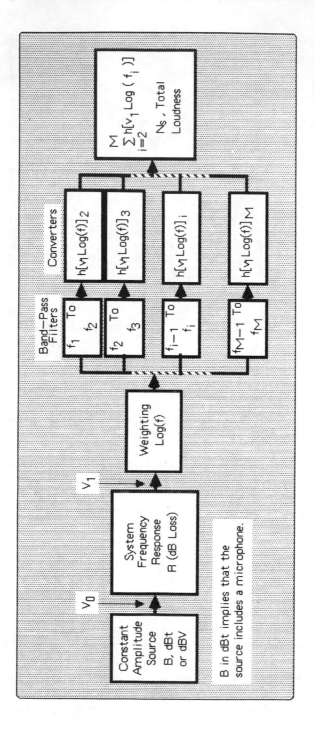

Fig. 2.26 Computation of Speech Loudness with Flat Spectrum Source

source is A and the gain of the system R is $r(f)$, then the voltage V_1 is $A \cdot r(f)$. The weighting, $\log(f)$ may have some proportionality constant, k_1, but this can be combined with the amplitude of V_0, i.e., $A \cdot k_1 = k_2$. The voltage out of the weighting block is then $k_2 r \cdot \log(f)$.

The filters on the right side of Figure 2.26 are selected to have equal $\log$(frequency) passbands. That is, $\log(f_{i+1}/f_i) = $ a constant.

All that is required now is to perform the conversions at the output of the filters and sum them as indicated. The relation $h(\cdot)$ is needed to change the amplitude of the output of each filter to a contribution to loudness. That is, the product of q_s by ΔS must be taken before the summation is performed. The values of q_s, except for a constant, can be obtained from the curve of Figure 2.22, loudness *versus* effective level, Z_s, for Z_s constant. This curve applies because:

1. the log-weighted voltage is divided by the filters into bands small enough to assume a constant spectral density within each of them;
2. the logarithmic weighting assures that each of these bands are equally important in comparing loudness to the reference spectrum.

The loudness scale of Figure 2.22 is linear on a logarithmic scale above the knee, that is, $Z_s = k_s \cdot \log(q_s)$ and the number of loudness units increase by a factor of 10 for a 44-dB change in level, or doubles every 13.2 dB. (This number compares with the 9.1 and 12.4 dB for the 50% reductions in loudness found for real speech in the Steinberg and Van Wynen filter tests, Figure 2.15.) The equation for q_s is

$$q_s = 10^{(1/44)Z_s} \tag{2.21a}$$

Now Z_s is in decibels, and for the configuration of Figure 2.26, is represented by $k_2 r \cdot \log(f)$. So, with Z_s given by

$$Z_s = 20 \log[k_2 r \cdot \log(f)] \text{ (dB)}$$

substituting in (2.21a) yields

$$q_s = [k_2 r \cdot \log(f)]^{1/2.2} \tag{2.21b}$$

This is the reason for writing the functions in the converter boxes as a function of the product of v_1 and the $\log(f)$ instead of simply as $h[v_2]$ as the voltage out of the filters might be labeled.

Referring to Figure 2.26, let

$$q_s = h[v_1 \cdot \log(f)]$$

Then each of the converter outputs, according to (2.14) will be

$$q_s = \int_{f_{i-1}}^{f_i} (v_1)^{1/2.2} d\, \frac{\log(f)}{df}\, df$$

But by choice of bandwidths, all of the v_1 are assumed constant within each filter, so

$$q_s = (v_1)^{1/2.2} \int_{f_{i-1}}^{f_i} \log'(f)\, df$$

or

$$q_s = (v_1)^{1/2.2} \log(f_i/f_{i-1})$$

and, in computational form, total loudness, N_s, is

$$N_s = \sum_{i=1}^{i=M} \log\left(\frac{f_i}{f_{i-1}}\right) (v_i)^{1/2.2}$$

but the v_i are the average values of v within each filter band so the sum becomes

$$N_s = \sum_{i=1}^{i=M} \log\left(\frac{f_i}{f_{i-1}}\right) \left[\frac{v(f_i)^{1/2.2} + v(f_{i-1})^{1/2.2}}{2}\right] \tag{2.22}$$

The voltages in this sum comprise the v_1 of Figure 2.26 (or $k_2 r$), and the sum is a relative voltage representing the loudness loss introduced by the R of the system. We would like to express the loss in decibels and we need a reference value or N_0. This can be found if R is removed, the measurement repeated (calculate a new N called N_0), and the ratio of N_s to this result is formed. In this case, the q_s become

$$q_0 = k_3(v_0)^{1/2.2} \log(f_i/f_{i-1}),$$

where k_3 is k_2/r raised to the power $1/2.2$. In the sum of these, the voltage is constant for all terms because there is no shaping of the flat spectral source, and so can be brought out to the front. The sum is then a sum of logarithms of ratios of successive frequency bands of the type $\log (f_2/f_1) + \log(f_3/f_2) + \cdots$, from f_1 to f_M. This reduces to $\log(f_M/f_1)$, and the total loudness N_0 is

$$N_0 = (v_0)^{1/2.2} k_3 \log(f_M/f_1)$$

Dividing this into (2.22) for the desired ratio makes the voltage terms dependent upon R only. That is, the constant voltage term from the last measurement, k_3, raised to the power 1/2.2 can be divided, under the exponent into each voltage, $k_2 r$, (v_1 of (2.22)). But $k_2 = k_3$, so the remaining voltage is proportional to the gain or loss of the system, r. The logarithm terms in the expressions are dimensionless, so the units for the ratio of N_s to N_0 are volts raised to the power 1/2.2. To recover volts before converting to a decibel expression, the entire result must be raised to the inverse power, 2.2. A decibel term is obtained by taking the logarithm of the result and multiplying by -20 so that loss will be positive. The result is called *objective loudness loss* (OLL), thus

$$ \text{OLL} = -20 \log \left\{ \frac{\sum_{i=2}^{i=M} \log(f_i/f_{i-1})[(v(f_i)^{1/2.2} + v(f_{i-1})^{1/2.2})/2]}{\log(f_M/f_1)} \right\}^{2.2} $$

objective loudness loss   (2.23)

with the voltage given by $v(f)_i = 10^{R(f_i)/20}$, where R is the response of the system at frequency i in decibels.

This is (7) of the IEEE standard, [5], without the identifier, OLL. Note that R of Figure 2.26 may be any linear device such as an electroacoustic transducer, and the $v(f)_i$ term may be the ratio of pascals to voltages, voltages to pascals, or pascals to pascals, with appropriate interpretations of the quantity OLL.

Finally, the implied restriction that equal log(frequency) intervals must be used for the f increments, can be removed. Equation (2.23) is a finite sum approximation to the underlying integral of the loudness equation and the restrictions on its use are dependent only on the approximation of area segments by rectangles. If the $v(f)_i$ are nearly constant over relatively wide frequency bands, use of the wide band does not increase the error. The required selection of *nearly flat* portions of measured or calculated spectra may not be obvious, however, because the data are usually in decibels rather than voltages. The term $\log(f_i/f_{i-1})$ in (2.23) automatically applies the correct weight to a term of any bandwidth. Essentially, constant voltage within any increment is the only real restriction.

2.6 THE CCITT OR UNITED KINGDOM LOUDNESS MEASURE

Until recently, the CCITT recognized only subjective testing as the method of measurement of loudness losses. In spite of increasing problems in the application of such measurements, explained in detail in Chapter 3, it was necessary to acquire a great deal of experience with methods of calculating a subjective quantity before one of them could be recommended for international use. With the issue of the Red Book, in 1984–1985, that body endorsed and recommended the use of one particular method.

The formulation adopted was developed at The British Post Office Research Centre in the United Kingdom, primarily by D. L. Richards. Extensive details of its derivation are given in [8]. The method is based on the same fundamental loudness equation (2.14), and is essentially the same as that described in the earlier sections of this chapter. The notation is different however.

In the CCITT approach, we begin with the loudness equation (2.14), rewritten here. Total loudness, N, is given by

$$N = \int Q(Z)S'(f) \, df \qquad (2.24)$$

To begin the development, consider a test in which two systems, each consisting of a microphone, a transmission system and a receiver, are compared for loudness by alternately listening to one and then the other. One of these systems is taken as a standard reference system, R; the other is an unknown system, u. The object of the comparison is to provide a loudness rating for the system, u, in terms of the standard system, R. The effective spectra delivered by these systems to the ear of a listener are Z_R and Z_u. Each of these spectra may be considered as being derived from some reference spectrum used as a source at the microphones. In the absence of either system, the source would deliver a reference effective spectrum, Z_0, to the ears of a listener. Consider this spectrum, Z_0, to be passed through the reference system loss, L_R, and the unknown system loss L_u. Thus,

$$Z_R = Z_0 - L_R \quad \text{and} \quad Z_u = Z_0 - L_u$$

The ratio of the loudness perceived through each, N_u/N_R, is

$$N_u/N_R = \frac{\int Q(Z_0 - L_u)S'(f) \, df}{\int Q(Z_0 - L_R)S'(f) \, df} \qquad (2.25)$$

Now assume that there is some constant, ΔX, such that when ΔX is added to L_u in the upper integrand of (2.25), the ratio will be unity. That is, the integral of $Z_0 - L_u$ shifted by ΔX and multiplied by $S'(f)$, will be equal to the integral of $Z_0 - L_R$ multiplied by $S'(f)$. The constant ΔX and the function $S'(f)$ must be determined to satisfy the equality:

$$\int Q(Z_0 - L_u + \Delta X)S'(f) \, df = \int Q(Z_0 - L_R)S'(f) \, df$$

As in the derivation of the EARS loudness equation, we take advantage of the fact that the relation between the logarithm of loudness and effective level, Z, is a line over much of the range of Z as in Figure 2.22. The form of the equation for such a line, with Q expressed in decibel-like terms, 10 log Q, is

$$10 \log Q = a + mZ$$

solving for Q yields

$$Q(Z) = C \cdot 10^{(1/10)mZ}, \qquad C = 10^{a/10} \tag{2.26}$$

(The slope of the upper portion of the line in Figure 2.22 is such that $m = 0.0228$ corresponding to a doubling in loudness for a 13.2-dB increase in Z.)

Substituting (2.26) and the equality condition in (2.25) yields)

$$1 = \frac{\int 10^{(1/10)m(Z_0 - L_u + \Delta X)} S'(f) \, df}{\int 10^{(1/10)m(Z_0 - L_R)} S'(f) \, df} \tag{2.27}$$

The integration is carried out over the frequency band of interest.

Now, because Z_0 is a fixed function, it can be combined with $S'(f)$ into a new function $G(f)$, thus,

$$G(f) = 10^{(m/10)Z_0} S'(f)$$

The quantity ΔX in (2.27) is to be a constant, so the term involving that is written separately outside the upper integral, and we have

$$1 = \frac{10^{(1/10)m\Delta X} \int 10^{(1/10)m(-L_u)} G(f) \, df}{\int 10^{(1/10)m(-L_R)} G(f) \, df} \tag{2.28}$$

and bringing the ΔX term to the left-hand side yields

$$10^{-(1/10)m\Delta X} = \frac{\int 10^{(1/10)m(-L_u)} G(f) \, df}{\int 10^{(1/10)m(-L_R)} G(f) \, df} \tag{2.29}$$

The upper integral in (2.29) is the loudness of the unknown system, the lower is that of the reference system. If this ratio is $\frac{1}{2}$, then the exponent on the left must be 0.3, or $m\Delta X = 3$.

Such a half-loudness point was found through tests similar to those of Van Wynen and Steinberg (data in Figure 2.15). Two identical reference systems, one with low-pass and high-pass filters installed, are called the *unknown*. Comparative loudness balance tests were conducted. For example, a high-pass filter removes some low frequencies. If the reference system is attenuated until the two systems are judged to be equally loud, then this attenuation, for a *balanced* condition, is a measure of the high-pass loudness loss. The high-pass and low-pass filter cut-off frequencies were kept equal to each other but varied across the band until their loudness losses were equal. For this frequency, the upper frequency loudness contribution is identical to the lower frequency contribution and their loudness must each be one half of the total. The loudness loss for this particular balance is the quantity ΔX, and m is $3/\Delta X$.

Once m and ΔX were found, a low-pass filter, with the cut-off stepped across the band, thus approximating a gradual increase in the upper limit of integration for the unknown, appropriate values of $G(f)$ could be determined by balance tests against the reference. For each setting, the L_u is known by the filter setting, ΔX is incremented by the change from the half-loudness value, and a $G(f)$ for that increment must be selected that will satisfy (2.29). $G(f)$ is a frequency weighting function representative of the growth of loudness as a function of frequency for the reference system. It is an $S(f)$ based on the speech of the test subjects. The $G(f)$ determined in this way is a cumulative function, but successive interval weights, needed in the computational forms of the integrals below, can be found by subtractions performed at each frequency interval (each change in the low pass filter setting).

The $G(f)$ appear in both the numerator and denominator terms of (2.29). In view of this, $G(f)$ can be scaled by a constant such that its integral over frequency is equal to one, and the ratio will not be affected. $G(f)$ is now a normalized weighting function and so performs an averaging operation on the power terms in the integrations.

Taking the logarithm of both sides of (2.29) and solving for ΔX yields

$$\Delta X = -10/m \log \frac{\int 10^{(1/10)m(-L_u)} G(f)\, df}{\int 10^{(1/10)m(-L_R)} G(f)\, df}$$

This can be written in computational form as

$$\Delta X = -10/m \log \frac{\sum_{i=1}^{i=M} 10^{(1/10)m(-L_{ui})} G_i \Delta f_i}{\sum_{i=1}^{i=M} 10^{(1/10)m(-L_{Ri})} G_i \Delta f_i} \tag{2.30}$$

where M is the number of intervals needed to cover the frequency band of interest. These will be discussed in Chapter 3.

Recall that L_R in the denominator represents the response of the reference system. Making use of the normalized weighting factors, $G_i \Delta f_i$, an average response value, $\overline{L_R}$, may be found from the expression:

$$\overline{L_R} = \sum_{i=1}^{i=M} 10^{(1/10)m(-L_{Ri})}G_i\Delta f_i \left(\frac{1}{M \displaystyle\sum_{i=1}^{i=M} G_i\Delta f_i}\right) \tag{2.31}$$

This constant average value may now be used in the denominator of (2.30) in place of the L_{Ri}, and the power term brought outside the summation leaving only $G_i\Delta f_i$ inside. But the sum on $G_i\Delta f_i$ is unity by design, and so may be dropped. The constant, $\overline{L_R}$, may now be divided into, and included in the numerator, yielding

$$\Delta X = -10/m \log \sum_{i=1}^{i=M} 10^{-(1/10)m(L_{ui}-\overline{L_R})}G_i\Delta f_i \tag{2.32}$$

If we now define

$$w_i = -(10/m) \log(G_i\Delta f_i) \tag{2.33a}$$

and

$$W_i = -\overline{L_R} + w_i \tag{2.33b}$$

then (2.32) can be written as

$$\Delta X = -10/m \log \sum_{i=1}^{i=M} 10^{-(1/10)m(L_{ui}+W_i)}$$

This is the basic formulation of the CCITT loudness measure. The ΔX is the average weighted loss difference between an unknown and a reference system. It is called *loudness rating* (LR):

$$LR = -10/m \log \sum_{i=1}^{i=M} 10^{-(1/10)m(L_{ui}+W_i)} \quad \text{CCITT loudness rating} \tag{2.34}$$

NOTES

1. An acoustic wave traveling in a reverberation free environment.
2. The subscript, s, will be used with acoustical functions, spectra, pressures, *etcetera* to specify speech as opposed to any general sound.
3. Complete results of the masking tests using the shaped noise are shown in Figure 2 of (3).

4.   These results were later refined to the concept of *critical band-widths,* exact maximum width noise bands for masking a tone. The width changes with frequency [4].

5.   The curve in Figure 2.12 was actually computed using values for k that differ slightly from those in Figure 2.10, see [4, Section 3.4].

6.   The researchers who reported the low values studied the details of the experiments conducted by the other four and concluded that different aspects of the speech levels were measured. Thus, there is lingering uncertainty about average speech level. However, in [1], Sullivan argues convincingly that the absolute value of the pressure has little effect on the loudness computation.

7.   The corresponding figure, numbered 11 in [1], includes data for the bandpass filter. The points for it scatter about the mean curve similarly to those shown here.

8.   The following description of the derivation of the S and q_s functions is not claimed to be accurate. The work was not published until about 30 years after it was completed, and the accounts are not totally clear. This explanation is thought to be reasonable and relatively easy to follow, while maintaining the essence of the original effort.

9.   There are other loudness measures considering spectral and temporal aspects of sounds (see [2]).

Appendix 2A

2A.1 RELATIONS BETWEEN SOME ACOUSTICAL MEASURES

Formal definitions of some acoustical measures encountered in communications are presented in the beginning of this chapter. This appendix should assist the inexperienced in their meaning and use. It also provides numerics related to common experience and a few useful conversion factors in the accompanying table.

Figure A2.1 is a pictorial representation of the relations between objective measures of acoustic pressure and their subjective derivatives. The heavy vertical arrow on the left is the reference scale, calibrated in pressure expressed in dBt. The reference pressure for dBt is $2.04 \cdot 10^{-5}$ newtons per square meter. There are other standard metrics in which this reference may be stated. Three of them are shown on the figure, in the box labeled reference pressure, P_0.

The two graphs to the right of the reference scale show accepted curves defining the thresholds of hearing, *versus* frequency, for pure sinusoids and for continuous spectra sounds. The ear is more sensitive to continuous spectra sounds; the reason is not clear but may relate to the multiple interconnection of nerve endings that individually respond essentially to single frequencies. They are positioned vertically so that their ordinate scales are the same as the reference scale.

The threshold for pure sinusoids has received most attention and been the subject of larger studies than that for continuous spectra sounds. The standard curve (See ISO R226-1967) represents a smooth fit to the median value of results of studies with 51 young people, average age of 20. The statistical nature of this curve is indicated by the width of the vertical space marked with a double ended arrow. The space is about 5 dB in height and shows the range of variations that can be obtained in a rapid series of measures with only one person. The range of variations within a large group of people is even greater than 5 dB.

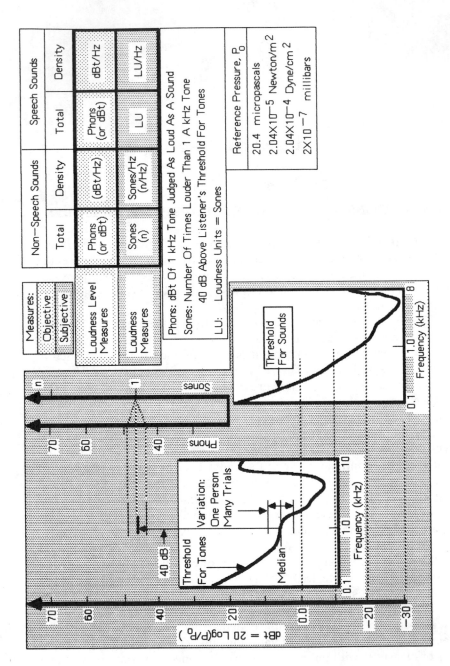

Fig. 2A.1 Objective and Subjective Acoustical Measures

The threshold for continuous spectrum sounds is shown for reference only; it has nothing to do with the quantities defined. It is important in the derivation of loudness computation procedures described in this chapter.

The quantity called the *phon* is the bridge between objective and subjective measures. The phon is a measure of loudness level, a pressure. Any sound that is judged to be equal in loudness to that of a 1000-Hz tone that is at a level 40 dB above *the listener's* threshold of hearing is said to have a loudness level equal to the pressure of the tone, in dBt. For example, starting at the 1-kHz point on the curve for the threshold of pure tones in the figure, move upward 40 dB. The arrow labeled 40 dB indicates this. The arrow terminates at about 46 dBt. If a sound were judged to be equally as loud as the 46-dBt tone, the loudness level of the sound would be 46 phons. The nature of the spectrum of the sound is of no consequence.[9] In this case, 46 dBt maps to and defines one sone, as indicated on the right-most scale. The converging dotted lines emphasize that, in any single measurement, any number of dBt within the range of about 44 to 50 might be defined as one sone ($n = 1$), reflecting the variability of this listener.

The sone scale is linear in loudness; that is, the value two sones ($n = 2$) is assigned to a sound that is judged to be twice as loud as the one of the last paragraph. Its corresponding value in dBt is of no interest for scaling purposes. In general, it will not be equal to the phon level plus 2 dB because loudness is not linear in decibels.

The table to the right of the dual phon-sone scale summarizes the metrics commonly used in acoustical work associated with communications. Total loudness is expressed in sones, or loudness units (LU) if one wishes to emphasize that the measure applies to speech. Loudness units and sones are numerically equal, but loudness units should not be used to describe nonspeech sounds. Total loudness level is expressed in either case in phons (dBt).

Loudness level density may be dBt/Hz for either tones or sounds and loudness density may be sones per hertz or loudness units per hertz.

2A.2 OTHER MEASURES

Some other acoustical measures that are either relatively common or apt to be encountered in communication related documents are defined below.

Decibel. With no reference, often seen in popular media relating to noise, usually means decibel with respect to the reference pressure *Po* ($2.04 \cdot 10^{-5}$ pascals), and also often implies a measurement made with a filter known as *A* weighting that attenuates frequencies below 1000 Hz (see 2, Figure 2.1, for example). The *A* weighting characteristic is rather similar to C-MESS weighting used in telephony, Figure 1.3, at the lower audio frequencies, but the upper cut-off frequency is above 10 kHz for the *A* filter.

dBSPL. General abbreviation for sound pressure level and the reference pressure should be specified. When encountered in communication literature, with no reference, decibels with respect to −94 dBPa (pascals), is assumed (0 dBPa = 94 dBSPL).

dBmB. 20 · log(milliBars) (reference one mB). Millibar is used in atmospheric pressure readings and is defined in the table of reference pressure measures in Figure A.1.

dBμB. 20 · log(p) (p in microBars).

Sound power. symbol, *J* (watts/cm²). $2.04 \cdot 10^{-3}p^2$ (p dynes/cm²) is equal to microwatts per cm².

Conversions among some of these metrics are given in Table 2A.1.

Table 2A.1
Conversion Factors

From:	dBPa	dBSPL	dBmB	dBμB
To:				
dBPa =	+0	−94	+40	−20
dBSPL =	+94	+0	+134	+74
dBmB =	−40	−134	+0	−60
dBμB =	+20	−74	+60	+0

For example to go to dBPa from dBSPL, subtract 94.

Some numerics for common sounds are:

- average speech at three feet: about 65 dBSPL, roughly equivalent to 0.0003 μW/cm², or $3 \cdot 10^{-10}$ W/cm² (at the ear drum);

- a soft whisper at five feet is about 20 dBSPL;
- subway station with express train passing: about 104 dBSPL;
- an office with many typists at work: 75 to 80 dBSPL.

REFERENCES

1. Sullivan, J.L., "A Laboratory System for Measuring Loudness Loss of Telephone Connections," *Bell System Technical Journal,* Vol. 50, No. 8., October 1971.
2. *Handbook of Noise Measurement,* Ninth Edition, GenRad, Inc., Concord, Massachusetts.
3. Fletcher, Harvey and Munson, W.A., "Relation Between Loudness and Masking," *J. Acoust. Soc. Am.,* Vol. 29, No. 5, May 1957.
4. French, N.I. and Steinberg, J.C., "Factors Governing the Intelligibility of Speech Sounds," *J. Acoust. Soc. Am.,* Vol. 19, No. 1, January 1947.
5. IEEE, "IEEE Standard Method for Determining Objective Loudness Ratings for Telephone Connections," *IEEE Std. 661-1979,* November 29, 1979.
6. IEEE, "IEEE Standard Method for Measuring Transmission Performance of Telephone Sets," *IEEE Std. 269-1971,* February 4, 1971.
7. Steinberg, J.C., "Positions of Stimulation in the Cochlea by Pure Tones," *J. Acoust. Soc. Am.,* Vol. 8, No. 176, 1937.
8. CCITT, "Calculation of Loudness Ratings," *Red Book, Volume V,* Rec. P.79, ITU, Geneva, 1984.

Chapter 3
Computing Loudness Ratings

The measurement of loudness is a difficult subject because of insufficient knowledge of the mechanisms translating acoustic waves into the perceived phenomenon. As a result, there is a large number of loudness metrics, each confined to a specific area of application. In this chapter we consider five of them that are used in telephony:

- IEEE loudness loss[1];
- CCITT loudness loss;
- Reference equivalent;
- Corrected reference equivalent;
- OREM—Objective Reference Equivalent Measurement.

We show why five metrics exist in this one area, and how they are related. The material of Chapter 2 is used as a starting point.

In Chapter 2 we introduced the fundamental loudness equation and traced the development of two computational variations repeated below.

With the parameters:

$$f = \text{frequency}$$

$$R(f) = \text{system frequency response in decibels}$$

$$v(f) = 10^{R(f)/20}$$

$$i = \text{an interval of the frequency band defined by } 1 \le i \le M$$

then, objective loudness loss, (2.23), is

$$\text{OLL} = -20 \log \left\{ \frac{\sum_{i=2}^{i=M} \log(f_i/f_{i-1})[(v(f_i)^{1/2.2} + v(f_{i-1})^{1/2.2})/2]}{\log(f_M/f_1)} \right\}^{2.2}$$

(3.1)

and with parameters:

m = slope of curve relating loudness, N_s, to effective spectral level, Z_s

$L_u = R(f)$ above

W = tabulated weighting factors

the CCITT loudness rating, (2.34), is

$$LR = -10/m \log \sum_{i=1}^{i=M} 10^{-(1/10)m(L_{ui}+W_i)}$$
(3.2)

These equations provide two different estimates of the same physical phenomenon, the loudness attenuation effected by spectral shaping of a speech waveform in telecommunication systems. Equation (3.1) evaluates attenuation relative to a band-limited free air path with a bandwidth defined by f_1 and f_M of (3.1). It is based on a defined reference spectrum for speech, and uses the logarithm of frequency as a weighting function. As we will see in Section 3.4.4, (3.2) is based directly on subjective tests between a system accepted as ideal and a reference *very good* communication channel. The characteristics of the reference channel are embedded in the weighting factors, W_i. Both equations are general to the extent that the independent variable, $R(f)$, may represent any form of spectral shaping, whether due to transducers, filters, or any other device. For engineering applications, both equations are further refined into a number of relations for the evaluation of specific portions of a communication channel. These equations, and those derived from them, form a significant part of the group of metrics known as objective measures of loudness. The other group is known as subjective measures.

3.1 OBJECTIVE AND SUBJECTIVE METRICS

Objective measures are those that do not depend on human judgment. Subjective measures are simply human judgments. Equations (3.1) and (3.2) are, of course, based on judgments, as described in Chapter 2. However, those judgments have been accepted as final and translated into objective measures for on-going application. The distinction is then that subjective measures require human judgment for every application.

Objective measures are relative newcomers in telephony, much of the fundamental work leading to (3.1) and (3.2) was done in the 1930s. Subjective measures of loudness were used much earlier than that, some are still in use and, more importantly, some objective measures are closely tied to the earlier subjective ones for purposes of continuity.

3.2 TERMINOLOGY

As one might anticipate, five different metrics give rise to a rather large vocabulary of terms and acronyms. They are defined in the text as they are introduced, but the uninitiated reader is in danger of becoming overwhelmed. This section contains a summary of them intended to ease the burden and provide a convenient reference. Several of the metrics are historically related in the sense that each is a variation of its predecessor. This chronological sequence is also presented as an additional familiarization aid.

Except for the first, the terms are presented in alphabetical order under two headings: test facilities and transmission ratings (measurement metrics) because many transmission ratings use terminology that refers to components or usage of a particular facility. These relations between facilities and transmission ratings are explained in the paragraphs between the two listings. The first term, AEN, is a transmission rating that will be of little interest in this book but is defined because the acronym forms a part of the names of some test facilities.

The test facilities are of two types, subjective and objective. Subjective facilities use people as test subjects to listen to the effects of various transmission conditions and rate them. Objective facilities make electro-acoustic measurements and the results are translated to subjective evaluations through various models. The two loudness equations above are examples of such models.

Subjective testing for telephony began early in this century with simple comparisons of the relative loudness loss of various types of common transmission cable assemblies. The arrangements used commercial telephone transmitters and receivers that introduced considerable spectral shaping of their own. Thus, observed loudness loss had to be attributed to the entire system and the relative contributions of the cables and components tested could not be strictly isolated. The *master reference system* (MRS) was the first facility that solved this problem through the use of high quality low distortion transducers. Many of the subjective test facilities in use today are evolutions of the MRS. Figure 3.1 shows the names of some major facilities, the approximate dates that they were introduced, and illustrates, via the arrows, the evolution of the subjective ones. Except for the one called SIBYL, all those introduced since 1950 are still in use. (NOSFER-84 is a technological revision of NOSFER.)

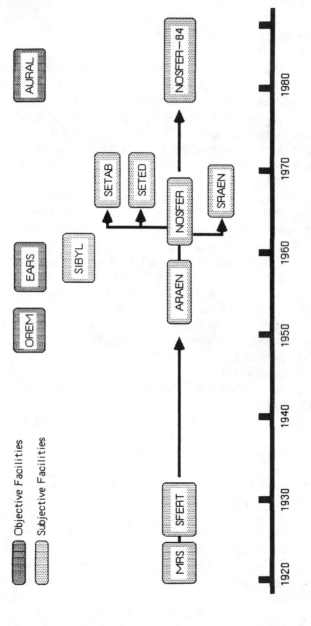

Fig. 3.1 Relations Between and Approximate Times of Introduction of Test Facilities

Test Facilities

In the following, names marked with an asterisk (*) are described in Sections 3.3 and 3.4. Others are listed for reference only because they are encountered in CCITT literature pertinent to those of interest in this book.

AEN*. An abbreviation for the French name *affaiblissement équivalent pour la netteté*, which is translated to *articulation reference equivalent* in United Kingdom terminology, or *equivalent articulation loss*, in U.S. terms. It is a measure of speech intelligibility and is described in Section 3.4.1.

ARAEN*. French abbreviation for *appareil de référence pour la détermination des affaiblissements équivalents pour la netteté*, which means *reference apparatus for the determination of AEN*, but is translated to *reference apparatus for the determination of transmission performance ratings*. It is a subjective test facility in Geneva, Switzerland, introduced in the 1960s. It comprises three major components:

- a transmission path subdivided into a transmitting or *sending end*, a *junction* which is a means for introducing attenuators and filters, or interconnecting with other systems, and a *receiving end*,
- a means for introducing room noise, an important consideration in subjective testing;
- required calibration and maintenance equipment.

The sending end includes a microphone placed at a standard distance from the talker's lips. The bandwidth of the sending and receiving ends is 8 kHz. The junction portion contains a bandpass filter with a nominal width of 4 kHz. ARAEN is unique in that it uses four identical receivers to allow four testers to simultaneously listen to transmitted speech.

AURAL*. *automatic speech quality rating system based on loudness* (AURAL) is an automated objective measuring system introduced by the *Nippon Telegraph Telephone Public Corporation* (NTT) in the early 1980s. It is equipped to yield results in one or more of several different metrics. It has been tested and approved for use by the CCITT.

EARS. electro-acoustic rating system (EARS) is an objective measuring apparatus designed and constructed at Bell Telephone Laboratories to measure telephones and telephone systems in terms of a specific loudness metric with the same name.

MRS. The *master reference system* (MRS) was the first reference system for subjective testing of loudness using very low distortion transducers and active devices to compensate for their poor efficiency. This innovation effectively removed the characteristics of the transmitter and receiver from being an intrinsic part of all measurements.

NOSFER*. Abbreviation from the French name *nouveau système fondamental pour la détermination des équivalents de référence* which is translated as *new standard system for the determination of reference equivalents*. Reference equivalent (RE) is one of the metrics to be described. NOSFER is basically the ARAEN system modified by the inclusion of sending-end and receiving-end equalizers, a change in the standard distance of the input microphone from the talker, and the disabling of three of the four receivers.

NOSFER-84*. The name given to the NOSFER system after it underwent a technological updating.

OREM*. *Objective reference equivalent measurement system* (OREM) is a laboratory instrument comprising a level meter with particular electrical and ballistic characteristics, an artificial mouth (microphone), an artificial ear (receiver), and an arrangement for supporting telephone handsets, or telephone transmitters alone, during measurements. When used to measure handsets, the result is called an *OREMA measurement*. When used to measure transmitters, the result is called an *OREMB measurement*. OREMB is mentioned here for completeness only. It is intended primarily as a manufacturing test measurement and so has little to do directly with transmission work. The specialized instrumentation of the OREM system is said to yield results in agreement with the subjective SFERT (defined below) system, at least for certain types of handsets. The system is widely used for routine measurements domestically and internationally.

SETAB. Abbreviation from the French name *Système-etalons de travail avec appareils d'abonné*. Known in English as *working standards with subscriber's equipment*. The name SETAB refers to any one of a number of standard subjective test apparatus, patterned after and calibrated with the NOSFER system, and for use with telephones using carbon microphones. A SETAB system may be used for measuring reference equivalents if it has been calibrated directly against NOSFER. Such a calibration designates it as a secondary or *working* standard.

SETED. Abbreviation from the French name *système-etalon de travail avec microphone et récepteur électrodynamique*. Known in English as *working standard having an electro-dynamic microphone and receiver*. The name SETED refers to any one of a number of standard subjective test apparatus, patterned after and calibrated with the NOSFER system, and for use with telephones using electro-dynamic microphones and receivers. A SETED system may be used for measuring reference equivalents if it has been calibrated directly against NOSFER. Such a calibration it designates it as a secondary or *working* standard.

SFERT. Abbreviation from the French name *le système fondamental de référence pour la transmission téléphonique*. Known in English as *European master reference system for telephone transmission*. SFERT is a subjective measurement facility that was in use before the introduction of the ARAEN.

SIBYL. A name, not an acronym, for a subjective test facility used at Bell Telephone Laboratories from the 1950s through the late 1970s. SIBYL enabled experimenters to introduce transmission impairments to a subjects' everyday business telephone and collect votes from the subjects relevant to their opinion of quality of the resulting transmission.

SRAEN*. Abbreviation from the French name *système de référence pour le détermination des affaiblissements équivalents pour la netteté*. Known in English as *Reference system for the determination of AEN*. SRAEN is an adaptation of ARAEN expressly for use in articulation or intelligibility tests. The bandwidth of the system is 300 to 3400 Hz and provision is made for the introduction of background noise via electrical sources within the transmission system.

Relations between Test Facilities and Metrics

A generalized test facility for telephones is illustrated at the top of Figure 3.2. The transmitting and receiving systems normally include a subscriber loop, the wire from a telephone location to the local telephone office or switching center (*Center Office* or (CO)), and provision for supplying current to a telephone through an impedance representative of the CO. Such a power source is known as a battery feed circuit. The junction shown in the center of the figure usually comprises an arrangement for switching between portions of the test facility and equivalent portions of an apparatus being testing. It also provides measurement access and standard 600-Ω terminations for the subsystems. The term junction is used internationally to refer to connecting facilities (trunks) between local switching machines. Such facilities could, in principle, be tested by inserting them in the test arrangement at this point. Physical size usually prohibits this but the concept is important.

The figure is representative of either subjective or objective measurement facilities. In these laboratory testing facilities, subjective measurements are made with trained operators speaking into microphones while trained listeners use the receiver(s) to compare the loudness of systems under test with that of standard systems. Objective measurements are made using instruments to measure voltage and acoustic pressure, and artificial ears and mouths (microphones and loudspeakers) are used to couple to telephone handsets.

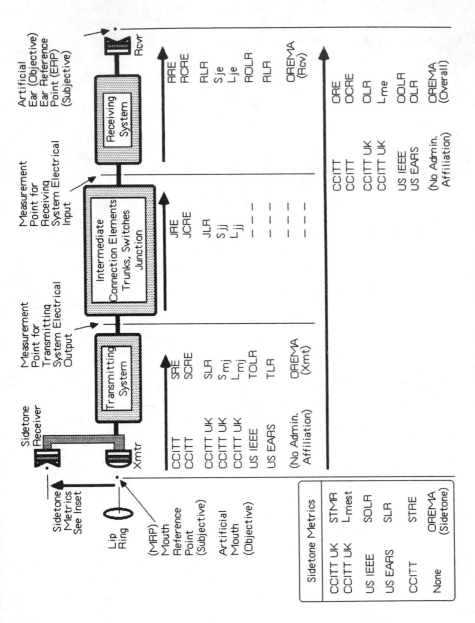

Fig. 3.2 Most Commonly Encountered Metrics

For subjective tests, a lip ring defines the location for a speaker's lips during tests. Acoustic pressures, measured during a calibration procedure, are referred to a point between the lip ring and microphone, 25 mm from the lip ring, known as the *mouth reference point* (MRP), and another at the output of a receiver, the *ear reference point* (ERP). The ERP is formally defined as the center of the entrance to the listener's ear canal. It is assumed that the distance between the handset receiver and the MRP is zero when the receiver is held close to the ear.

For objective tests, an artificial ear, a microphone assembly that clamps and seals tightly to the handset receiver, is used. An artificial mouth, a calibrated speaker, is positioned with a defined spatial relation to the handset transmitter. A variable frequency oscillator or other standard electrical source such as shaped noise or a defined artificial speech is used to generate signals:

- to drive the speaker to produce an acoustic field picked up by the transmitter; or
- to drive the handset receiver to supply an acoustic field for the microphone.

The objective transmission ratings in the upcoming list, except for the one called *OREM,* are determined by applications of the loudness equations of Chapter 2. The OREM system ratings are read directly from a specialized level meter designed to emulate subjective ratings. In the general case there are five kinds of ratings:

- acoustical to electrical efficiency for a transmitting system;
- electrical to electrical loss across a junction element;
- electrical to acoustical efficiency for a receiving system;
- acoustical to acoustical loss for a complete system;
- acoustical to acoustical loss for the sidetone path.

The IEEE ratings do not distinguish a junction element. Such losses may be treated in a number of ways, one of which is illustrated in Chapter 7. The OREM system also makes no provision to account for a junction element. In principle, junction-like devices could certainly be included in an OREM test arrangement as part of the transmitting or receiving system, but this is not usually done.

The various transmission ratings are shown in rows and columns in the lower part of Figure 3.2. Columns are positioned below the portion of the test facility to which they apply. The elements in any row are common to the metric named at the beginning of the row. Subscripts are used with the metric designated CCITT UK (United Kingdom) in the figure. The letters used in the subscript notation are: m for mouth, j for junction, e for ear, and ST for sidetone. The subscript pairs indicate endpoints, and their ordering follows the direction of the arrows in the figure.

The ratings shown in rows beginning with *CCITT* are subjective measures; the remainder are objective. The identifiers *CCITT UK* refer to objective ratings developed in the United Kingdom and subsequently adopted by the CCITT. The U.S. metrics were developed at Bell Telephone Laboratories using the electro-acoustic rating system. These metrics were adopted, with slight modification, by the IEEE and published as standard in the 1970s.

The ratings are listed below, in the order shown on the figure, with the definitions for the acronyms, and are defined in detail as they appear in the text. An asterisk indicates a subjective measure. By way of explanation, the word *send* is commonly used instead of *transmit* in CCITT literature. The wording and double notation of junction to junction and *JJ* are part of accepted CCITT terminology. The meaning is *across the junction*. Junction equipment is almost always electrically symmetrical so direction is seldom of interest. In any nonsymmetrical situation the intended direction would be obvious from the context.

Transmission Ratings

Transmit

SRE:*	Send Reference Equivalent
SCRE:*	Send Corrected Reference Equivalent
SLR:*	Send Loudness Rating
S_{MJ}:	Send, mouth to junction
L_{MJ}:	Loss, mouth to junction
TOLR:	Transmit Objective Loudness Rating
TLR:	Transmit Loudness Rating
OREMA Transmit:	Objective Reference Equivalent Measurement "A" for a transmitting system

Receive

JRE:*	Junction Reference Equivalent
JCRE:*	Junction Corrected Reference Equivalent

Transmission Ratings (cont'd)

JLR:	Junction Loudness Rating
S_{JJ}:	Send, junction to junction
L_{JJ}:	Loss, junction to junction
RRE:*	Receive Reference Equivalent
RCRE:*	Receive Corrected Reference Equivalent
RLR:*	Receive Loudness Rating (CCITT UK)
S_{JE}:	Send, junction to mouth
L_{JE}:	Loss, junction to mouth
ROLR:	Receive Objective Loudness Rating
RLR:	Receive Loudness Rating (EARS)
OREMA Receive:	Objective Reference Equivalent Measurement "A" for a receiving system

Overall

ORE:*	Overall Reference Equivalent
OCRE:*	Overall Corrected Reference Equivalent
OLR:	Overall Loudness Rating (CCITT UK)
L_{me}:	Loss, mouth to ear
OOLR:	Overall Objective Loudness Rating
OLR:	Overall Loudness Rating (EARS)
OREMA Overall:	Objective Reference Equivalent Measurement "A" for an overall system (transmit plus receive)

Sidetone

STRE:	Sidetone Reference Equivalent
STMR:	Sidetone Masking Rating
L_{meST}:	Loss, mouth to ear, Sidetone
SOLR:	Sidetone Objective Loudness Rating
SLR:	Sidetone Loudness Rating
OREMA Sidetone:	Objective Reference Equivalent Measurement "A" for a sidetone path

---

In the following sections, these metrics are described qualitatively and, when appropriate, quantitatively. The reader should be aware that the several types of metrics are independent and, because they are either subjective or involve nonlinear functions of nonlinear parameters, there can be no accurate conversion factors relating them. Furthermore, none of the metrics permit accurate algebraic addition of ratings when elements are placed in tandem. Nevertheless, useful approximations are possible to circumvent both of these problems. The additivity issues are addressed as they arise, and the question of conversion factors is discussed in Section 3.4.7.

3.3 SUBJECTIVE AND OBJECTIVE TEST FACILITIES

Subjective testing, to enable fairly reproducible quantitative results, is a science unto itself. Intrinsic differences between, and long-term drift of bias within people, cause test results to have variances that may be large compared to mean values. For results to be meaningful and useful, it is necessary to be confident that both the mean and variance found are due to normal responses to the stimuli presented and not influenced significantly by other factors. Great care must be taken to identify all the variables that might affect the outcome of a test, closely regulate those that are subject to control, and attempt to monitor those that are not. The need for close control dictates a laboratory environment which in itself can be a factor affecting the response of a test subject. It is reasonable to assert that any testing environment will bias subjects' opinions and that valid results are obtained only under normal day-to-day conditions and without the subject's knowledge. Such tests can be devised but the opportunity for control of variables is then minimal. The practical answer to this apparent dilemma is that certain kinds of tests are suited to the laboratory and others must be conducted in the natural environment. There is, of course, any number of intermediate situations between the two extremes of total laboratory control and the completely natural environment. The appropriate blend of the two for any test depends upon the type of information desired or simply the question that is being asked. For example, if one wished to know which of two sounds was judged to be louder, a laboratory would be an appropriate test location. A question with more unknown variables such as "Which of these do you prefer?" might better be asked in the subject's normal environment. Many of the fundamental studies leading to the results discussed in this chapter and the next were carried out with one of two subjective test facilities. The first is representative of a pure laboratory, the second is one that enables collection of data in the subjects' normal working situation with a minimum disturbance.

The most widely recognized subjective test facility for telephonometric[2] measurements is operated by the CCITT in Geneva, Switzerland. A complete description of the facilities and their use is given in Volume V, P Series, of the CCITT Yellow Book [1], and the CCITT Red Book [2]. As indicated in Figure 3.1, the facility that is now in use has evolved over several decades with numerous modifications and additions to the equipment and procedures.

3.3.1 MRS

As mentioned earlier, the master reference system may be considered to be the evolutionary starting point for subjective facilities currently in use. An adequate description for present purposes was given in the definition section, Section 3.1.1.

3.3.2 ARAEN

ARAEN, the reference apparatus for the determination of transmission performance ratings, is a subjective measurement system [3]. It is the direct descendent of the MRS and SFERT and comprises three elements: a transmission path, a source for room noise, and calibration equipment. Of interest here is the transmission path, which in turn has three components: sending portion, junction, and receiving portion. The sending portion has a microphone and amplifier; the receiving portion, an amplifier and four receivers for simultaneous listening by four subjects. The nominal bandwidth of these portions, 60 to 8000 Hz, is probably a reflection of its ancestral cable test facilities. The junction is an arrangement to connect attenuators, filters, *etcetera* between the sending and receiving portions, or to interconnect these with portions of other systems to be evaluated.

In this, and later revised versions, the person talking into the microphone places his lips at a fixed position defined by a guard or lip ring. The listener holds a receiver *closely* to his ear to judge the quality of the speech received from the talker. With no attenuators in the junction, the path from the lip ring to the ear of a listener (the ERP) is designed to have zero loss. With an attenuator in the junction, set to 30 dB, the path has the same loss as the free field air path between two people at a distance of one meter if monaural listening is assumed. ARAEN forms the basic set of equipment in the constitution of both SRAEN and NOSFER.

3.3.3 SRAEN

SRAEN, the reference system for the determination of AEN (AEN is described in Section 3.4.1) is a particular arrangement of the AREAN equipment. A bandpass filter, 300 to 3400 Hz, is placed in the junction and provision is made for the introduction of noise with a Hoth spectrum[3] at the receiving portion input. It is used for the talking and listening tests to determine AEN, and as a calibration system for similar arrangements used in various countries.

3.3.4 NOSFER

NOSFER, the new standard system for the determination of reference equivalents, comprises the ARAEN equipment modified by the addition of transmitting and receiving equalizers and with three of the four receivers replaced with terminating resistors. This configuration is the accepted standard reference transmission system, against which, other systems are evaluated. A simplified block diagram, general enough to represent NOSFER or NOSFER-1984, the recently revised version, is shown in Figure 3.3.

NOSFER consists of a transmission system, known as the sending end, and a receiving system, known as the receiving end, and a section known as the junction. The junction is a means for interconnecting the sending and receiving ends through attenuators, filters, or other devices. It also provides for substituting an unknown receiving or transmission system for its NOSFER counterpart and switching between the two. This arrangement is used to determine the loudness of an unknown system relative to the NOSFER reference. Most often a simple attenuator is used in the junction to adjust the receiver output during tests. The lip ring and microphone in Figure 3.3 are shown placed 140 mm apart with a point between, 25 mm from the lip ring, designated the mouth reference point (MRP). The MRP is the point at which acoustic pressure from the talker is referenced. The pressure at the output is measured at the center of the receiver, the point called the ear reference point (ERP). The speech voltmeter is used to monitor the talker's speech level. The sending and receiving portions have a nominal 8-kHz bandwidth. The channel characteristics are published in the CCITT Red Book because they enter into some calculations as described later in this chapter. The points labeled JS and JR (J is for junction) define the output of the sending end and the input of the receiving end. The impedance at these points is 600 Ω, resistive. This value is a standard for measurements and calculations of power. Use of the system is described in Section 3.4.2.

3.3.5 SIBYL

The system used for much of the work relating to the model, discussed in Chapter 4, was done using a system known as *SIBYL* [5].

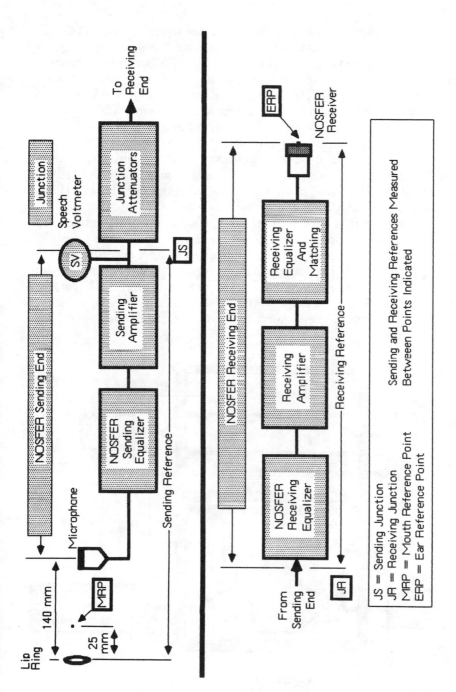

Fig. 3.3 NOSFER System

JS = Sending Junction
JR = Receiving Junction
MRP = Mouth Reference Point
ERP = Ear Reference Point

Sending and Receiving References Measured
Between Points Indicated

(SIBYL is a name, not an acronym.) The construction of SIBYL was begun in the late 1950s, and was used until the late 1970s, at which time its use was discontinued due to incompatibilities with electronic switching machines. SIBYL permitted transmission impairments of known magnitude to be inserted in the telephone lines of employees at Bell Telephone Laboratories. A functional block diagram of the system is shown in Figure 3.4. The subject employees were selected in a random manner and then asked to participate in an experiment. If they agreed, SIBYL was connected to their lines for a period of many months. Impairments were inserted from time to time, and changed in a sequence known only to the experimenters. When one of the subjects used his or her phone, SIBYL recognized the event. When the phone was returned to on-hook, SIBYL called the user with a special alerting ringing pattern. This notified the user to pick up the phone, think about the *acceptability* of the transmission quality of the call just completed, and dial a number corresponding to a rating for that call. The rating was a five-point scale, carrying the descriptive terms: excellent, good, fair, poor, unsatisfactory. SIBYL sometimes called back even if there were no inserted impairments. Initially, SIBYL could introduce only noise and loss; the capabilities were later expanded to include echo.

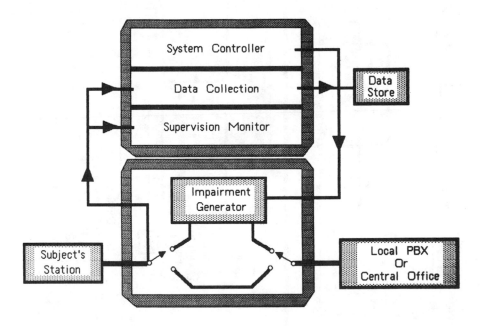

Fig. 3.4 SIBYL Functional Diagram

It is appropriate to mention here that the subjective opinion models in use today and discussed in chapter 4, are very elaborate and include the effects of room noise, frequency response, sidetone, frequency response of the sidetone path, and quantizing noise. Not all of these impairments were studied *via* the use of SIBYL; as the work progressed, and more insight to the modeling of subjective opinion was gained, methods for incorporating the results of smaller scale tests for other effects were developed. These did not require the SIBYL facility because, with improved knowledge of effects of more closely controlled impairments, relatively small but valid experiments could be designed. Results were incorporated as modifiers to the basic SIBYL results as will be seen in Chapter 4.

3.3.6 OREM

A system that is popular in many countries, and used by some telephone handset manufacturers in the U.S., is the *objective reference equivalent measurement system* (OREM). OREM was developed in the 1940s and 1950s in Europe, and was the first objective measuring equipment demonstrated to yield results in close agreement with subjective RE measures, at least for a few types of telephones. The OREM system can be used for two kinds of measurements: one, called OREMA, for complete handsets, and the other for carbon transmitter units only, called OREMB. When OREMA is used, the portion of the system rated is not designated with a letter prefix, such as S for Send, as in the other systems, but the complete phrase, *i.e., OREMA for sending,* is spelled out.

3.3.7 AURAL

AURAL, the automatic speech quality rating System based on loudness, is not another metric system but an automated measuring apparatus with considerable flexibility. It was designed by the Nippon Telegraph Telephone Public Corporation. AURAL is programmed to provide transmission ratings in one or more of several metrics from objective measurements. Through comparative tests, AURAL measures have been shown to be in close agreement with subjective results from the CCITT Laboratory [6]. Such a system is very desirable because it removes the relatively large variability of (even trained) human listeners. There are several similar systems in use, for example, the British Telecom operates one known as TIGGER (see [7]). These other systems do not appear to have the flexibility of AURAL in terms of the number of metrics in which results may be given.

3.4 TRANSMISSION RATINGS

3.4.1 AEN

Several kinds of measurements may be made with subjective test facilities such as NOSFER and exactly what it is to be measured must be clearly defined. The object of one kind of measurement, AEN or equivalent articulation loss, is to determine a value, to be assigned to a collection of telephone equipment, that defines the intelligibility of speech as heard by a user of that equipment. It is referred to simply as an articulation test and is accomplished, for telephony purposes, by use of the SRAEN arrangement together with the system that is to be rated ([8] and [9]).

An articulation test is made by simply recording the percentage of words spoken by one person that are understood by another. For measurements of AEN, two systems, a reference and the unknown, are used. Attenuators are placed in the transmission path of each and adjusted to a number of different values during each test. Talkers and listeners conduct tests at these various settings, and the results for each system plotted as the percent of words understood *versus* the attenuator setting. That is, the attenuator in the reference system is adjusted to a number of values and an articulation test is made at each setting. The procedure is repeated using the system being rated. Actually, the tests alternate between the two systems in order to balance listener fatigue effects over the total test time. From the plots of these data, the attenuator setting in each test, that corresponds to the value at which 80% of the test words were understood, is determined. If the attenuator value, in decibels, for the reference system is called $A1$, and that for the system being rated, $A2$, then AEN is defined as AEN $= (A2 - A1)$. AEN then is a decibel-loss value relating the relative *articulation* quality of the two systems. The emphasis is to point out that the parameter being measured here is based on the relative ease with which spoken words are understood. It is an *articulation test* and does not imply anything more about the general *quality* of the received speech. Other measures are intrinsic to other tests and other systems.

3.4.2 Reference Equivalents, REs

Another type of subjective test is made to determine transmission ratings known as *reference equivalents* (REs). There are five REs:

- SRE (send reference equivalent): for describing the acoustic to electric transfer characteristic of a transmission system;
- RRE (receive reference equivalent): for describing the electric to acoustic transfer properties of a receiving system;
- STRE (sidetone RE): for relating acoustic pressure at the transmitter to that at the receiver of the same handset;
- JRE (junction RE): for describing the electric-to-electric loss across the *junction* or intervening connection between the transmitting and receiving systems;
- ORE (overall RE): an overall connection acoustic-to-acoustic loss.

REs are determined in a procedure similar to that for determining the AEN, but the test has properties which, one might agree, allow REs to be thought of as quality ratings of broader scope than articulation ratings. An important difference between articulation and reference equivalent tests is that for an RE determination, short sentences are listened to, first on the reference system, and then on the system being rated.[4] The listener turns the attenuator controlling the loss in the system being rated until the two systems are judged to be *balanced*. As above, the RE is the decibel difference in the attenuator settings. Figure 3.5 illustrates the SRE and RRE test arrangements.

Notice from the formulas in the figure that if the system under test has more loss than the reference system, then the rating number is positive. For example, in Figure 3.5(a) if the *Send System Under Test* introduces more loss than the *Send Reference System,* then the attenuator setting called $X1$ must be greater than the one called $X2$ in order for the levels at JR to be comparable. The difference, SRE $= X1 - X2$, is then positive. All ratings are loss values, in this case, loss with respect to the reference. Rating numerics may be positive or negative but, within any one metric, a larger (more positive) number indicates greater loss.

3.4.3 EARS and IEEE Metrics

Objective loudness loss, (3.1), is the basis for a set of objectively measured metrics, similar in concept to the CCITT REs, known as the electro-acoustic rating system. The EARS system has only four metrics, TLR, RLR, SLR, and OLR, transmit, receive, sidetone, and overall loudness ratings. There is no designator for intermediate connection elements such as the junction. When such elements are considered, they may be included in the calculation of TLR or RLR. Alternatively, a loudness loss may be computed for these elements, (3.4) below, and added to the overall loss, OLR, (3.8). In 1979, the IEEE adopted a modified version of the EARS system of measuring loudness ([10]). The IEEE rating metrics are

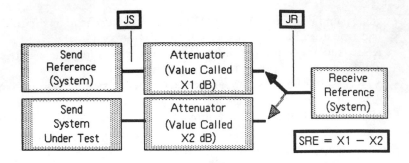

(a) Arrangement For Measuring Send Reference Equivalent, SRE

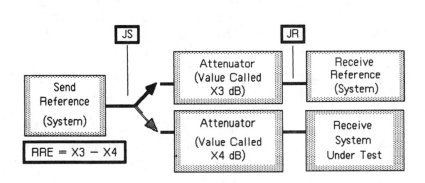

(b) Arrangement For Measuring Receive Reference Equivalent, RRE

Fig. 3.5 NOSFER Testing Arrangements

numerically different from those of the EARS system, so to eliminate confusion, the word objective is added to the names. The differences are principally due to changes in the test arrangements, primarily in the distance between the transmitter and the point of measurement of the acoustic pressure derived from a calibrated loudspeaker.

The IEEE metrics are defined below in (3.5)–(3.8). The notation, $f(\cdot)$, in these definitions, such as $f(v/p)$ means that the ratio specified is to be used for the variable in (3.1) for objective loudness loss repeated below as (3.4).

The specified ratios are all frequency dependent, and this is expressed in the following summation by the index, i. The ratios to be used for X in the X_i are usually in decibels and so must be converted by the inverse decibel operator.

$$Y_i = (10^{X_i/20}) \tag{3.3}$$

$$\text{OLL}(X) = -20 \log$$

$$\left\{ \frac{\Sigma_{i=2}^{i=M} \log(f_i/f_{i-1})[(Y(f_i)^{1/2.2} + Y(f_{i-1})^{1/2.2})/2]}{\log(f_M/f_1)} \right\}^{2.2} \quad \text{(dB)} \tag{3.4}$$

The M and i are chosen to properly cover the frequency band, 300–3300 Hz, the frequency band used with this system. Refer to the end of Section 3.5.4, *the EARS Equation,* for a discussion of how to properly cover the frequency band.

The four IEEE Loudness Ratings are:

- TOLR: *Transmit objective loudness rating* is a measure that relates acoustic pressure at the transmitter to voltage across the load of a transmitting system. TOLR is a function of the ratio of voltage delivered at the output of a transmitting system to the acoustic pressure at the input microphone:

$$\text{TOLR} = f(v/p) \quad \text{(dB)} \tag{3.5}$$

  The quantity, v, is the load voltage in millivolts, and p is the acoustic pressure in Pascals. Note that for a constant output voltage, v, if the microphone is more sensitive and less pressure, p, is required, by virtue of the minus signal of (3.4), TOLR becomes more negative. Larger negative numbers for TOLR means a *louder* transmission system. This is in keeping with the fact that ratings are loss values, as mentioned earlier. TOLR values may range from about -30 to -60 dB and depend upon many variables; typical values for installed telephones found in North America are around -45 dB, including the loss of the cable to the central office. Comparable values common in other administrations are usually louder by up to 15 dB in equivalent OOLR, the sum of the transmitting and receiving losses, (3.8) below. The differences are proportioned in various ways, determined by the administration, between equivalent TOLR and ROLR.

- ROLR: *Receive objective loudness rating* is a measure that relates voltage at the input of a receiving system to acoustic pressure at the earpiece:

$$ROLR = f[p/(v/2)] \quad \text{(dB)} \tag{3.6}$$

The voltage, v, is the open circuit value of the source in millivolts. The division by two translates the open circuit voltage to the actual closed circuit voltage delivered to a matched impedance receiving system. In the standard test configuration, described in [11], the source impedance is specified as 900 Ω. This must be taken into account in station design and system analyses. Because the function (3.4), applied to the ratio in (3.6) has a negative sign in front of the logarithm, as with TOLR, a smaller positive (larger negative) number for ROLR means more gain in the system. For installed systems, ROLR values may range from +35 to +60 dB, but typically (in the U.S.) are on the order of +48.0 dB.

• SOLR: *Sidetone objective loudness rating* is a measure that relates acoustic pressure at the transmitter to that at the receiver:

$$SOLR = f(p_e/p_m) \quad \text{(dB)} \tag{3.7}$$

where the subscripts e and m on the p terms refer to the acoustic pressure at the ear and mouth, respectively. Greater sidetone loss is indicated by a larger positive number. A desirable value of SOLR for U.S. telephones is about 6 to 14 dB, and may range from about zero to perhaps 18. As will be shown later (Chapter 7), SOLR is very difficult to control.

• OOLR: *Overall objective loudness rating* is the acoustic loss (or gain) from the transmitter of a talker to the receiver of a listener over a complete connection:

$$OOLR = TOLR + ROLR \quad \text{(dB)} \tag{3.8}$$

The measurement procedures ([10] and [11]) specify the acoustic field delivered by the loudspeaker to the telephone handset for TOLR measurements, and the source voltage to be applied to the handset receiver during receive measurements, as well appropriate source and load impedances. The values of voltage and pressure are chosen to facilitate measuring transmission ratings in the typical ranges mentioned above: An acoustic pressure of −6 dBPa (0.5 Pa) at a distance of 25 mm (one inch) from the transmitter, and a receiver open circuit driving voltage of 50 dBmv (316 mv). In the case of carbon microphone transmitters, measurements are difficult to repeat because of the nature of the device (see Chapter 5). Special conditioning procedures are detailed in order to make measurements as nearly repeatable as possible.

3.4.4 CCITT UK Loudness Ratings

Equation (3.2), CCITT loudness rating (LR), is the basis for a set of equations applicable to portions of a system, in a manner similar to (3.1). The derivations of these is a little more complex however.

The W_i in (3.2) incorporate the characteristics of a standard reference system, against which other systems are to be evaluated. A special modification of the NOSFER system was used in their determination. The frequency response characteristics of channels in use in many administrations were examined and culled by quality. Based on this work, a set of filters was designed for each portion of the NOSFER. The resulting sending and receiving portions are taken to be representative of very good quality transmission and receiving systems actually in use. Similarly, the junction, with the special filter in place, is taken as representative of a very good analog carrier system. The NOSFER, in this configuration, is known as the *intermediate reference system* (IRS) ([12]). The IRS is used as a reference for the evaluation of systems and in network design, particularly, international networks.

To use (3.2), the parameter m must be predetermined. This was done using low-pass and high-pass filters in balancing tests, similar to those for finding the half-loudness point as described in Section 2.6, with the unmodified NOSFER system (8-kHz bandwidth) as the reference. The value for m was determined to be 0.175. The quantity $(1/10)m$ in (3.2) then becomes 0.0175. This is commonly written in the expressions for the various loudness ratings as its reciprocal, $1/57.1$.

Recall that the relation between log(frequency) and loudness contribution is taken as linear (Figure 2.16, for example). Therefore, any set of equal logarithmic intervals on the frequency axis will yield equal contributions to loudness if the effective spectrum, Z_0, is uniform with frequency. The increments of frequency chosen for the summation in the LR equation (3.2) are the ISO preferred $\frac{1}{3}$-octave intervals.[5] The frequency range is the NOSFER bandwidth, 100 to 8000 Hz, which is covered by 20 of these $\frac{1}{3}$-octave bands. They are numbered sequentially, 1 to 20, so a band number specifies a frequency interval.

The L_{ui} in (3.2) refer to the loss-frequency characteristic of the overall response, mouth to ear, of the unknown system. Similarly, the weighting factors, W_i, are derived from and apply to the overall response of a reference system, the IRS described above. For the evaluation of portions of a system, sending, receiving, junction, different weights must be used. For example, if the unknown is a sending portion, L_{uMJ} (loss unknown—mouth to junction), a balance test as in Figure 3.5(a), would compare the unknown sending portion in tandem with the reference receiving portion, to the overall reference system. The comparison, written in terms of losses, and using the additional subscript notation, R = reference and O = overall, to identify components, is

$$(L_{uMJ} + L_{RJE}) - (L_O - W_O) \tag{3.9}$$

The term, $L_{uMJ} + L_{RJE}$, in (3.9) would then be used to form the L_{ui} in (3.2) to compute the loudness rating for the unknown sending system. The L_{RJE} represent part of the IRS and so are not subject to change. Therefore, they may be included as part of a new set of weightings applicable in 3.2 when the unknown is the sending portion. This set of weights is labeled W_S for *send* weighting. Three other sets of weights are defined in this manner: W_R for receive, W_J for junction, and W_O for overall.

The loss functions, L_{UMJ}, *etcetera,* are obtained from measurements of the input-output characteristic of the portion of a system to be rated, and expressed in decibels. These are referred to as sensitivities, and designated with an S. For example, for a sending system, the input is acoustic pressure in Pascals (Pa), and the output is a voltage (v) at the junction. The send sensitivity, S_{MJ} is given by

$$S_{MJ} = 20 \log(v_j/Pa_m), \quad v \text{ in volts (dB)} \tag{3.10}$$

Other sensitivities are defined below in (3.12). Notice that volts are used instead of millivolts as in the IEEE formulation. This difference alone gives rise to a numerical difference of 60 dB in the two types of ratings. Different weightings applied in the two loudness equations account for further differences. Note that the definition of the sensitivities are such that more efficient devices have greater sensitivity. This is a gain-like expression, greater gain yields a larger positive number of decibels. The loudness equation requires a loss value so the sensitivities must be entered with a negative sign. The negative sign preceding the exponent for the power term in (3.2) is used to absorb this and change the sign associated with W_i.

Equation (3.2) may be written for each particular portion of a system. The ratings are named according to the portion, *i.e., send loudness rating* is designated SLR, *receive loudness rating* is RLR, *etcetera*. Thus,

$$\text{SLR} = -57.1 \log \Sigma_i \, 10^{(1/57.1)[S_{uMJ}(i) - W_S(i)]} \tag{3.11a}$$

$$\text{RLR} = -57.1 \log \Sigma_i \, 10^{(1/57.1)[S_{uJE}(i) - W_R(i)]} \tag{3.11b}$$

$$OLR = -57.1 \log \Sigma_i \, 10^{(1/57.1)[S_{uMJ}(i)+S_{uJE}(i)-W_O(i)]} \tag{3.11c}$$

$$JLR = -57.1 \log \Sigma_i \, 10^{(1/57.1)[x_{uJJ}(i)-W_J(i)]} \tag{3.11d}$$

Note the change in notation in the expression for JLR, x_{uJJ} instead of S_{uJJ}. The junction does not involve a transducer so sensitivity, S, is not appropriate. The letter, L, could be used to indicate loss, but the CCITT notation is as shown.

The theory behind the sidetone calculation is slightly different. It treats sidetone heard through the speaker's internal acoustic coupling between throat and ear, plus that through the air from mouth to ear, plus the mechanical coupling through the handset, as a loss through which a masking sidetone is heard. The handset sidetone loss is evaluated against this masking loss. The reader is referred to [13] for details. The rating is called the *sidetone masking rating* (STMR), and is given by an expression similar to the others, except that the value for m in this case was found to be 0.225, yielding $(1/10)m = 44.44$. Thus,

$$STMR = -44.4 \log \Sigma_i \, 10^{(1/44.4)[L_{meST}(i)-L_E(i)-W_M(i)]} \tag{3.11e}$$

where $L_{meST}(i) = 20 \log[P_m(i)/P_e(i)]$ the sidetone loss, and L_E is an acoustic leakage factor accounting for the imperfect seal between the earcap of a handset and the listener's ear. The weights W_M are broken into two subsets for use as representations of the situations in which a person presses the receiver closely to the ear, called *sealed,* or loosely, called *unsealed.*

The weighting factors, taken from [13], are listed in Table 3.1 and 3.2. The column identifiers used in the tables have the following meanings:

- Band No: identifies an ISO preferred $\frac{1}{3}$-octave band;
- Midfrequency: the defined center of the $\frac{1}{3}$-octave band;
- Send W_S: weighting factors for calculating SLR;
- Receive W_R: weighting factors for calculating RLR;
- Junction W_J: weighting factors for calculating JLR;
- Overall W_O: weighting factors for calculating OLR;
- W_{MS}: weighting factors for calculating STMR, sealed;
- W_{ML}: weighting factors for calculating STMR, unsealed;
- L_E: weighting factors for earcap acoustic leakage.

Table 3.1
Weighting Factors For Computing Loudness Ratings

Band No.	Midfrequency (Hz)	Send W_S	Receive W_R	Junction W_J	Overall W_O
1	100	154.5	152.8	200.3	107.0
2	125	115.4	116.2	151.5	80.1
3	160	89.0	91.3	114.6	65.7
4	200	77.2	85.3	96.4	66.1
5	250	62.9	75.0	77.2	60.7
6	315	62.3	79.3	73.1	68.5
7	400	45.0	64.0	53.4	55.6
8	500	53.4	73.8	60.3	66.9
9	630	48.8	69.4	54.9	63.3
10	800	47.9	68.3	52.8	63.4
11	1000	50.4	69.0	54.1	65.3
12	1250	59.4	75.4	61.7	73.1
13	1600	57.0	70.7	57.6	70.1
14	2000	72.5	81.7	72.2	82.0
15	2500	72.9	76.8	71.1	78.6
16	3150	89.5	93.6	87.7	95.4
17	4000	117.3	114.1	154.5	76.9
18	5000	157.3	144.6	209.5	92.4
19	6300	172.2	165.8	245.8	92.2
20	8000	181.7	166.7	271.7	76.7

(Source: Red Book, Volume V, CCITT, Geneva, 1985)

Table 3.2
Weighting Factors For Computing STMR

Band No.	L_E	W_{MS}	W_{ML}
1	20.0	110.4	94.0
2	16.5	107.7	91.0
3	12.5	104.6	90.1
4	8.4	98.4	86.0
5	4.9	94.0	81.8
6	1.0	89.8	79.1
7	−0.7	84.8	78.5
8	−2.2	75.5	72.8
9	−2.6	66.0	68.3
10	−3.2	57.1	58.7
11	−2.3	49.1	49.4
12	−1.2	50.6	48.6
13	−0.1	51.0	48.9
14	3.6	51.9	49.8
15	7.4	51.3	49.3
16	6.7	50.6	48.5
17	8.8	51.0	49.0
18	10.0	49.7	47.7
19	12.5	50.0	48.0
20	15.0	52.8	50.7

(Source: Red Book, Volume V, CCITT, Geneva, 1985)

For some applications of (3.11) there are other small adjustments that may be needed, bandwidth corrections when calculating OLR, for example. Some of the basic ones are discussed below in Section 3.4.6, and examples for a number of special cases are given in [14].

The sensitivities in (3.11) are defined as follows, p is pressure in pascals and v is in volts:

For SLR, $S_{mj} = 20 \log(v_j/p_m)$ (3.12a)

For RLR, $S_{je} = 20 \log P_e/\tfrac{1}{2}v_{je}$ (3.12b)

For JLR, $S_{jj} = 20 \log(v_j/v_j)$ (3.12c)

For OLR, $S_{me} = 20 \log(p_e/p_m)$ (3.12d)

For STMR, $S_{meST} = 20 \log(p_e/p_m)$ (3.12e)

In (3.12c), the ambiguous notation, jj, means loss across the junction. The direction must be self-evident or stated for nonsymmetrical junction devices.

The measurements of sensitivities and the calculations should be carried out over the 8-kHz NOSFER band of frequencies. This is done using the ISO standard $\tfrac{1}{3}$-octave bands for the frequency intervals, and the weighting factors listed in Tables 3.1 and 3.2. In practice, data over the full 8 kHz is not often available so the calculation includes only the band of interest. Because the weighting factors were computed for the full band, a penalty appears in the narrow-band result that can be subtracted out as a correction factor. The size of the penalty is determined by the bandwidth ignored in the calculation. Factors for some of the more frequently used bands are listed in Table 3.3. The correction factors for the penalties are tabulated according to the number of ISO bands included in the calculation. These are identified by band number (see Table 3.1). These factors are to be subtracted from the LR calculated over the narrower band to make the result consistent with the NOSFER standard reference system.

3.4.5 Reference Equivalent (RE)

Reference equivalents, REs, date back to the SFERT system and are currently measured by means of the NOSFER, described in Section 3.3.4 and illustrated in Figure 3.3. It was designed to provide a measure of loudness for telephone transmitters and receivers, alone or in conjunction with subscriber loops and battery feed circuits. Being a direct evolutionary product of the MRS, a cable-oriented test arrangement, NOSFER has

Table 3.3
Bandwidth Correction Factors

Bands Included	Factor
3–18	0.1 dB
3–17	0.1 dB
4–18	0.3 dB
4–17	0.3 dB
6–16	2.1 dB

(Source: Red Book, Volume V, CCITT, Geneva, 1985)

a bandwidth extending from 100 to 8000 Hz, referred to as the 8-kHz band. The transmission characteristics of the NOSFER system are available in [1, Volume V, Recommendation P41].

The arrangements of NOSFER for particular measurements can be diagrammed as shown in Figure 3.6. The complete NOSFER system is illustrated in 3.6(a). The subjective test procedure involves comparing a part of the reference system with a corresponding unknown arranged as in Figure 3.6(b) or 3.6(c). Switches at the junction points, JR or JS, enable the unknown to be interchanged with the reference. The attenuator, $x1$ or $x2$ depending on the configuration, is set to an arbitrary value within a restricted range (about $\pm$ 15) around 25 dB. A talker, on the transmitting (left) side, speaks into a microphone on the left and the switch is used to alternately select one of the paths. A listener, using a receiver on the right side, adjusts the attenuator, $x0$, for equal loudness or *balance* between the reference system and the system under test. When the listener decides that the two systems are comparable, the reference equivalent assigned to the unknown is

$$SRE = x0 - x1 \quad \text{(sending RE)} \tag{3.13a}$$

for sending systems, and

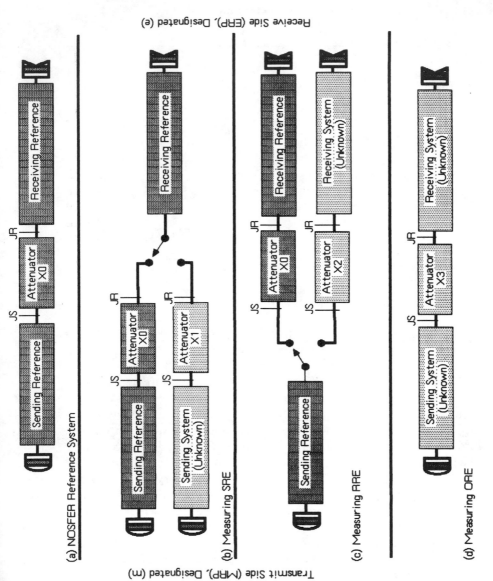

Fig. 3.6 NOSFER Test Configurations

RRE $= x0 - x2$ (receiving RE)   (3.13b)

for receiving systems. The procedure is repeated for several controlled combinations of talkers and listeners and results averaged for the final value.

Figure 3.6(d) shows an arrangement for evaluating ORE on a complete system. In this case, the comparison would be made between the complete NOSFER system, Figure 3.6(a), and the unknown 3.6(d).

3.4.6 REs, CREs, and Additivity

Using these arrangements, a great deal of data have been compiled assigning accepted values of SRE and RRE to frequently used pieces of equipment, telephones used by the various administrations in particular. As interest in ORE increased, it became apparent that the results of measurements of ORE did not always agree with those of SRE and RRE for the same unknown system. The problem was traced to two aspects of the testing procedure.

First, the arbitrary setting of $x1$ (or $x2$) at the start of each test, causes the absolute received levels to vary and, it was discovered, subjective loudness is not linear with level defined in decibels. Thus, in general, it was not true that ORE $-$ SRE $+$ RRE for a given system because all three measurements may not have been made at the same absolute received level.

The problem could have been eliminated by changing the test procedure. If, instead of arbitrarily setting attenuator $x1$ and adjusting $x0$ for balance, $x0$ could be left fixed and $x1$ adjusted. This would have the effect of making all tests at the same absolute listening pressure. However, because of the large number of systems that had been measured and assigned ratings, it was decided not to change the test procedure but to make adjustments in calculations using the measured quantities. Studies of this aspect of the problem resulted in a conversion equation to compensate for the level changes; the term *corrected* was prefixed to the converted numbers, and a C included in the identifiers. Hence, SCRE is the corrected value of SRE, *etcetera*. The correlation equation, usable for converting any reference equivalent to a corrected reference equivalent is

$$y = 0.0082(q + 70)^2 - 39.7$$ (3.14)

where q is the value of the RE, and y is the corrected quantity, CRE. The same equation is used for any RE: SRE, RRE, *etcetera*. For example, given an RRE value of -2.5, then $q = -2.5$ and $y = -2.3 =$ RCRE. If occasion should arise to compute q from y, then the positive result of the square root operation must be used.

The second aspect of the problem with additivity has to do with the 8-kHz bandwidth of NOSFER. Most commercial systems tested have bandwidths much less than 8 kHz and part of the loss in loudness expressed in a measured SRE or RRE is, therefore, a bandwidth penalty. If the sending and receiving portions of a commercial system have the same bandwidth, the penalty is measured and included in the REs twice, once when SRE is determined and once when RRE is determined. If the ORE is then measured, the penalty is included only once. (If the send and receive bandwidths are not identical, similar unequal errors occur, of course.) This error in the direct addition of SRE and RRE is handled by introducing a factor D_0, defined by

$$D_0 = \text{ORE} - (\text{SRE} + \text{RRE}) \tag{3.15}$$

Note that D_0 compensates for the double bandwidth penalty incurred if SRE and RRE are measured separately and then summed to form ORE. (This is the explanation often provided but, D_0 actually includes corrections for errors arising from the linear addition of nonlinear functions. See Section 3.4.8 on conversion factors.) Tests made in the CCITT Laboratory have found that D_0 is typically about -2 dB for a number of analog systems and about -6 dB for digital systems. Results for a number of such tests, with values for D_0 from -0.58 to -6.7, are reported in Annex B of [14]. The reference also states that, in the absence of complete measurements, the value -3.9 may be used for D_0.

3.4.7 CCITT Loudness Insertion Loss (LIL)

Using the notation introduced in Section 3.4.4 to identify portions of a system, the overall, mouth to ear (CCITT) loudness rating may be written as the sum of the component LRs from mouth to junction to ear:

$$L_{OME} = L_{MJ} + L_{JJ} + L_{JE} \tag{3.16}$$

Solving this for L_{JJ}, we have

$$L_{JJ} = L_{OME} - L_{JE} - L_{MJ} \tag{3.17}$$

and the now isolated L_{JJ} is seen to be the only term void of acoustic components. This is the essence of the *loudness insertion loss* (LIL) concept. The term may be used for any junction device and so applies to network elements, *i.e.,* trunks. Loudness insertion loss is defined for use with either loudness ratings, SLR, *etcetera* or with CREs. In the case of loudness ratings, it is calculated as the junction loss, (3.11d). As in the determination of any insertion loss, the impedances at the interfaces must be known and be the same when the x_{JJ} terms for the junction device are measured, or suitable corrections made. Junction impedances are usually assumed to be 600 Ω, resistive.

The use of the LIL concept with the subjective metrics CREs, is explained as follows. Recall that the need for CREs was the fact that subjective loudness is not linear in decibels, and the standard NOSFER test procedure allowed measurements of REs at many different acoustic pressures because the attenuator $X0$ (Figure 3.6) is set to a number of different values during the test sequence.

The NOSFER system is designed to simulate the orthotelephonic situation described in Chapter 2. In the orthotelephonic situation, the acoustic loss from a point 25 mm in front of the talker's lips to the position of a listener's ear, is 25 dB. This is reflected in the NOSFER calibration so that when $X0$ is set to 25 dB, the overall acoustic loss, MRP to ERP, is 25 dB. Thus when $X0$ is set to 25 dB, NOSFER is assumed to be representative of the ideal speaking situation, bandlimited to 8 kHz.

The weighting factors used in the loudness rating equation were obtained with $X0$ in the NOSFER system adjusted to 25 dB. This setting thus ties CREs, LRs, and REs (through (3.14)), together. One thus finds reference to R25 REs, referring to the setting of $X0$. (Equation (3.14) does not convert a q of 25 to a y of 25, however, because it was derived using components with typical telecommunication bandwidths. The equality point is RE = CRE = -4.3.) Loudness insertion loss is, therefore, a valid concept to use with CREs and an objective determination of LIL may be used with CREs in network planning. The factor(s), D_0 of (3.15), must be applied as appropriate for any network connections considered.

3.4.8 Relations between Metrics

The two loudness equations discussed in this chapter are different nonlinear functions of frequency dependent variables and, the subjective measures of loudness are of unknown form. Strictly speaking then, there is no simple way to convert between ratings in different metrics. If the spectra or the talkers can be duplicated; then one must re-compute or re-measure the rating in terms of the desired metric.

Nevertheless, there is a real engineering need to make conversions and, as it turns out, reasonable approximations are available. Before presenting them, it is desirable to look at the range of variations that may be expected in errors incurred by their use. In other words, "How good are the conversion factors?" To answer this we compare three channels, each comprised of a sending end and a receiving end. One is the CCITT defined IRS, mentioned earlier, one has idealized bandpass filters in both portions, and the third is a system with a flat response and zero gain in both the sending and receiving portions. The first two are illustrated in Figure 3.7. The third is just a constant, 0.0 dB in the figure.

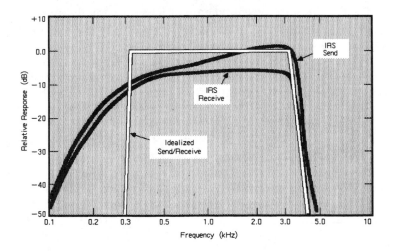

Fig. 3.7 Response of Systems Used for Rating Comparisons
Source: (IRS Channel Data) *Red Book, Volume V*, CCITT, Geneva, 1985

For the IEEE system, (3.8) states that OOLR = TOLR + ROLR; the comparable sum of CCITT loudness loss is qualified by the D_0 factor of (3.15). It was mentioned, when the D_0 was introduced, that only a part of the correction was attributable to the double bandwidth penalty. In either system of metrics, the overall loudness loss can be computed exactly only if the overall frequency response is known. Both formulations of loudness loss involve taking the logarithm of sums of (weighted) voltages or gain factors and for systems in tandem, such as the sending and receiving portions, the overall response is the product of the individual gain factors, which must then be weighted, summed, and changed to decibels. Adding the loudness losses for the two is similar to claiming that the log of a sum is equal to the sum of the logs. Some error is expected.

Computing the errors that obtain for the three channels considered here provides some indication of the magnitudes of inconsistency that may be encountered. These numbers establish a frame of reference when considering errors in the conversion factors.

Sending, receiving, and overall transmission ratings for the three channels are computed in three different ways: the IEEE method, the CCITT method using the full 8-kHz bandwidth formulation, and the CCITT method assuming limited data, large attenuation and no information outside the band between 200 and 4000 Hz. Reference [13] recommends a correction factor of +0.3 dB for this case.

The frequency responses for the IRS channel, from [12], are shown in Table 3.4 for the center frequencies of the 20 ISO bands and values obtained by linear interpolation at the lower and upper IEEE defined frequency limits, 300 and 3300 Hz. The reference values for the responses in dB are one volt per pascal, one pascal per volt, and one pascal per pascal for the sending, receiving, and overall segments. These are shown in columns two, three, and four. Those involving voltages are adjusted by 60 dB for the IEEE use of millivolts and are shown in columns five and six. The last column is a repeat of the overall response (no adjustment is required), and includes the interpolated values. In the table, the notations for pressure are: Pe = pressure in Pascals at the ERP, and Pm = pressure in Pascals at the MRP.

The transmission response for the idealized shape is zero in-band and −50 out-of-band for both sending and receiving for the CCITT metrics. It is taken as +60 in the flat portion and +10 out-of-band for the IEEE metrics. The receiving system is taken as −60 in-band and −90 elsewhere for the IEEE case. These values yield a through loss of zero in-band, and the correct sending and receiving IEEE ratings relative to the IRS system.

The ratings for the systems are as given in Table 3.5. The type of rating, or loudness loss, is listed in the left-most columns. The corresponding values are presented in the following three columns for the channels: IRS, ideal, and flat.

The transmission, or sending ratings appear in the first three rows, receiving ratings in the next three, followed by computed overall. The fourth set of three rows is the approximation to overall obtained from the sum of send and receive. The last set of three rows contain the correction factors, D_0, obtained as the corresponding differences between the two methods of obtaining the overall, using (3.15). A D_0 is calculated for the IEEE metrics even though this system does not recognize it.

From the Table it is seen that the ratings of the IRS channel are nearly exactly as desired. The sending, receiving, and overall ratings, calculated for data over the full 8 kHz, or the narrower bandwidth, are

Table 3.4
Response of IRS Channel in Figure 3.7

Center Freq.	CCITT Metrics			IEEE Metrics		
	Send dB $[v/Pm]$	*Receive dB* $[Pe/v]$	*Total dB* $[Pe/Pm]$	*Send dB* $[mv/Pm]$	*Receive dB* $[Pm/mv]$	*Total dB* $[Pe/Pm]$
100	−45.8	−27.5	−73.3			
125	−36.1	−18.8	−54.9			
160	−25.6	−10.8	−36.4			
200	−19.2	−2.7	−21.9			
250	−14.3	2.7	−11.6			
300*				48.4	−53.8	−5.4
315	−10.8	7.2	−3.6	49.2	−52.8	−3.6
400	−8.4	9.9	1.5	51.6	−50.1	1.5
500	−6.9	11.3	4.4	53.1	−48.7	4.4
630	−6.1	11.9	5.8	53.9	−48.1	5.8
800	−4.9	12.3	7.4	55.1	−47.7	7.4
1000	−3.7	12.6	8.9	56.3	−47.4	8.9
1250	−2.3	12.5	10.2	57.7	−47.5	10.2
1600	−0.6	13.0	12.4	59.4	−47.0	12.4
2000	0.3	13.1	13.4	60.3	−46.9	13.4
2500	1.8	13.1	14.9	61.8	−46.9	14.9
3150	1.8	12.6	14.4	61.8	−47.4	14.4
3300*				54.9	−55.2	−0.3
4000	−37.2	−31.6	−68.8			
5000	−52.2	−54.9	−107.1			
6300	−73.6	−67.5	−141.1			
8000	−90.0	−90.0	−108.0			

* IEEE loudness loss frequency bounds.

essentially zero. The weighting factors were chosen to achieve zero for this reference channel. The largest departure occurs in the wideband receive calculation (RLR = −0.4 dB). Also note that D_0 is zero for this case, again because of the method of choosing weights.

Table 3.5
Transmission Ratings for Three Channels

Channel:		*IRS*	*Ideal*	*Flat*
Rating:				
8 kHz	SLR	0.0	−4.4	−6.0
Narrow	SLR	+0.1	−4.6	−5.5
IEEE	TOLR	−56.7	−58.4	−60.0
8 kHz	RLR	−0.4	11.7	8.9
Narrow	RLR	−0.1	11.6	10.2
IEEE	ROLR	48.2	61.6	60.0
8 kHz	OLR	−0.4	5.3	−2.2
Narrow	OLR	−0.1	5.9	3.0
IEEE	OOLR	−8.0	1.7	0.0
Sums:				
8 kHz	S + R	−0.4	7.3	2.9
Narrow	S + R	0.0	7.0	4.7
IEEE	T + R	−8.5	3.2	0.0
D_0:				
8 kHz	D_0	0.0	−2.0	−5.1
Narrow	D_0	−0.1	−1.1	−1.7
IEEE	D_0	−0.3	−1.5	0.0

The next item to look at is the IEEE evaluation of the flat channel (last column). There is no distortion and it is so rated.

The idealized channel (middle column), evaluated by the IEEE equation, appears nearly distortionless, a departure from the flat channel of 1.7 dB overall, and 1.6 dB each for the sending and receiving portions. Note that the ideal used in this case has nonperfect edges (Figure 3.3) because of the arbitrary use of the nearest ISO center frequencies for the first measure of in-band loss values. If the width of the sloping band edges is made small by specifying a one-hertz difference between in-band and out-of-band frequencies, the calculation duplicates the result of that for the flat channel.

Because the reference system for the IEEE metric is a free air path, the IRS channel shows modest distortion in the transmission path, about 3 dB, and about 12 dB in the receiving path. Overall it exhibits a gain of 8 dB (the minus sign indicates a negative loss).

The range of channels in this example is from zero distortion for the flat channel, through a nearly ideal band-limited telephone channel, to one that is considered typical of excellent real channels. The interesting thing to observe is that the "log-addition" approximation is not bad. The errors are measured by the quantity D_0. This is seen to be small, 2 dB or less, except for the channel with no bandlimiting whatsoever, in which case the CCITT 8-kHz rating is affected significantly, $D_0 = -5.1$ dB. That rating system was designed to evaluate real channels, however, not free air paths.

This exercise is presented as a demonstration that the linear addition is good enough in practice for approximations in engineering problems. Errors can be expected to be less than 2 dB. Now we are ready to look at conversions between the various metrics.

Some approximate, empirically derived, conversion factors between the metrics are provided in Table 3.6.[6] These conversion factors are statistical means taken over many observations. They are useful in obtaining quick comparisons and in early design stages. Experience has indicated that one can expect agreement of a conversion with measured values within about 1.5 dB (plus or minus) depending upon the frequency response and actual impedances realized in specific system designs.

The values given are for two specific cases. The first is that in which the measurement arrangement is identical for any metric; that is, the same source and load impedances are used, no cable (simulated loop) changes are permitted, and the same acoustic environment exists for all measurements. In practice, the CCITT and OREM measurement standards call for different kinds of cable, different feeding bridges, and different spatial relations between the handset mouthpiece and the artificial voice, than does the IEEE standard. The simplest important difference between the IEEE standard and the others, is the use of 900-Ω source and load impedances instead of 600. This difference is the basis for the second set of table entries labeled 900/600 Adj. These second factors are to be used when the impedances in the measurement systems are as the standards specify. The conversions apply for an IEEE measurement with 900-Ω terminations and the other measurements with 600-Ω terminations. Other differences must be accounted for as required *i.e.,* different types of cable.

Some of the conversion factors in the table may be compared with the calculations for the channels of Table 3.5. For example, the first three rows of Table 3.5 can be used to obtain six estimates of the conversion from TOLR to SLR by subtracting the TOLR values from the SLR values within each column. The results range from 53.8 to 56.8 which compare with the value 56.1 of Table 3.6. Similarly, observed ROLR to RLR conversions in Table 3.5 range from -48.3 to -51.1 compared to -47.5 in Table 3.6. The calculated OOLR to OLR differences range from -2.2 to 7.9, with 6.0 in Table 3.6. Note that the -2.2 value is associated with an 8-kHz CCITT rating of the flat channel. This is the case that exhibited the largest D_0 value as noted above.

Table 3.6
Loudness Rating Conversions

Metric	IEEE	CCITT	Corrected CCITT	UK (CCITT)	OREMA
Transmit	TOLR	(S)RE	(S)CRE	SLR	Transmit
Equal Z	T	*	T + 52.5	T + 56.1	T + 46.9
900/600 Adj	T	*	T + 54.1	T + 57.7	T + 48.5
Receive	ROLR	(R)RE	(R)CRE	RLR	Receive
Equal Z	R	*	R − 45.6	R − 47.5	R − 50.3
900/600 Adj	R	*	R − 47.6	R − 49.5	R − 48.3
Sidetone	SOLR	(ST)RE	(ST)CRE	STMR	Sidetone
	S	*	S + 6.0	S + 8.0	S + 5.1
Overall	OOLR	(O)RE	(O)CRE	OLR	Overall
900/600 Adj	O	*	O + 6.0	O + 1.0	O + 0.2

* The RE, q, is given in terms of original CRE value, y, by $q = |[(y + 39.7)/.0082]^{1/2}| - 70$ (from (3.14)).

NOTES

1. IEEE: Institute of Electrical and Electronics Engineers.
2. A comparison of two voice-transmission systems by voice and ear is referred to as a *telephonometric measurement*.
3. Noise with a frequency content taken to be representative of that in common indoor environments (see [4]).
4. Usually five trained people listen, one at a time, and their results are averaged.
5. This is a standard set of frequency intervals recommended for general acoustical analysis. The frequencies defining the band centers are based on the formula $f = 10^{n/10}$, where n takes on integer values. The usual starting frequency is 100 Hz, $n = 20$, and the frequencies given by the formula are rounded, *i.e.*, 100, 125, 160, . . . , instead of 100, 125.9, 158.5, *etcetera*. See the standard: ANSI A1.11-1966.
6. The table is based on data used, with permission, from internal AT&T memoranda. Some of the original numbers have been modified by the author.

REFERENCES

1. CCITT, *Yellow Book,* VIIth Plenary Assembly, ITU, Geneva, 10–21 November 1980.
2. CCITT, *Red Book,* VIIIth Plenary Assembly, ITU, Málaga-Torremolinos, 8–19 October 1985.
3. CCITT, "Description of the ARAEN," *Yellow Book, Volume V,* Rec. P.41, ITU, Geneva, 1980.
4. CCITT, "Noise Spectra," *Red Book, Volume V,* Supplement No. 13, ITU, Málaga-Torremolinos, 1985.
5. CCITT, "The SIBYL Method of Subjective Testing," *Red Book, Volume V,* Supplement No. 5, ITU, Geneva, 1985.
6. Irii, H., and Kakehi, K., "Instrumentation for Objectively Measuring Loudness Ratings of Telephone Systems," *Review of the Electrical Communications Laboratories,* Research and Development Headquarters, Nippon Telegraph and Telephone Corporation, Tokyo, Vol. 34, No. 4, 1986, pp. 429–436.
7. Ward H.F., and Cross, R.C., "TIGGER—An Automatic Test System for Measuring the Transmission Performance of Telephones," *British Telecommunications Engineering,* Vol. 2, July 1983, pp. 70–77.

8. CCITT, "Description and Adjustment of the Reference System for the Determination of SRAEN," *Yellow Book, Volume V,* Rec. P.44, ITU, Geneva, 1980.

9. CCITT, "Articulation Reference Equivalent," *Yellow Book, Volume V,* Rec. P.12, ITU, Geneva, 1980.

10. IEEE, "IEEE Standard Method for Determining Objective Loudness Ratings for Telephone Connections," *IEEE Std. 661-1979,* November 29, 1979.

11. IEEE, "IEEE Standard Method for Measuring Transmission Performance of Telephone Sets," *IEEE Std. 269-1971,* February 4, 1971.

12. CCITT, "Specification for an Intermediate Reference System," *Red Book, Volume V,* Rec. P.48, ITU, Geneva, 1985.

13. CCITT, "Calculation of Loudness Rating," *Red Book, Volume V,* Rec. P.79, ITU, Geneva, 1985.

14. CCITT, "Corrected Reference Equivalent (CREs) and Loudness Ratings (LRs) in an International Connection," *Red Book, Volume III,* Rec. G.111, ITU, Geneva, 1985.

Chapter 4
Measures of Quality

Chapter 1 defined the transmission impairments that affect the quality of speech and data on telecommunication facilities. In this chapter, methods to model users' perception of the quality of the received signal will be described, and some of the relations that map transmission impairments into quantitative quality measures presented. Applications of these relations in the design of systems are illustrated in Chapter 7 after the introductory material on terminals and networks in Chapters 5 and 6.

4.1 VOICE QUALITY

4.1.1 Introduction to Voice Quality

In the design of a network, the expected magnitude of each of the parameters that relate to voice transmission must be examined to assure that it falls within acceptable bounds. These bounds determine the probability that a user will be satisfied with the quality of a connection through the network. The subject of this section is the development of mathematical tools needed to estimate that probability. This is generally a two-step process: first, subjective assessment of the relative quality produced by various transmission conditions is quantified; then expressions that provide a mapping between the transmission conditions and these assessments are determined. We begin our discussion with the development of one particular model of user reaction to transmission conditions. This leads to the concept of transmission *grade-of-service* (GoS), an overall quality measure, computed through a set of transmission-quality mapping relations.

GoS is a voice transmission evaluation tool. For data transmission, no such equivalent has been developed but a number of quality-oriented measures exist. These are considered in the latter sections of the chapter.

4.1.2 A Grade-of-Service Model

The object of grade-of-service modeling is the development of expressions relating impairments to a user's assessment of transmission quality. It requires knowledge of human factors, subjective testing, acoustics, and statistical analysis, as well as telecommunications. The construction of such a model requires years of subjective testing, analysis, and verification. A detailed description of the procedures used to develop a very useful model for three impairments is given in [1]. Extensions to it for additional parameters are described in [2]. The model as extended, includes:

- loss
- noise
- talker echo
- room noise
- channel frequency response
- sidetone level
- sidetone frequency response
- quantizing noise
- listener echo

The model derivation in [1] is in terms of the EARS metric, the extensions in [2] are in CCITT *corrected reference equivalents* (CRE). For completeness, and direct use in the U.S., the development in the following is in the IEEE metric; the CCITT formulation and factors required for conversion between the CCITT and IEEE relations are given in Section 4.1.7. The general procedure used in the construction of the model is presented below.

The modeling methods and the concepts embodied in the results are important in understanding the general rationale behind transmission objectives. This does not mean that all requirements encountered are based on extensive models. The intent of good requirements is the same, even if some do not have the benefit of elaborate models behind them. An understanding of the models provides a framework for thinking about and evaluating any set of requirements that may be encountered. Some applications of the results discussed below are illustrated in Chapter 5, Section 5.2, Optimizing GoS.

The modeling proceeds, in general, in the following manner: Consider results of a subjective test, obtained by use of a facility such as SIBYL, plotted in histogram form as in Figure 4.1(a), then accumulated, smoothed, and fit with a normal cumulative probability function or CDF, $G(\mu, \sigma)$,[1] as in (b). Such a plot represents the distribution of subjective evaluations of a connection with a fixed set of transmission impairments.

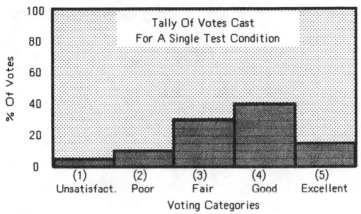

(a) Vote Histogram

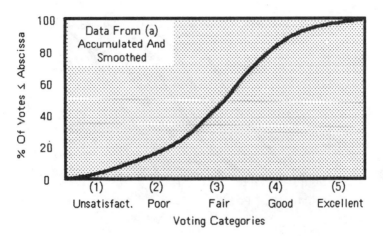

(b) Cumulative Distribution Function (CDF)

Fig. 4.1 Presentations of Subjective Test Data

The average value of such a CDF is called the *mean opinion score* (MOS). For analysis purposes, a scale is created by assigning the values 1 through 5 to the midpoints of the histograms of votes (opinions unsatisfactory to excellent). In the case illustrated in Figure 4.1(b), the MOS would be about 3.1.

If one parameter, say the amount of noise on the connection, is adjusted to different values, and the experiment repeated for each of them, a set of CDFs is obtained, as illustrated in Figure 4.2(a). Note that, in general, the standard deviation as well as the MOS can be expected to

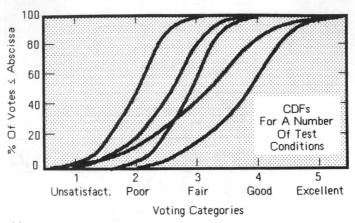

(a) Raw Data

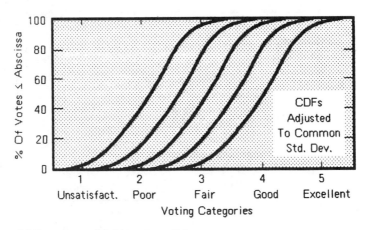

(b) Data From (a) After Smoothing

Fig. 4.2 Smoothing and Fitting Data to Best Common Standard Deviation

change with changes in a parameter, as indicated by the different slopes of the linear portions of the curves. (The change in standard deviations is exaggerated in the illustration.) The data resulting in plots as shown here, are smoothed and fit in a special manner. The procedure yields a set of normal distribution functions of (best fit) constant standard deviation, and with mean values that are best fits to the MOSs for each test condition. In other words, the data are smoothed to the extent that all the CDFs have the same slope. The mean values of the distributions are called *fitted means* (fm) and correspond very closely to the original MOSs. The procedures for doing this are outlined in [1]. Graphic comparisons of results of the fitting process with the original data are also provided there. The results of the fitting process appear as in Figure 4.2(b). The single (constant) standard deviation, derived in the smoothing, recognizes typical differences of opinion among the test subjects, but forces the spread of opinions for each condition to be the same even though the mean opinion changes from condition to condition. Thus, the smoothing forces the variation in opinion to be constant, but allows the fitted means to change as the transmission conditions change.

If the set of derived CDFs is a reasonable representation of user's opinion of quality, then, if a fitted mean can be estimated from a given set of transmission conditions, the entire distribution of users' reaction to that set of conditions will be known. What is required then is the determination of an empirical relation between fitted means and the transmission conditions for each test. Thus, for impairments x and y:

$$fm = f(x, y) \tag{4.1}$$

Equation (4.1) is written for two impairments, but, as shown by the list given earlier, fm, in general, is a function of many parameters. This is the essence of the subjective modeling, but the question of stability of opinion distributions with time, groups of people (test subjects), and perhaps other unknown factors needs to be addressed. This complicates the procedure.

It is found that subject group opinions tend to change with time (years) and perhaps other unknown factors, even though the laboratory or other test conditions are maintained as nearly constant as possible. The change with time appears to be consistent in that subjects become increasingly critical (as tracked in at least four tests over about one decade.) MOS appears to decrease over the years. Modest changes in the variance or standard deviation of overall best fits have also been found. This implies that separate equations relating impairments to fitted means are required for each test, making application of the models cumbersome at best.

Each test, in the sense of this section, is unique and comprises a large amount of data. It has become the practice to distinguish these by referring to each with a name, reflecting some characteristic of the test, followed by the words *data base*. Thus, one finds reference to the *long toll data base*, the *CCITT data base, etcetera*. The first complete test, using loss, noise, and talker echo, was conducted at the Murray Hill location of the Bell Telephone Laboratories, so these data are referred to as the *Murray Hill* (MH) *data base*. A number of tests of the same type were conducted later at the Holmdel location. These are designated *Holmdel No. 1* (HO1), *Holmdel No. 2* (HO2), *etcetera*.

A scatter plot of fitted means for a number of transmission conditions from the MH and HO1 data bases is shown in Figure 4.3 together with the results of a linear regression analysis of the MH means on the HO1 means:

$$fm_{MH} = -0.206 + 1.372fm_{HO1} \qquad (4.2)$$

The downward bias of opinion of the HO1 subjects compared to the MH subjects is reflected in the negative constant, or offset of -0.206. The fact that the slope, 1.372, exceeds one indicates a tighter overall clustering of votes within the HO1 data base. The spread of opinions was larger in the MH scores and, on average, about 0.2 of an opinion category higher than those from the HO1 data base.

This problem of shifting opinion, or biases between tests, was treated in the following way. Consider the Murray Hill data base as the reference, and let each of the others be designated data base i, $i = 1, 2,$ . . . . Linear regressions were performed between the MH data base, and each data base i, of interest. The resulting relations were used to translate fitted means of the other data bases into equivalent Murray Hill fitted means by application of the regression equations. Thus, a regression of the form:

$$fm_{MH} = c_i + d_i \cdot (fm_i) \quad \text{(Regression Equation)} \qquad (4.3)$$

applied to the data from base i yields equivalent points in the MH data base. The resulting larger base was then used to find fm_{MH} as a function of the transmission parameters. For example, if the transmission variables in a test are noise, n, and loss, l, then the empirical relation between these and the fitted means can be represented as

$$fm_{MH} = f(n, l) \qquad (4.4)$$

Now, given a noise-loss pair, (n, l), these can be substituted in the expression for fm_{MH}, (4.4), and the result used in the regression, (4.2) with HO1 replaced with i, to solve for the fm_i for any data base, i, of interest. The resulting estimate of fm_i will be accurate to within the errors represented by the departure of the actual fitted means from the regression lines.

If one knows the fitted mean for an (n, l) pair, and the standard deviation for the data base, then a CDF such as one of those in Figure 4.2(b) is specified. These curves provide estimates of the percent of all votes that were cast in any particular voting category or in a selected set of categories. For example, consider the left-most curve in Figure 4.2b.

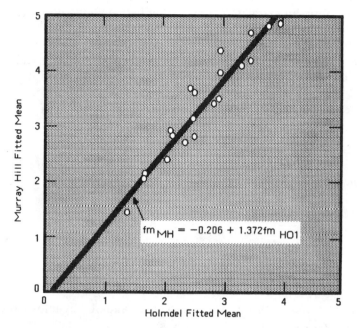

Fig. 4.3 A Sample of the Regression Analysis
Source: Cavanaugh, *et al.*, "Models for the Subjective Effects of Loss, Noise, and Talker Echo on Telephone Connections," *Bell System. Tech. J.*, Vol. 55, No. 9, November 1976. Reprinted with permission from the Bell System Technical Journal. Copyright 1976 AT&T

Of the curves shown, this one represents the worst transmission conditions used in the tests, it has the lowest mean value, *fm*. A fairly large percent of all voters rated this condition either unsatisfactory or poor. The voting category poor is bounded by opinion scores of 1.5 and 2.5. Unsatisfactory is bounded by 1.5 on the right, includes all values less than that, and so extends all the way to $-\infty$. The percentage of all voters who rated this condition unsatisfactory or poor is the value of the CDF corresponding to the abscissa value 2.5, about 40%. The exact percentage can be found by integrating the density function (pdf) from $-\infty$ to 2.5.

Now, transmission grade-of-service (GoS) is typically specified by the values of two such integrals. The first is called *percent good* or *better* (percent GoB) and is 100 times the integral of the pdf from 3.5, the lower bound of the good category, to infinity, and represents the upper tail of the CDF. The second is called *percent poor* or *worse* (%PoW), and is 100 times the integral of the pdf from $-\infty$ to 2.5, the lower tail of the CDF. The 100 is needed because a pdf is expressed in fractional measure rather than percent. Percent good or better is an estimate of the percentage of users who would be relatively pleased with the transmission quality; percent poor or worse is an estimate of the percentage who would be relatively unhappy with it.

For the example above, the mean (also the 50% point) for the leftmost curve of Figure 4.2(b) occurs at about 2.2 on the opinion scale. If the standard deviation for that distribution were equal to s, then the density function, $g(\mu, \sigma)$, is $g(2.2,s)$. The percentage of votes in the poor and unsatisfactory categories, P_{pow}, would be

$$P_{pow} = 100 \int_{-\infty}^{2.5} g(2.2,s) \, dx \qquad (4.5)$$

where the function g is an exponential given by

$$[1/(2\pi)^{1/2}]\exp[-(x - \mu)^2/(2s^2)] \qquad (4.6)$$

The integrations can be simplified with a normalizing change of variable:

$$t = (x - \mu)/\sigma \qquad (4.7)$$

This results in a pdf with zero mean and unit standard deviation called the unit normal pdf, $g(0, 1)$:

$$g(0, 1) = [1/(2\pi)^{1/2}]\exp(-t^2/2) \qquad (4.8)$$

Tables and computer algorithms for the integral of this function are readily available.

If the fitted means of data base i are designated fm_i, the finite limits needed for GoS calculations are, by reference to (4.7):

$$(3.5 - fm_i)/\sigma_i \quad \text{limit for \%GoB} \qquad (4.9a)$$

and

$$\%GoB = \int_{(3.5 - fm_i)/\sigma_i}^{\infty} g(0, 1) \, dt$$

Similarly,

$$(2.5 - fm_i)/\sigma_i, \quad \text{limit for \%PoW} \qquad (4.9b)$$

and

$$\%PoW = \int_{-\infty}^{(2.5 - fm_i)/\sigma_i} g(0, 1) \, dt$$

The values for fm_i for use in these relations may be found by the inverse of the regression, (4.3). Thus,

$$fm_i = (fm_{MH} - c_i)/d_i \qquad (4.10)$$

and the several different empirical relations mapping impairments to fm_i for each data base, i, are no longer necessary. Only the regression coefficients, c_i and d_i, need be preserved.

Thus, for any combination of noise and loss, (4.4) will yield an fm_{MH} for the Murray Hill data base that can be translated by (4.10) to an fm for the selected data base. The desired grade-of-service is found by integrating the unit normal pdf, (4.8), with the finite limit given by (4.9a) or (4.9b).

Now, (4.4) is written for the two parameters, noise and loss, and we mentioned that many parameters may be of interest. This indicates that many different functions corresponding to (4.4) are needed and, for each case of interest, the steps above must be repeated to determine the grade-of-service. Let the m in the notation fm of (4.4) stand for the particular parameter combination, noise and loss, then $f_j, j = 1, 2, \ldots$, can represent any of the numerous other combinations, for each case a specific j must be stated. Each j then leads to a new integration for GoS. As will be shown later, each function of j, for j not equal to m, can be formulated as a modification of fm. Because all transmission involves some noise and loss, fm may be considered the basic relation and other parameters, represented by j, merely perturb the base value given by fm. This is equivalent to accepting a shift in mean opinion score for noise and loss by other parameters, but assuming that the standard deviation of the CDFs is unaffected. Now the function that must be integrated is fixed, and only the limits of integration change as other parameters are introduced.

We now need a procedure to relate or map the shifted means, f_j, to equivalent original values, fm. This is accomplished by the introduction of a new variable called *transmission rating* (R). The R scale is a common denominator to which all other f_j relate. We would like R to assume values between zero and 100 with zero representing a very poor transmission condition and 100 a very good one.[2] These extremes may never be realizable, but all f_j from real conditions must map to this scale. R is simply a linear transformation, with constraints, of the fitted means, f_j. It can be thought of as being created from the Murray Hill data base for noise and loss, f_{MH}. The constraints are that, for two preselected test conditions, the f_{MH} values map to two preselected R values. If the preselected test conditions are properly chosen, and assigned appropriate R values, then the remaining f_{MH} values (for other loss and noise conditions) will map over the scale zero to 100. Transformations with constraints such as these are referred to as being anchored, and the preselected values are called *anchoring points*. Anchoring serves to force the mapping effected by the transformation into a desired range. The transformation is

$$R = a + b \cdot fm_{MH} \quad \text{(transformed } fm_{MH}) \tag{4.11}$$

where a and b are found by the solution of the two equations determined by substituting the two selected anchoring noise and loss values for fm_{MH} and setting the equations equal to the corresponding preselected R values. The anchoring points were chosen to be well separated, one representing relatively good transmission, and one relatively poor, but within the range of realistic conditions:

$$\text{noise} = 25 \text{ dBrnC}, \quad \text{OOLR} = 10 \text{ dB}, \quad R = 80 \tag{4.12a}$$
$$\text{noise} = 40 \text{ dBrnC}, \quad \text{OOLR} = 25 \text{ dB}, \quad R = 40 \tag{4.12b}$$

The inverse of (4.11) is needed below, so

$$fm_{MH} = (R - a)/b \tag{4.13}$$

The anchoring points chosen correspond to specific grades-of-service, %PoW and %GoB. These now serve as the link between test results for other parameters and the R scale. When other parameters are tested and the data analyzed, the new parameter conditions that result in the same grades-of-service must be assigned the same R values, 40 and 80. For other parameters, then, an empirical relation may be sought that maps their values directly into the R scale instead of f_j.

4.1.3 Data Base Descriptors A and B

The R scale, as defined above, is tied to one series of subjective tests, that summarized in the Murray Hill data base. This restriction can be removed and limits of integration for %GoB and %PoW, (4.9) expressed in terms of an R related to any data base.

Any noise-loss pair, used in tests for any data base i, can be substituted in (4.1), derived for that data base, to obtain an fm_i. The fm_i can be converted to an fm_{MH} by the use of the regression (4.3):

$$fm_{MH} = c_i + d_i \cdot fm_i$$

The fm_{MH} substituted in (4.11) yields an R value:

$$R = a + b(c_i + d_i \cdot fm_i) \tag{4.14}$$

A diagram illustrating these relations is shown in Figure 4.4. Various combinations of two parameters, X and Y, are illustrated as giving rise to the sets of mean opinion scores on the vertical bar labeled fm_i. The regression equation for data base i maps these scores to equivalent Murray Hill data base scores shown on the vertical scale labeled fm_{MH}. These, in turn, are mapped to the R scale on the right through the linear transform with parameters a and b. This flow from left to right is indicated by the expressions and arrows at the top of the figure. The reverse mapping is indicated similarly at the lower part of the drawing.

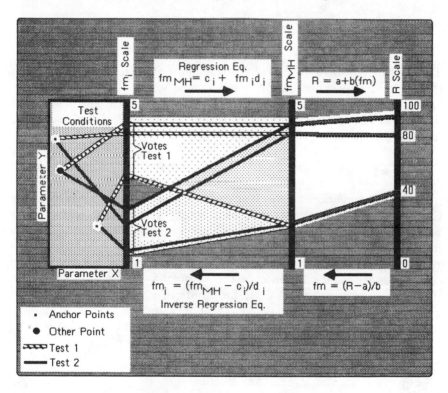

Fig. 4.4 Mapping of fm_i to and from the R Scale

The central point in the X, Y parameter plane is shown receiving a higher voting score than the apparently better upper anchoring point. Such a result is not uncommon because the parameters may exhibit trade-offs, or be compensatory over portions of their range, *i.e.*, the function mapping, (X, Y) to voting score, may be nonlinear. Some examples of this are shown later in Section 4.1.6.

Two separate tests, labeled 1 and 2, are hypothesized in Figure 4.4 to illustrate the effect of anchoring. It is assumed that the subjects in Test 2 cast votes consistently lower than those in Test 1. Note the corresponding votes for the two anchor conditions; dashed lines for the Test 1 group, and solid for the Test 2 participants. The bias is removed through the regression, and anchoring maps the R values for the anchor conditions to the selected values 80 and 40.

The equations for the limits of integration for the two measures of GoS can now be rewritten. Equation (4.14) can be solved for fm_i in terms of R:

$$fm_i = (R - a - c_i b)/bd_i \tag{4.15}$$

This replaces fm_i in (4.9a) and (4.9b), which then give the limits in terms of the linear transform coefficients a and b, and the regression coefficients c_i and d_i. If the vote values of 3.5 and 2.5 used for %GoB and %PoW are denoted as V_k, $k = 1, 2$, then the finite limit of integration, $\lim'_k$ for either measure is:

$$\lim'_k = -(R - V_k b_i d_i - a - c_i b)/(b_i d_i \sigma_i) \tag{4.16a}$$

This formulation for the limits of integration is consistent with the CDF structure of Figure 4.2(b) and a voting scale that assigns a larger number to better quality, i.e., 1 = bad and 5 = good. This arrangement is intuitively pleasing to the author and, it is hoped, to the reader. The tests were actually scored in the reverse fashion, 1 = good and 5 = bad. This is of little consequence because of the symmetry of the normal density function. For the reversed scale the limits of integration can be interchanged and their signs reversed. The limits $\lim'_k$ to infinity for %GoB become $-\infty$ to $-\lim'_k$ and similarly for %PoW. The new limits will then be given by the negative of (4.16a):

$$\lim_k = \frac{R - [a + b(c_i + V_k d_i)]}{\sigma_i bd_i} \tag{4.16b}$$

With $V_k = 3.5$, the value of (4.16b) is called A, and with $V_k = 2.5$, it is called B. With these definitions, the equations for the two grades-of-service become

$$\%\text{GoB} = (2\pi)^{-1/2} \int_{-\infty}^{A} \exp(-t^2/2)\, dt \quad \text{(definition of } A) \tag{4.17a}$$

and

$$\%\text{PoW} = (2\pi)^{-1/2} \int_B^\infty \exp(-t^2/2) \, dt \quad \text{(definition of } B) \tag{4.17b}$$

The results of different tests (different data bases) can then be reported simply in terms of the appropriate values for A and B. This is the way in which the various data bases are summarized. A and B are tabulated as functions of R, (4.16b), using the translation coefficients, a and b, with the appropriate regression coefficients, c_i and d_i. V_k is set to 3.5 to find A, and to 2.5 to find B. Some examples and applications of these test result descriptors are given in the following sections.

4.1.4 Classifying Transmission Ratings, R_y

Now recall that in Section 4.1.2 it was stated that an R value can be determined for any combination of parameters, j, and their corresponding f_j. Thus, R must be identified as to which ones were considered when the equation relating it to the j was derived. Hence, equations that map impairments to the R scale are labeled R_y, where the y describes the transmission parameters included. For example, R_{LN} means transmission rating when the impairments considered are loss and noise. The corresponding equations, $f(y)$, mapping impairment(s) y to the R scale are found empirically by seeking the best fit to the corresponding set of fitted means, $fm(s)$, taken from CDFs such as in Figure 4.2(b) for the y of interest. The independent variables are the values of the transmission impairments for each mean.

Some of the transmission ratings are R_{LN}, just mentioned, R_{E} for talker echo, R_{LE} for listener echo, R_{LNLE} for combined loss, noise, and listener echo, *etcetera*. For some of these ratings, many factors may enter into the computation. A partial list is:

- room noise
- sidetone loss
- sidetone frequency response
- channel frequency response
- quantizing noise
- echo path loss
- echo delay

Interaction of these factors make the empirical relations for the R functions somewhat lengthy. In addition, one may choose to include variations in handset efficiency or positioning, for example, as parameters. As a result, a complete evaluation of an entire telephone connection, including room noise at both ends, position of the handset, *etcetera*, is a tedious calculation, but is straightforward to program and carry out on a computer for studies of entire networks. Rather complete evaluation programs will fit into programmable calculators, so appearances can be deceiving in terms of the actual program size required.

4.1.5 Detail of R_{LN}

As an example of one of the more lengthy rating calculations, the transmission rating for loss and noise, R_{LN}, is detailed next. The noise used must be the total noise heard. It may include room noise at the near or far ends, and quantizing noise if digital facilities are used, as well as analog circuit noise. Including all of these is computationally lengthy.

The basic R_{LN} equation, derived by fitting the votes from the Murray Hill tests to the loss and noise values, is given in [1] in EARS metrics, and in [2] in CCITT CRE metrics. (The noise computations are also explained in [2].) In IEEE metrics, it is

$$R_{LN} = 147.76 - 2.257[(OOLR - 2.2)^2 + 1]^{1/2} - 1.907N$$
$$+ 0.02037 \cdot N(OOLR), \quad \text{(rating for loss and noise)} \quad (4.18)$$

where OOLR is the IEEE *objective overall loudness loss* found by use of (3.8), or (3.1) applied to the system acoustic-to-acoustic loss characteristic; and N is the total effective noise in dBrnC, referred to a receiving system with an ROLR of +46. With elements defined below, total effective noise is the power sum of all noise sources:

$N = N_c \oplus N_{Re} \oplus N_{Qe}$ ($\oplus$ indicates power addition)

N_c is circuit noise;

N_{Re} is effective room noise, and

N_{Qe} is effective quantizing noise.

N_c, the circuit noise, is in dBrnC referred to a receiving system with an ROLR of +46; N_{Re} is the effective circuit noise (dBrnC) equivalent of room noise in dBA referred to a receiving system with an ROLR of +46; and N_{Qe} is the effective circuit noise (dBrnC) equivalent of quantizing noise referred to a receiving system with an ROLR of +46. N_{Qe} is computed in a special manner, explained below, depending upon the type of quantizing and encoding used.

The effective circuit noise values for room and quantizing noise are found from the following expressions:

$$N_{Re} = N_R - 35 + 0.0078(N_R - 35)^2$$
$$+ 10 \log[1 + 10^{(1-SOLR)/10}] \text{ dBrnC} \qquad (4.19)$$

where N_R is the room noise in dBA at the receiving end, and SOLR is the sidetone loss of the listening end telephone. A default value for N_{Re} of 27.37 dBrnC is used if no other value is stated. (See the note below.)

Effective quantizing noise is expressed in terms of speech level, in VUs, and the resulting signal-to-noise ratio out of the quantizer:

$$N_{Qe} = Vo + 89 - SNR \text{ dBrnC} \qquad (4.20)$$

where Vo is the received speech level in VU referred to a receiving system with an ROLR of +46 (Vo can be taken as equal to $-8 - $ OOLR); SNR (signal-to-quantizing noise ratio) may be approximated as SNR = $2.36Q - 8.34$, in decibels; and Q is the ratio of speech-to-speech correlated noise in decibels (see [2] and [3]); a single PCM codec pair has a Q of about $0.78B - 12.9$ dB; B is the system bit rate in kilobits per second; in [2], L is used instead of B for bit rate, but we are using L for loss here.

Note on room noise. The default value of 27.37 dBrnC for N_{RE} corresponds roughly to values obtained using (4.19) with sidetone loss values, SOLR, and room noise values, N_R, that are typical:

$5 < SOLR < 10$ dB   (found in practice)

$55 < N_R < 57$ dBA   (average office environment)

Plots of (4.18), R_{LN}, for loss in terms of OOLR and with total noise, labeled in both dBrnC and dBmp, as a parameter are shown in Figure 4.5.

For a sample calculation, one of the anchor conditions shown in the figure, OOLR = 10 dB and N_c = 25 dBrnC, can be used in (4.18):

Power addition of the 25 dBrnC with the default N_{RE} of 27.37 yields

$$N = 10 \log(10^{27.37/10} + 10^{25/10}) = 29.36 \text{ dBrnC}$$

With this value for N, and 10 for OOLR in (4.18), we find R_{LN} = 80.0, as expected.

It can be seen from Figure 4.5 that, for realistic noise values, the best possible condition for transmission is with an OOLR of 2.5 and noise

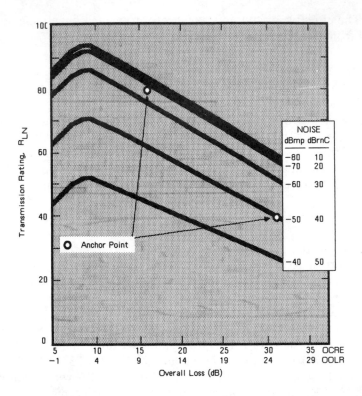

Fig. 4.5 Transmission Rating for Loss and Noise, R_{LN}
Source: Red Book, Volume V, CCITT, Geneva, 1985. Reprinted with permission

of 10 dBrnC which, used in (4.18), yields an R_{LN} of about 94.5. As we will see in Chapter 7, this can be improved. To develop some feeling for R values and their meaning it is necessary to compute some GoS numbers.

The GoS parameters, A and B, defined by (4.17a) and (4.17b) and used to describe the results of subjective tests, are given in Table 4.1 for three data bases. They are presented in terms of the variable, R, without any subscript indicating that any R_y may be used. The formulas for A and B in the table are simply (4.16b) evaluated with the transformation coefficients, a and b, and appropriate regression coefficients, c_i and d_i, for each data base. The first pair is that for the original Murray Hill test. The second, called *long toll,* is from a test in which subjects were asked to rate the quality of long distance connections. The third, CCITT, is from data collected by the CCITT from a number of different countries and represents a very nonhomogeneous group of subjects, at least in terms of nationality.

Table 4.1
Integration Limits, A and B, for Some Data Bases

Data Base	A	B
Murray Hill, 1965	(R − 64.07)/17.57	(R − 51.87)/17.57
Long Toll	(R − 51.5)/15.71	(R − 40.98)/15.71
CCITT tests	(R − 62)/15	(R − 43)/15

A word of caution is in order at this point. Grade-of-service is not an absolute scale. It should not be used as an evaluation of a single entity. To say that 80% of listeners would rate a system as *good or better* is meaningless. But after using these techniques to compare two entities with a single parameter difference, it is quite proper to say that 20% more people would be expected to prefer entity X over entity Y, if the calculated GoSs differed in that direction by 20 points. Grade-of-service analysis provides an estimate of the percent of users that will notice transmission design changes. The phrase *will notice* means that their opinion is shifted by a measurable amount.

Figure 4.6 compares opinion ratings calculated using each of the three data bases in Table 4.1. The drawing shows percent GoB and percent PoW, as a function of transmission rating. The ordinate scale is that for the normal probability function, so a straight line in the figure represents a normal CDF. These curves are interpreted as follows: For a transmission rating of 80, well over 95% of subjects in the long toll tests rated the connection as good or better as compared to only about 80% of the subjects in the Murray Hill tests. For this same rating of 80, about 5% of the Murray Hill subjects rated the connection poor or worse. The percentage of subjects in the long toll tests who rated the connection poor or worse was less than one. The percent GoB curve has a positive slope because more people will give a high rating as the transmission rating increases, reflecting a better quality connection. Conversely, if the quality diminishes, R decreases and more people vote in the poor or worse categories.

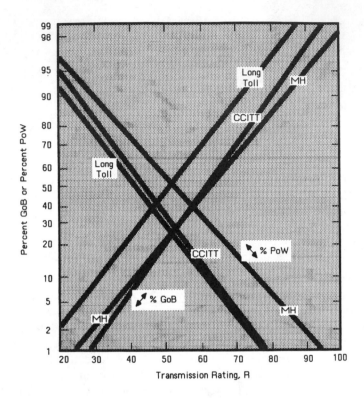

Fig. 4.6 GoS as Function of Transmission Rating for Three Data Bases
Source: Red Book, Volume V, CCITT, Geneva, 1985. Reprinted with permission

Notice that the plot for long toll calls shows a greater tolerance for impairments than the other two. This may reflect a user expectation of poorer quality transmission on long distance calls, but that conjecture has not been verified. It is interesting to note that the differences between the Murray Hill data, derived by using Bell Laboratories employees as subjects, and the CCITT data, collected from tests in a number of countries, are small compared to those between either of these and the long toll data.

Equation (4.18) for transmission rating R_{LN}, and the A, B limits for a give database, can be used to derive contours of constant grade-of-service for combinations of loss and noise. Such a plot for a percent good or better grade-of-service is presented in Figure 4.7 for the CCITT data base. These curves show the trade-off between noise and loss as the listeners perceived it. Notice that no matter what the noise on a circuit may be, there is a constant preferred loss value of about 2.5 dB OOLR (8.5 dB overall CRE). At this time, no explanation can be offered for this; it is simply an experimental observation. Remember that only loss and noise

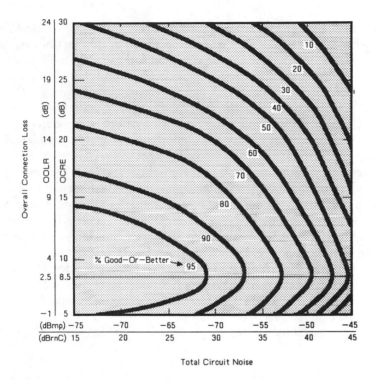

Fig. 4.7 Contours of Constant Grade-of-Service for Varying Noise and Loss Conditions
Source: Adapted from *Red Book Volume V*, CCITT, Geneva, 1985

are considered in this example. Analogous occurrences do not appear if other parameters are considered.

Another model, constructed by NTT using techniques that parallel those described here, is reported in [4]. In this case, preferred loss was found to decrease slightly as noise increased. The only obvious difference in the subjective procedures is that room noise was considerably higher (about 10 dB) in the tests conducted by NTT than it was in the Murray Hill tests. Also, the frequency response and sidetone characteristics of the telephones used have a strong effect on such results and these are not reported in the Reference.

4.1.6 Models for Other Parameters

Expressions to include many other parameters in a composite transmission rating are given in [2]. The number and complexity of the equations for a complete calculation dictate the use of a computer as mentioned earlier. Plots of most of the individual transmission ratings as functions of the parameters are also provided in [2]. Some of the R_y and corresponding plots are presented and discussed in this section. A study of these and the others in [2] provides a good feel for acceptable ranges of the various impairments and the trade-offs between them. The reader is reminded that all of the functions defining transmission ratings, R_y, are simply best fits to the test data. There is no theory underlying them. In many instances, realistic limits were imposed as bounds in deriving constants. Values for minimal room noise or circuit noise are examples. Such considerations also aided in the curve fitting.

When other parameters are added to noise and loss, the tests show that a scaling of the basic R_{LN} relation, (4.18), is sufficient to describe subjective reaction to the composite. In fact, that is what one would most likely expect to happen. The scaling operations involve two types of factors: impairment related constants such as echo delay, and derived weighting factors that are empirical functions of impairments. As one might expect, there is a large number of such factors needed to describe and weight the numerous impairments that have been considered.

The following is a list of transmission parameters and factors involved in the calculations of the many possible R_y. The numbers in brackets, following the factors, refer to the equations further on in which they are used.

Echo Factors:

D   Talker echo path round trip delay in milliseconds.   [4.31]

E   Talker echo path round trip loudness loss in decibels. Calculated using (3.3) and (3.4) with frequency response in dB as the variable X in (3.3).
($X = 0$ at 1 kHz)   [4.31]

LEPD   Listener echo path delay in milliseconds.   [4.29]

LEPL   WEPL computed for the listener echo path;
WEPL is defined in (1.15).   [4.29]

Channel Bandwidth Factors:

f_l Frequency, in hertz, at which the lower end channel frequency response is 10 dB below the 1-kHz value.   [4.21]

f_u Frequency, in hertz, at which the upper end channel frequency response is 10 dB below the 1-kHz value. If $f_u > 3200$, then 3200 is used.   [4.22][3]

Bandwidth Weighting Factors:

$$K_1 = 1 - 0.00148(f_l - 310) \quad [4.25] \tag{4.21}$$

$$K_2 = 1 + 0.000429(f_u - 3200) \quad [4.25] \tag{4.22}$$

Slope Weighting Factors:

$$K_3 = 1 + 0.0372(S_l - 2) + 0.00215(S_l - 2)^2 \quad [4.25] \qquad (4.23)$$

$$K_4 = 1 + 0.0119(S_u - 3) - 0.000532(S_u - 3)^2$$
$$- 0.00336(S_u - 3)(S_l - 2) \quad [4.25] \qquad (4.24)$$

Slope Factors:

S_l Low frequency channel response, slope,
from f_l to 1 kHz in decibels per octave.   [4.23]

S_u High frequency channel response, slope,
from 1 kHz to f_u in decibels per octave.   [4.24]

Note. S_l and S_u are computed, by trial and error, using (3.1) to yield the same loudness loss as the actual channel frequency response.

Composite Channel Shape Factor:

$$K_{BW} = K_1 K_2 K_3 K_4 \quad [4.27] \qquad (4.25)$$

Sidetone Weighting Factor:

K_{ST} Computation given below, (4.26).   [4.28]

Sidetone Factors:

SR   Approximate sidetone frequency response type based on the low-end and high-end response slopes in decibels per octave: [4.26] [4.32]

$$SR = 0 \text{ if low end slope} = 0$$
$$SR = 3 \text{ if low end slope} = +3$$
$$SR = 6 \text{ if low end slope} = +6$$

Intermediate values may be obtained by linear interpolation. The high-end slope is assumed to be 1.5 times the low-end slope. Slope is defined as for channel response, below and above 1 kHz. (This imposes some constraints on the overall response shape.)

The quantity K_{ST} is a sidetone weighting factor that is a function of the sidetone loss and the sidetone frequency response characteristic, SR. It is computed by the expression:

$$K_{ST} = 1.021 - 0.002(SOLR - 10)^2$$
$$+ 0.001(SOLR - 10)(SR - 2)^2 \quad [4.28] \qquad (4.26)$$

These factors appear in the following equations to compute the various R_y as the parameters of interest increase in number.

As the y in R_y includes more impairments, the calculation of R becomes lengthy and the y are dependent upon preceding ones. The equations for them are presented here in order of increasing complexity. The notation used for y is a string of letters of the form $a(b)(c)d$, where each letter identifies a parameter, and a parameter enclosed in parentheses means that the expression is valid with or without its inclusion.

The expression for R_{LN} was given earlier as (4.18). The first extension considered here is the effect of channel frequency response expressed in terms of the bandwidth and slope factors K_1 through K_4, and their composite, K_{BW}. The bandwidth pair, K_1, K_2, and the slope pair, K_3, K_4, are independent, and, if one is not interested in the effect, either pair may be omitted in the computation of K_{BW}. On real channels, the product of either the bandwidth pair or the slope pair may range from about 0.8 to 1.2, so K_{BW} may vary from 0.64 to 1.44.

The channel frequency response (bandwidth) transmission rating is $R_{LN(BW)}$:

$$R_{LN(BW)} = (R_{LN} - 22.8)K_{(BW)} + 22.8 \tag{4.27}$$

Notice that $R_{LN(BW)}$ is a scaling of R_{LN}. As stated above, ratings for most other parameters are computed as scalings of the basic one, R_{LN}. Also note that the (BW) notation shows that channel frequency response is an optional parameter.

The contribution to R due to sidetone, R_{ST}, is taken into account through the sidetone weighting, K_{ST}, (4.26), as a direct multiplier:

$$R_{LN(BW)(ST)} = K_{ST}R_{LN(BW)} \tag{4.28}$$

K_{ST} can range from about 0.6 to 1.1 and so has the potential of shifting R by plus or minus 33 percent about a midrange value. As will be seen below, the impact of sidetone can be even greater if talker echo is a factor.

Listener echo factors, determining R_{LE} and $R_{LN(LE)}$, are taken into account next:

$$R_{(LE)} = 9.3(LEPL + 7)(LEPD - 0.4)^{-0.229} \tag{4.29}$$

is the listener echo factor that may be combined with $R_{LN(BW)(ST)}$:

$$R_{\text{LN(LE)}} = \frac{R_{\text{LN(BW)(ST)}} + R_{\text{(LE)}}}{2}$$

$$- \left[\left(\frac{R_{\text{(LN)(BW)(ST)}} - R_{\text{(LE)}}}{2} \right)^2 + 169 \right]^{1/2} \qquad (4.30)$$

Notice the simplification of notation here; (BW)(ST) has been dropped in $R_{\text{LN(LE)}}$. This is arbitrary and for convenience only.

The R for talker echo is designated R_E and is a function of both echo path loss and delay, E and D, in the listing above. Echo effects may be evaluated independently as

$$R_E = 106.36 - 53.45 \log \left[\frac{1 + D}{(1 + (D/480)^2)^{1/2}} \right] + 2.277E \qquad (4.31)$$

Some plots of R_E *versus* echo path loss with delay, D, as a parameter are shown in Figure 4.8. The plots vividly demonstrate the need for increased loss in the echo path to maintain a constant R_E as the delay increases. For example, to keep R_E at 60 as the delay increases from 5 to 20 ms, one would have to add about 13 dB of loss to the circuit. It is apparent that the required loss will rapidly become excessive as delay increases. How this problem is handled is the subject of the section on loss plans in Chapter 5.

Talker echo may be combined with sidetone effects through the relation:

$$R_{E\text{(ST)}} = R_E + 2.6(7 - \text{SOLR}) - 1.5(4.5 - \text{SR})^2 + 3.38 \qquad (4.32)$$

and all of the parameters considered here may be used to generate a composite transmission rating, $R_{\text{LN(LE)}}$:

$$R_{\text{LN(BW)(E)(ST)(LE)}} = \frac{R_{\text{LN(LE)}} + R_{\text{(E)(ST)}}}{2}$$

$$- \left[\left(\frac{R_{\text{LN(LE)}} - R_{\text{(E)(ST)}}}{2} \right)^2 + 100 \right]^{1/2} \qquad (4.33)$$

For completeness, the subscript notation has been expanded in the result to delineate all the parameters. Note that if listener echo is ignored, (4.30) reverts to (4.28) and $R_{\text{LN(LE)}}$ in (4.33) is replaced with $R_{\text{(LN)(BW)(ST)}}$. Plots of (4.33), with BW, LE, and ST ignored,

$$(R_{LN} + R_E)/2 - \{[(R_{LN} - R_E)/2]^2 + 100\}^{1/2}$$

and using echo path loss as a variable and delay as a parameter are shown in Figure 4.9. The asymptotic behavior of the curves at larger echo path loss values is due to noise and talking path loss becoming controlling. (R_{LN} was taken equal to 74 for these plots, and it is assumed that echo path loss can be increased without increasing talker path loss.)

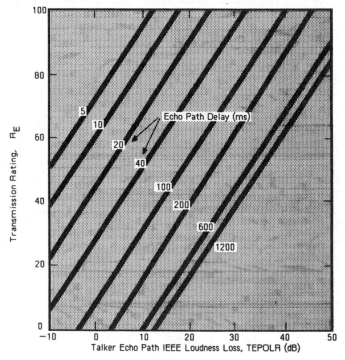

Fig. 4.8 Transmission Rating for Talker Echo
Source: Red Book, Volume V, CCITT, Geneva, 1985. Reprinted with permission

    The increased impact of sidetone combined with talker echo is illustrated in Figure 4.10. The independent variable is SOLR with SR and delay as parameters. Though not explicit in the figure, for these plots:

$f_l = 300$

$f_u = 3000$

S_l is $+2$

S_u is -4

    These yield a K_{BW} (see (4.25)) of 0.8263. In the worst case shown, the lower curve, $R_{LN(BW)(E)(ST)}$ is forced monotonically over the range from

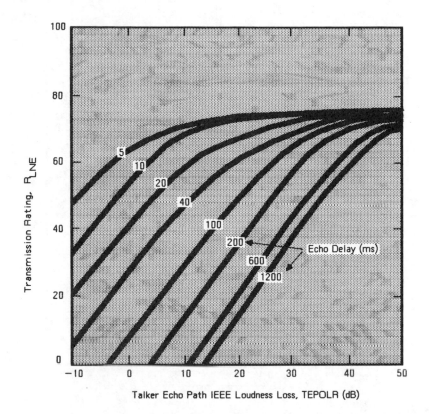

Fig. 4.9 Transmission Rating for Overall Loss, Noise, and Talker Echo
Source: Red Book, Volume V, CCITT, Geneva, 1985. Reprinted with permission

40 to 5.6 as SOLR increases from 5 to 20 dB. Substituting these two end values for R in the relations from Table 4.1, for A and B for the long toll data base, we have

$$A_{40} = (40 - 51.5)/15.71 = -0.73$$

$$A_{5.6} = (5.6 - 51.5)/15.71 = -2.92$$

$$B_{40} = (40 - 40.98)/15.71 = -0.062$$

$$B_{5.6} = (5.6 - 40.98)/15.71 = -2.25$$

Using these as limits in the integrals of (4.17a) and (4.17b), it is found (through tables or algorithms) that %GoB moves from 23.28% to 0.17%, while %PoW changes from 52.49% to 98.78% by simply controlling the sidetone loss (SOLR).

If the sidetone frequency response is now improved, from SR = 0 to SR = 6, then, as illustrated in Figure 4.10, R can rise from 40 to 52.0,

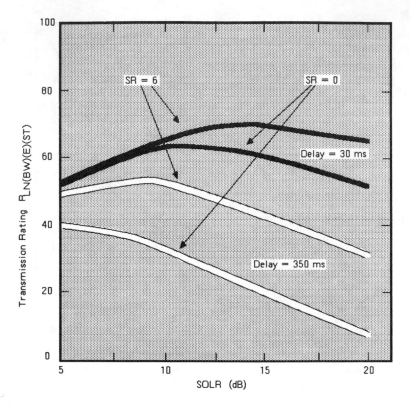

Fig. 4.10 Effect of Sidetone Changes When Talker Echo is Significant

%GoB increase to 48.73, and %PoW drop to 24.15.[4] Notice that in this case optimum values for SOLR occur at about 9 dB for the long delay, and 11 or 12 dB for the shorter one. Implied trade-offs are discussed in Chapter 7.

A study of these curves illustrates the interplay between loss, delay, and sidetone. For the delay of 30 ms, the black curves, with a rising sidetone frequency response, SR = 6, the optimum SOLR is about 14 dB. When the response is flattened, SR = 0, the relative masking of echo by the higher frequency sidetone is diminished and more sidetone is desired, preferred SOLR decreases. The optimum now results in a lower transmission rating; however, the best R value is smaller than was possible with the preferred larger SR. For the long delay, 350 ms, the echo is much more annoying, the entire curves are low compared to the first case, and even more sidetone (less sidetone loss) is needed to try to mask it. The effect of the high frequency sidetone components is even greater in this case as noted by the relative spread within the two pairs of curves. A delay of 350 ms is very long and unrealistic in this context. The example merely illustrates the nature of the role that sidetone loss may play.

This section has presented an outline of a methodology for modeling listener opinion of quality and relating it to transmission impairments on telephone connections. Use of these results in a practical design situation is illustrated in Chapter 7.

It was stated at the beginning of the discussion of grade-of-service modeling, Section 4.1.2, that the models presented are described in [2] but in terms of the CCITT CRE metrics. The functional forms are identical to the IEEE formulations and the differences are numerical only. They are reproduced in the next section together with the factors used to convert between metrics.

4.1.7 Model Equations in the CCITT Metrics

The equation for R_{LN}, as given in Section 4.1.5, (4.18), is in terms of the IEEE metrics, OOLR for loss and dBrnC for noise. The conversion to CCITT metrics may be accomplished by using the relations:

$$OOLR = OCRE - 6 \tag{4.34}$$

and

$$dBmp = dBrnC - 90 \tag{4.35}$$

Substituting these in (4.18) yields:

$$R_{LN} = -34.88 - 2.257[(L'_e - 8.2)^2 + 1]^{1/2}$$
$$- 2.0294N'_F + 1.833L'_e + 0.02037L'_eN'_F \tag{4.36}$$
$$(R_{LN} \text{ in CCITT metrics})$$

with L'_e = overall connection CRE in dB, and N'_F in dBmp. In these metrics, the noise must be referred to that on a receiving system with a plus one dB receiving CRE.[5]

Equations for other parameters (Section 4.1.6) that use sidetone loss, SOLR, are rewritten in terms of SL, the EARS sidetone loss by use of the relation

$$SL = SOLR + 5 \tag{4.37}$$

For ease of reference, the equations in Section 4.1.6 are rewritten below in the CCITT notation.

Echo Factors:

D Talker echo path round trip delay in milliseconds.

E' CRE of talker echo path round trip loss, in decibels, $E' =$ OOLR + 10 (OOLR computed for the *talker echo path overall loss,* also called TEPOLR).

LEPD   Listener echo path delay in milliseconds.

LEPL   WEPL computed for the listener echo path; WEPL is defined in (1.15).

Channel Bandwidth Factors:

f_l Frequency, in hertz, at which the lower-end channel frequency response is 10 dB below the 1-kHz value.

f_u Frequency, in hertz, at which the upper-end channel frequency response is 10 dB below the 1-kHz value.
      If $f_u > 3200$, then 3200 is used.

Bandwidth Weighting Factors:

$$K_1 = 1 - 0.00148(f_l - 310) \tag{4.38}$$

$$K_2 = 1 + 0.000429(f_u - 3200) \tag{4.39}$$

Slope Weighting Factors:

$$K_3 = 1 + 0.0372(S_l - 2) + 0.00215(S_l - 2)^2 \tag{4.40}$$

$$K_4 = 1 + 0.0119(S_u - 3) - 0.000532(S_u - 3)^2$$
$$+ (S_u - 3)(S_l - 2) \tag{4.41}$$

Slope Factors:

S_l Low frequency channel response, from f_l to 1 kHz in decibels per octave

S_u High frequency channel response, from 1 kHz to f_u in decibels per octave

Note. S_l and S_u are computed using (3.1) to yield the same loudness loss as the actual channel frequency response. The same IEEE method may be used in lieu of extensive subjective testing to find an equivalent sidetone CRE.

SR   Approximate sidetone frequency response type based on the low-end and high-end response slopes in decibels per octave:

$$SR = 0 \text{ if low-end slope} = 0$$
$$SR = 3 \text{ if low-end slope} = +3$$
$$SR = 6 \text{ if low-end slope} = +6$$

Intermediate values may be obtained by linear interpolation. The high-end slope is assumed to be 1.5 times the low-end slope. Slope is defined as for channel response, below and above 1 kHz. (This imposes some constraints on the overall response shape.)

Composite Channel Shape Factor:

$$K_{BW} = K_1 K_2 K_3 K_4 \tag{4.42}$$

Sidetone Weighting Factor:

$$K_{ST} = 1.021 - 0.002(SL - 15)^2 + 0.001(SL - 15)(SR - 2)^2 \tag{4.43}$$

$$R_{LN(BW)} = 22.8 + K_{BW}(R_{LN} - 22.8) \tag{4.44}$$

$$R_{LN(BW)(ST)} = K_{ST} R_{LN(BW)} \tag{4.45}$$

$$R_{(LE)} = 9.3(LEPL + 7)(LEPD - 0.4)^{-0.229} \tag{4.46}$$

$$R_{LN(LE)} = \left(\frac{R_{LN(BW)(ST)} + R_{(LE)}}{2} \right)$$
$$- \left[\left(\frac{R_{(LN)(BW)(ST)} - R_{(LE)}}{2} \right)^2 + 169 \right]^{1/2} \tag{4.47}$$

$$R_E = 83.62 - 53.45 \log \left[\frac{1 + D}{(1 + (D/480)^2)^{1/2}} \right] + 2.277E' \tag{4.48}$$

$$R_{E(ST)} = R_E + 2.6(12 - SL) - 1.5(4.5 - SR)^2 + 3.38 \tag{4.49}$$

$$R_{LN(BW)(E)(ST)} = \left(\frac{R_{LN(LE)} + R_{E(ST)}}{2} \right)$$
$$- \left[\left(\frac{R_{LN(LE)} - R_{E(ST)}}{2} \right)^2 + 100 \right]^{1/2} \tag{4.50}$$

If listener echo is ignored, (4.47) reverts to (4.45) and $R_{LN(LE)}$ in (4.50) is replaced with $R_{(LN)(BW)(ST)}$.

These relations are identical to those in Section 4.1.6 except for the few changes of constants to adjust for the CCITT metrics.

4.2 DATA QUALITY

The definition of quality would appear to be an easier problem for data transmission than voice transmission because the human element can be removed from the assessment of the received signal. An error is an error and readily shows up as such or at least can be detected and counted. Obviously, error-free transmission is good and anything other than that is undesirable. However, because noise in any real channel can never be zero, information theory tells us that error-free transmission is impossible. Nevertheless, for a given channel, through coding or signal-to-noise improvement, the error rate may be made arbitrarily small. The trade-offs are, respectively, time and economic cost. This brings the human aspect back to us and a grade-of-service model for data transmission might describe the subjective balance of incorrect information against time and money. No such model exists.

One of the reasons that there is no model against which data transmission performance can be evaluated is the lack of a meaningful measure of acceptability. Many measures of performance are available and several of them can be monitored on an on-going or routine basis but relating these to acceptability is still a subject for research.

Another factor, that is really distinct from the acceptability issue but complicates it, is the practical one of measurement. Data transmission error rates are quite small, commonly less than one in one million. Popular voiceband transmission rates are on the order of 1000 to 10,000 bits per second, and so measuring such small quantities with reasonable accuracy is time consuming.

We will return to the problem of data quality measures in a later section, but first we will describe some of the tools used to predict data transmission performance, and present statistics describing performance in networks.

The first commercial long-distance transmission systems were designed for telegraphy and so were digital in nature. With the invention of the telephone, the impetus changed to telephony and analog transmission. Because of the economies and flexibility provided by modern digital transmission and switching, networks are becoming digital once more. However, data transmission in the current meaning, evolved on analog systems through the use of modems to convert digital information to analog form. The need for these devices will continue for some time and data transmission performance is determined by their operation. Modems are tested, and performance predicted, with transmission line simulators.

4.2.1 Transmission Line Simulators

This section and Section 4.2.2 apply to data transmission using modems as network terminating devices. The remaining ones, except for the data shown in Section 4.2.3, are relevant to data transmission in general. In data transmission, the tools equivalent to NOSFER and SIBYL for voice are transmission line simulators. They have the capability of approximating all of the transmission impairments discussed in Chapter 1. A general block diagram for one is shown in Figure 4.11.

A modem to be tested is connected as indicated by the two darker shaded blocks at the bottom of the drawing. A data generator drives the transmitter with selected digital patterns or a random bit stream. The generator output is also delivered, after a suitable delay, to one input of a comparator that compares transmitted and received data and records errors.

The carrier simulation unit shown in the upper portion of the drawing simulates carrier system impairments, frequency offset, jitter, *etcetera*. It may be inserted, by means of connectors *A* and *B*, either before or after the linear and nonlinear baseband channel elements in the lower part of the figure. Switches S1 and S2, used to introduce hits are implemented with variable length gates with adjustable switching times and duty cycles to control rise time, length, and repetition rate of the disturbances.

The baseband channel in the lower portion of the figure has a linear element placed between two nonlinear ones. The nonlinear elements are often implemented as third-order polynomials with adjustable coefficients. The linear channel has provision for introducing variable amplitude and phase distortions. The additive impairments, noise, *etcetera,* are shown introduced at the output of the channel, but the system may be configured to place the summing node elsewhere. The calibration and control unit may be any mix of manual and automatic devices. In digital realizations it can be programmable to facilitate running a sequence of tests.

4.2.2 Modem Testing

Modem testing is done for one of two reasons. There is verification testing in the case of new designs, and performance testing for transmission objectives and requirements studies.

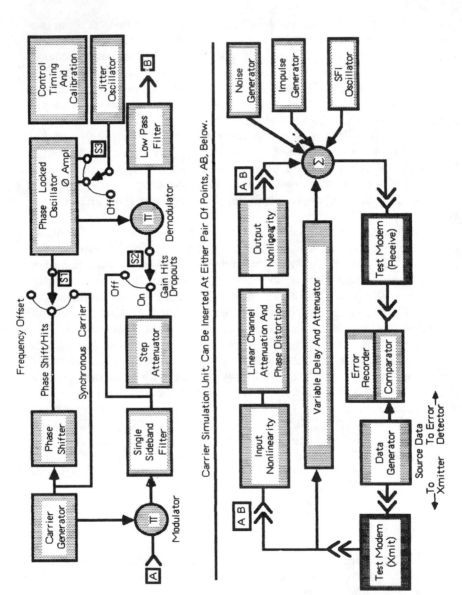

Fig. 4.11 Elements of a Transmission Line Simulator

Because all the measurements are objective, modem testing is faster than the subjective testing necessary to study voice transmission impairments. The number of impairments of interest is larger, however, and this increases testing time because of the need to study interactions or trade-offs in the multiplicity of possible combinations. Several months may be needed to completely characterize a new modem. Good experiment design is important in these cases, and a programmable simulator is a real asset.

Modem test results typically take the form of plots showing: bit error rate, bits received in error per total bits received; or, for transient phenomena, average errors per event as a function of event magnitude. These data may be used to:

1. Compare performance of modems.
2. Evaluate new designs.
3. Predict performance on real connections.
4. Assist in establishing network performance criteria.

4.2.3 Performance Statistics

Bit error rate, defined above, is only one of many measures describing data transmission quality. A stream of data bits may be segmented into successive blocks of some arbitrary number and, it has been observed that errors tend to occur in clusters or bursts. Viewing data transmission with these ideas in mind gives rise to other measures of quality such as:

1. Block error rate, Bler:

   Bler = (no. of blocks with one or more errors)/(total blocks)

2. Error burst rate:

   (no. of error bursts)/(unit time)

3. Average number of intervals between error bursts.
4. Percent of intervals of time that contain errors, %EI:

   %EI = %(no. intervals with errors)/(total no. of intervals)

5. Percent of error-free intervals, %EFI:

   %EFI = %(no. error free intervals)/(total no. of intervals)

6. Reliability measures based on items 4 and 5, *i.e.*, mean time between error events.

It is informative to look at some real data using these measures. The results of two extensive data transmission surveys are presented in [5] and [6]. Some of those reported are summarized here in Figures 4.12, 4.13, and 4.14. It is important to note that error statistics are contingent upon the bit transmission rate, and the modulation technique[6] used in modems, as well as the transmission characteristics of a connection. In general, error rates increase as transmission speed increases because faster systems are more sensitive to noise and other impairments. (Consider the eye patterns of Figure 1.17 with the sampling intervals closer together.) The data shown are for a bit rate of 1200 per second. Data in Figures 4.12 and 4.14 were obtained in 1969, those in Figure 4.13 were obtained in 1982. (Block sizes other than 1000 bits and the structure of errors within bursts were not published for the 1982 survey.) In this latter study, two different modem implementations were evaluated. These are designated Set 1 and Set 2 in Figure 4.13. The improvement in performance observed in the later survey, Figure 4.13 versus 4.12(a), is due predominantly to the appearance of newer technology in the network.

On the linear ordinate used, the distributions of bit error rate (Figures 4.12(a) and 4.13) show significant curvature. They have been found to be more nearly linear on a logarithmic scale, and are shown in that manner in [6]. (They were replotted on the linear scale here to facilitate the comparisons with the data from [5].) The distributions were constructed by calculating the error rate from the total number of errors measured on a large number of calls, each of length approximately two million bits (1969), and one million bits (1982). Therefore, they represent distributions across calls and do not necessarily indicate performance on a long single call that is segmented into a series of observational periods.

For block error rates the calls were segmented into successive blocks of bits according to the sizes shown. The resulting data were then accumulated for all calls. These distributions thus represent a mix of performance on single calls and across calls.

As mentioned earlier, a characteristic of errors in data transmission is the clustering they exhibit. A measure of clustering is provided by burst statistics. A burst, for purposes of the 1969 survey, was defined as any sequence of bits that contained at least one error and was bounded by error free intervals of at least 50 bits in length. The length of a burst is the total number of bits in the group in which errors occurred (excluding the 50 error-free bits on either end). Distributions of number of errors per burst and burst lengths are shown in Figure 4.12(b). The abscissa is a dual scale: a measure of length in bits, and a number of bits. The lower curve

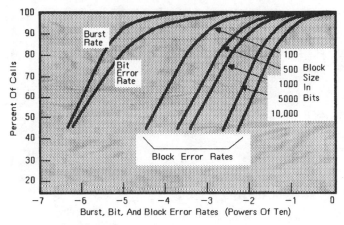

(a) Distributions Of Error Rates

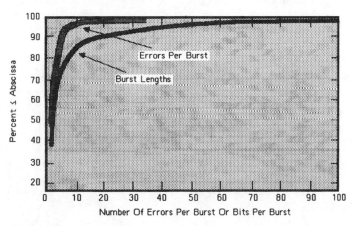

(b) Errors Per Burst And Burst Lengths

Fig. 4.12 Measured Distributions of Data Transmission Errors (1200 b/s, 1969)
Adapted from: Balkovic, *et al.*, "High Speed Voiceband Data Transmission Performance on the Switched Telecommunications Network," *Bell System Tech. J.*, Vol. 50, No. 4, April 1971

shows the distribution of the length of error bursts, in bits; the upper one shows the number of errors in a burst. (All bits in a burst are not in error.) It is seen that bursts may be as long as 100 bits and contain over 30 errors. With some effort, it is possible to estimate the mean and median errors per burst from Figure 4.12. The mean is found by estimating the number of intervals that span small count increments on the abscissa, multiplying by the midpoint of the count increment and summing. The median is about two, and the average is about nine. The average burst length is also about nine bits.

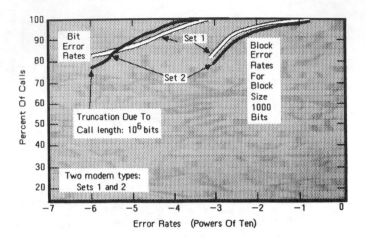

Fig. 4.13 Measured Bit and Block Error Rates, (1200 b/s, 1982)
Source: Adapted from: Carey, *et al.,* "1982/83 End Office Connection Survey: Analog Voice and Voiceband Data Transmission Performance of the Public Switched Network," *AT&T Bell Laboratories Tech. J.,* Vol. 63, No. 9, November 1984

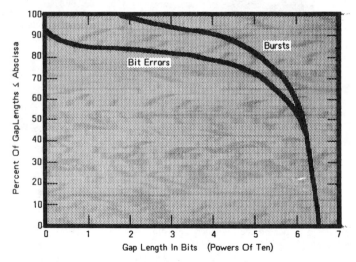

Fig. 4.14 Measured Burst and Bit Error Gap Distributions (1200 b/s, 1969)
Source: Adapted from: Balkovic, *et al.,* "High Speed Voiceband Data Transmission Performance on the Switched Telecommunications Network," *Bell System Tech. J.,* Vol. 50, No. 4, April 1971

Figure 4.14 shows the distributions of the lengths of intervals between bits in error and between bursts, measured in numbers of bits. These error-free intervals are called gaps. The total length of the test calls in the survey for the 1200 bit per second modems was 30 minutes or 2.16 million bits. About 33% of all calls were error free so the curve below 33%

(gap lengths greater than two million) is an extrapolation. The curves show the percent of all gaps that were shorter than the value indicated on the abscissa. For example, 50% of either bit error or burst gaps were shorter than about one million bits. The definition of burst length imposed a shortest inter-burst gap length of 50 bits so all burst gaps must be longer than that and the burst length gap distribution intersects 100% at about $10^{1.5}$. (The abscissa is in powers of ten, a logarithmic scale.)

The average burst rate, as estimated from Figure 4.12(a), yields an expected number of bursts per 2.16 million bits of about 24. Their average length, as noted above, is about nine bits. If these were uniformly distributed over the 30 minute test interval, the average gap length would be 90,000 bits. The average gap length, as estimated from Figure 4.14, is about 1.2 million, more than ten times the uniform value. This is another indication of the clustering, not only of bit errors but of bursts of errors.

While it is not obvious from these plots, analysis of data of this type reveals that the mean and standard deviation of error rate distributions are often of the same order of magnitude. This is significant in terms of verification of performance. For example, one might use simple random sampling to estimate the average error rate in a particular situation. The number of measurements, n, needed to estimate the average is given by the equation

$$n = (ts/w)^2 \qquad (4.51)$$

In this relation, s is the estimated standard deviation of the variable that is being sampled, w is the width of the allowed limit of precision for the estimated mean. That is, we would like the estimate to be correct within plus or minus $w/2$. Estimates of the mean are assumed to be normally distributed. We want the number n to be large enough to assure that our estimate will fall sufficiently near the mean of such a distribution to satisfy our precision criterion. Equation (4.51) is formulated so that we can force the probability that our estimate (plus or minus $w/2$) will be wrong, to be arbitrarily small. The multiplier t accomplishes this. It is the variable in a normal distribution (the normal deviate) corresponding to the probability that our estimate is as good as we want. If we take $t = 2$ (for plus or minus 2 standard deviations) we will have a probability of error of 5%. This is equivalent to assuring that our estimate will fall within plus or minus two standard deviations (95%) of the true mean. (See [7, Section 4.5], for example.) For a precision of 20%, approximate equality of the mean and standard deviation yields $w = 0.2s$, and 100 measurements will be needed. If the anticipated mean is one in 100,000 bits, a reasonable measurement may consist of transmitting 200,000 bits. At 1200 bits per second this estimate of the average requires nearly three minutes per measurement or about four and one half hours of actual transmission time to make 100 measurements.

The example just given of time required to estimate an average error rate is rather simplistic, but introduces a significant difference between voice transmission and data transmission quality estimates: measurement time. A noise and a loss measurement will usually suffice to determine voice quality. These can be completed in less than a half a minute, or about six times faster than the data measurement.[7] Furthermore, the ratio of standard deviations to means for these parameters (around 1 : 4 to 1 : 8) yield sample sizes less than ten as opposed to 100, for comparable precision. Relatively speaking, assessing data quality takes a great deal of time. Partly because of this difference, and partly because of another time related variable discussed next, the results of data quality measurements have different meanings depending upon why they were made. One meaning applies to network survey results; a second is needed for measurements made in the resolution of trouble reports. These two are considered briefly in the following section.

4.2.4 Error Rate Estimation

A network survey may take months to complete so the averages are taken over a relatively long time interval compared to the few days it may take to measure performance as viewed from a single, or small number of, customer locations. This difference in the time bases underlying the statistics collected can be significant because some transmission factors contributing to errors are time dependent. Such things as fading in microwave systems, changing traffic patterns affecting network routing, and periodic maintenance activities all influence performance. These events may exhibit daily, weekly, or seasonal cycles, thus limiting the interpretation of short term measurements. For a single customer then, it may be true that the measurements do not reflect his experience.

In this discussion of error rate we have implicitly defined it to be error rate observed over a specific time interval, the length of one test call. With the burst and gap structures illustrated in Figures 4.12 and 4.14, it should not be surprising that error rate, observed on a continuous basis, must wander around, and only gradually approach a long term average. (If long term were defined as 12 hours, there might even be different day time and night time averages.)

Because error rate is so highly variable, other measures are often used. These measures are in terms taken from reliability theory. They do not avoid the time variability problem but cast it in a different light. Instead of being concerned with error rates, one may view a data transmission session as broken into a series of contiguous one-second intervals, each of which is classified as containing at least one error or being error free. The terms *errored seconds* (ES) and *error-free seconds* (EFS) are used in describing channel performance. One published network objective now states that 99.5% of transmission seconds should be error free. The CCITT has extended the concept to cover grades of degradation within errored intervals.[8] These metrics have more intuitive appeal than error rate. They are availability measures, and convey more information and a better understanding to a user. They are also metrics that are easily measured and stored by automatic surveillance equipment such as might be used for digital network maintenance. In this last case, the interval of second may be replaced by a coding reference such as a framing interval. In any event, the measurement and data collection processes are more easily defined than they are for error rate. These metrics are also time variable. That is, the percentage of error-free intervals is dependent upon the observation time. Nevertheless, observation times are usually not stated. The CCITT has suggested one month as an interim guideline for an averaging time until standard time periods can be defined.

The task of verifying customer complaints, for noncatastrophic failures, in a reasonable and satisfactory manner is yet to be solved. C.A.W. McCalla has presented an approach to this that seems promising but needs further development, [9]. By studying data on error rates in digital networks, he has determined that the discrete log-normal distribution can be used to describe their properties. In particular, he has focused on the distributions of EFS and shown that, if his choice of model is valid, good performance estimates with high confidence levels can be achieved with a small number of measurements. There is a possibility that as few as two or three measurements, of perhaps five or fifteen minutes duration, can yield useful information. This is an area that requires more study.

For quality in data transmission then, we are left with an unresolved technical problem, that of defining and determining what level of quality is actually being experienced by a user. Another one, taken up in the next section, is subjective: just how error free received data should be. Cost resolves this one for now, but the question remains. Is there an *error rate* below which no one would perceive a difference, and therefore be foolish to try to achieve even at costs approaching zero?

4.2.5 Defining Acceptability

The question at the close of the last section can be rephrased: *How many errors are acceptable?* Zero is not a realizable answer. Furthermore, the premise here is that zero errors is a conceptual fiction that is never realized, and in fact never needed. In support of this premise, the following argument is presented. Note that we will use both bit and character error rates interchangeably because the important receiver is a human being. If an error is perceived by this receiver, it does not distinguish type. The fact that more than one bit error may be absorbed in one character error results in a trivial numeric difference.

A little known study by a human factors psychologist, E.T. Klemmer, while at the IBM Corporation, provides a framework useful in understanding the question about acceptable errors. The study was done with the assistance of a banking company. Bank records with alphanumeric data were routinely put in machine format by keypunch operators. The keypunch operators were classified as trainees, experienced, or highly skilled. Keypunched records, generated by these three groups were examined for errors. Trainees were found to make errors at a character error rate between 10^{-3} and 10^{-4}; experienced operators achieved 10^{-4} to 10^{-5}; and the rates for highly skilled people approached 10^{-6}.

Similar studies have been made in other fields. Reported error rates associated with familiar things are: periodicals and newspapers, 10^{-2} to 10^{-4} character error rate; and microprocessors, 10^{-12} to 10^{-13} bit error rate. The numbers for microprocessors have appeared in various journals but it is not clear whether these are estimates based on some statistical estimate associated with the physics of chip technology or experimental data. Network designers have reported design goals ranging from 10^{-9} to 10^{-14}. This last number must be taken in a light hearted manner at best. Even if a designer chose to use such a value, there is no practical way that system performance could be verified. At the time of this publication, the most advanced bit rate available as a single user channel over any significant distance was on the order of ten megabits per second. The system would have to be monitored for 116 days before 10^{14} bits could be observed. Any sensible verification procedure would require years of testing even if one could devise a measurement apparatus of sufficient reliability to do the testing. We will ignore that number and arbitrarily decrease it one order of magnitude. These results are summarized in Table 4.2.

The range of error rates in human experiences seems to span eleven orders of magnitude and we live, rather contentedly, with the entire gamut. Table 4.2 and this comment are offered as evidence that the question at the beginning of this section, *How may errors are acceptable?*, can be answered, and that any quest for zero errors is a philosophic one. Until such time as this question is resolved, the answer will continue to be simply *Whatever is economic*.

Table 4.2
Some Published Error Rates

Source	Error Rates From	To
Newspapers–Magazines	10^{-2}	10^{-4}
Keypunch Trainee	10^{-3}	10^{-4}
Keypunch, Skilled	10^{-4}	10^{-5}
Keypunch, Expert	10^{-5}	10^{-6}
Network Design	10^{-9}	10^{-13}
Microprocessors	10^{-12}	10^{-13}

NOTES

1. The notation, $G(\mu, \sigma)$, stands for a cumulative normal probability function with mean μ and standard deviation σ. The most common notation for this function is $N(\mu, \sigma)$ but N and n are reserved here to represent noise. We adopt G (Gaussian) to represent the cumulative, and g the corresponding density function, as an alternative. Then $G(X) = \int_{-\infty}^{X} g(x)\, dx$.

2. We will see in Chapter 7 that R may exceed 100, it is not restricted to zero and 100.

3. The weighting factor K_4 achieves its maximum, 1.0, at $f_u = 3200$. No similar restriction on f_l is stated in [2].

4. Two decimal places are carried in this example only to prevent the sum %GoB and %PoW from equaling 100, leaving no room for the *fair* category. This degree of precision is not warranted for other purposes.

5. See examples in Chapter 7.

6. Frequency, phase, and various quadrature modulation techniques are common.

7.   This is measurement time only. Set-up time is not considered.
8.   See Rec. G.821 of [8].

REFERENCES

1.   Cavanaugh, J.R., Hatch, R.W., and Sullivan, J.L., "Models for the Subjective Effects of Loss, Noise, and Talker Echo on Telephone Connections," *Bell System Technical Journal,* Vol. 55, No. 9, November 1976, pp. 1319–1371.
2.   CCITT, "Transmission Rating Models," *Red Book, Volume V,* Supplement No. 3, ITU, Geneva, 1985.
3.   CCITT, "Modulated Noise Reference Unit (MNRU)," *Red Book, Volume V,* Rec. P.70, ITU, Geneva, 1984.
4.   Osaka, N., and Kakehi, K., "Objective Model for Evaluating Telephone Transmission Performance," *Review of the Electrical Communications Laboratories,* Research and Development Headquarters, Nippon Telegraph and Telephone Corporation, Tokyo, Vol. 34, No. 4, 1986, pp. 437–444.
5.   AT&T, "1969–70 Switched Telecommunication Network Connection Survey," *Bell System Technical Reference,* AT&T Pub., April 1971.
6.   Carey, M.B., *et al.,* "1982/83 End Office Connection Study: Analog Voice and Voiceband Data Transmission Performance Characterization of the Public Switched Network," *AT&T Bell Laboratories Technical Journal,* Vol. 63, No. 9, November 1984.
7.   Cochran, W.G., *Sampling Techniques,* John Wiley and Sons, New York, 1964.
8.   CCITT, *Red Book,* VIIIth Plenary Assembly, ITU, Málaga-Torremolinos, 8–19 October 1984.
9.   McCalla, C.A.W., "Error Performance Metrics and the Generic Design of Digital Data Circuits," *GLOBECOM '85 Convention Record,* Vol. 1, December 1985, pp. 36–39.

Chapter 5
Voice Terminals

For transmission and grade-of-service considerations, the term *voice terminals* is just a new name for plain old telephones. The topic is multidisciplinary, including acoustics and electromagnetics as well as electrical engineering. In this chapter we describe the basic operation of a telephone, the acoustical and mechanical characteristics of handsets that are important in design, and significant characteristics of the environment in which it operates. We then show how these factors affect quality and how grade-of-service can be used to optimize performance.

5.1 STATION SETS

5.1.1 Transmitters

The devices that convert sound waves into electrical energy in a telephone handset are called *transmitters*. Most transmitters in service today consist of a small amount, perhaps half a teaspoon, of carbon granules resting loosely in a capsule between two electrodes. When a dc voltage is applied to the electrodes, the amount of current that flows through the granules is determined, in part, by how tightly the granules are packed. The capsule is arranged in the handset so that when sound waves pass it the granules are compressed more during the high pressure part of the wave than during the low pressure part. This compression and relaxation causes the current flowing through the device to vary, superimposing on the dc, an ac component that represents the sound. This current at some point passes through a load resistor where the current changes are sensed as voltage changes. The changing voltage is called the *transmitter output*.

The carbon transmitter is popular because it is economical and functions well. It is not an ideal device, however; it has two major nonlinear characteristics. First, for a fixed available dc, the output does not change linearly in magnitude with changes in the sound level. Over the range of normal speech volumes (about 40 to 60 dBA at conversational distances or 80 to 100 at modal distances), this problem is not usually important. The other nonlinearity has to do with the output variations for a constant sound level when the available dc changes. This is important because the dc is supplied from batteries in a central office over the copper wire loop. For a fixed size of wire, as the loop increases in length with distance from the central office, its resistance increases, and the available current for the transmitter decreases. When the dc decreases the ac output decreases, which is just the reverse from what we would like to have happen. The longer copper loops not only reduce the dc, they also cause greater attenuation of the voice signal output of the transmitter on its path to the CO. The amount of this attenuation depends on the gauge of the copper conductors and design of the cable. For wire gauges commonly used, at a frequency of 1000 Hz, it ranges from about 1.1 decibels per mile for 19 gauge, to about 2.9 decibels per mile for 26 gauge [1, p. 823].

Many telephones use varistors (devices with resistance inversely proportional to current) in parallel with the transmitter. As the loop length increases and current diminishes, the resistance of the varistor increases and more of the available current goes through the transmitter. This limits the variability in transmitter output. Figure 5.1 shows the output variation of a complete telephone for constant sound level as the current changes. The abscissa (current axis) is annotated to indicate the magnitude of current expected on short and long loops. The dc resistance of most U.S. telephones ranges from about 100 ohms at 60 ma, to more than 200 ohms at 20 ma. The relative contribution of the transmitter itself to the total resistance is small, however. Central office batteries have an available voltage of about 50 volts. The voltage drop across the telephone is only about 4 to 10, most of the remainder of the 50 is dropped across the loop. (There is a small drop across some equipment in the central office.)

Uniform content and structure of the carbon granules used in transmitters is very important for uniformity of performance and great care is taken in their preparation. Even so, the variability within and between manufactured lots is large, and there are aging effects that do not seem to be well understood. The total variation in output for transmitters of identical type may be as large as 10 or 12 dB.

In recent years there have been sufficient reductions in the costs of manufacturing a type of microphone known as the *foil-electret* to make it economically competitive with the carbon transmitter. A functional diagram of such a microphone is shown in Figure 5.2. The foil-electret microphone derives a voltage from sound pressure by causing the pressure variations to move an electrostatically charged foil ([2], [3], [4]). The foil

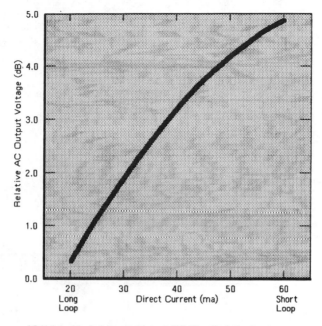

AC Values Are As Measured Into A 600 Ohm Resistive Load.
Normal Variations From Set To Set May Be As Great As ± 6.0 dB.

Fig. 5.1 Typical Output of a Carbon Transmitter Type Telephone Set

is supported by a flexible grid that serves as one plate of an air capacitor. The bottom cover serves as the other plate. An acoustic wave passing over the device flexes the foil and grid, thus changing the distance between the charge on the lower surface of the foil and the bottom cover. The voltage across the capacitor, being proportional to this distance, follows the acoustic signal. These devices have a linear relation between acoustic pressure and voltage output over a wide dynamic range. This characteristic makes them very desirable for use as telephone handset transmitters. The microphones themselves can be held to manufacturing tolerances of about plus or minus three decibels, another advantage over the carbon unit.

For reasons not yet understood, foil-electret, and other linear microphones, appear to have a subjective quality in the speech reproduced from their outputs that makes them superior to carbon. That is, for identical acoustic level output of receivers responding to signals from carbon and linear microphones, the one driven by the linear microphone is preferred. This preference has been measured in terms of equivalent loudness loss in dB. If a positive number of dB is used to express preference for the linear microphone, various subjective tests have reported results

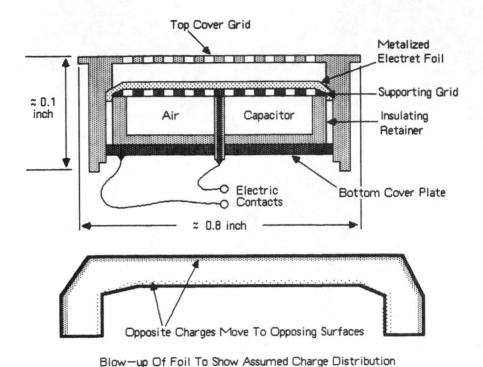

Fig. 5.2 Principle of Foil-Electret Microphone Construction
Source: Adapted from: G.M. Sessler and J.E. West, *J. Acoust. Soc. Am.*, Vol. 46, No. 5, (Pt. 1), November 1969

ranging from about −1 to 5 dB. It appears that the experimental differences may be dependent upon factors such as absolute talker volume and room noise levels. In general, however, a linear microphone is preferable [5].

Electrets operate from dc voltages of 1 to 20, require about 10 to 50 microamperes of current, and have a relatively high output impedance, on the order of 3000 ohms. Typical sensitivities are sufficient to produce an open circuit voltage of about 0.1 mv when driven by a sound pressure equal to average speech intensity about one inch from a talker's lips, about 92 dBSPL referred to 20 micro pascals.

5.1.2 Receivers

Most telephone receivers employ electromagnetic devices to convert electric signals to acoustic pressures. One type of construction is illustrated in Figure 5.3. The diaphragm is a thin disc of ferro-magnetic material. It is held suspended above the permanent magnet to which it is attracted. The edge of the disc is held rigid by the support, so the diaphragm is normally bent or cupped downward into the small gap by the *bias* of the magnet. The gap size may be on the order of 0.010 inches. The electric signal carrying speech information from a transmitter is applied to the coil. This creates a varying magnetic flux superimposed on the bias of the permanent magnet. As the applied current varies, the magnetic flux follows, resulting in a varying force on the diaphragm. The spring action of the cupped diaphragm, reacting with the changing magnetic force, causes the diaphragm to take on a vertical motion corresponding to changes in the applied signal. The diaphragm motion then sets up acoustic waves, regencrating a sound field similar to that impinging on the transmitter. Without the permanent magnet bias, the reference plane for the spring action of the diaphragm would be the undistorted position. The flux due to the coil current cannot push the diaphragm outward, a full wave rectifier action results, and the unbiased device acts as a frequency doubler. If constructed as shown in the simplified drawing (Figure 5.3), the receiver frequency response would be sharply peaked at the resonant frequency of the diaphragm. Acoustic damping and tuning techniques must be employed to achieve desired response characteristics. A detailed discussion of this is beyond the scope of this material.

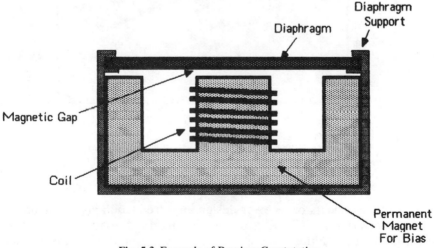

Fig. 5.3 Example of Receiver Construction

The bias also causes the rest position of the diaphragm to locate in a position of higher mechanical stress, compared to a nonbias situation, so motion consumes more energy; thus, receiver sensitivity can be controlled by the strength of the magnetic bias. If a part of the bias is due to a direct current in the coil supplied from the CO battery, sensitivity can be made proportional to loop length. In combination with varistors to modify loop current, a typical receiver output as a function of loop current is shown in Figure 5.4. Note that the change in sensitivity with current, or loop length, is inverse compared to that of the transmitter. This provides some compensation for the characteristic of the transmitter and the increased attenuation of long loops.

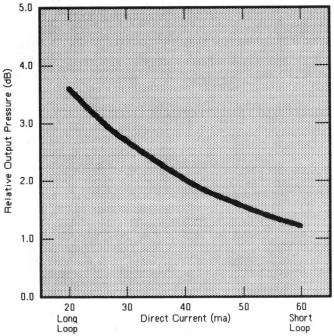

AC Values As Measured With A 600 Ohm Resistive Source Driver
Normal Variations From Set To Set May Be As Great As ± 3.0 Db

Fig. 5.4 Typical Output of a Telephone Receiver

Critical aspects of receiver design and production are the acoustic tuning mechanisms and the gap size. Production variations are not as great as for transmitters, however; a total range of about 6 dB may be expected in commercial receivers.

Other types of receivers are found in some handsets. In principle, any electroacoustic device may be used as either a transmitter or receiver. Small loud speakers with treated paper cones have recently been used successfully in both applications. These tend to be more linear in their transfer of electric energy to acoustic, and can be made to have very good frequency characteristics as measured within the enclosing handset. They are not as yet in wide use, perhaps because of concern over life expectancy and long-term stability.

5.1.3 Hybrid Devices

Hybrid devices are certainly one of the most cost-effective devices in telephony. They permit simultaneous two-way transmission over a single pair of wires, a two-wire trunk, by being able to distinguish and separate signals in the two directions. In the telephone station, hybrids prevent the transmitter from coupling directly to its companion receiver even though they are both coupled to the loop. In two-wire trunks, they allow *repeaters* (two-way amplifiers) to be inserted as needed to compensate for cable loss. In these two applications, by permitting one rather than two pairs of wires to be used they nearly halve the cost of transmission. Use of a hybrid is illustrated in Figure 5.5 where a four-wire trunk[1] is connected to two-wire trunks. The communications flow is indicated by the thin directional lines.

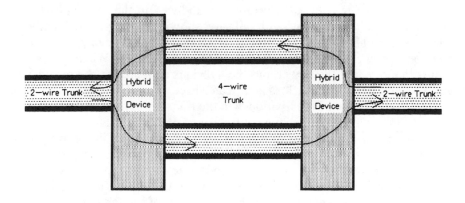

Fig. 5.5 An Application of Hybrids

Hybrid operation can be understood by examining Figure 5.6. Drawing A shows the basic arrangements with generalized impedances Za through Zd attached to the windings. The three coils have equal numbers of turns. Placement of the coils in a telephone set circuit is indicated parenthetically. Drawing B illustrates what happens when signals are coming from the line (loop) toward the set. These are represented by the

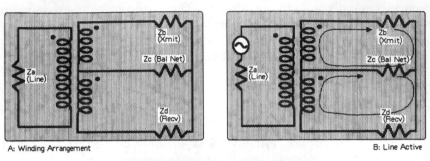

A: Winding Arrangement   B: Line Active

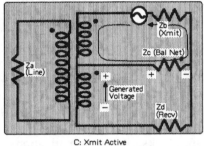

C: Xmit Active

Fig. 5.6 Functional States of a Hybrid Transformer

generator in series with Za. Currents are induced in the two right-hand circuit loops by equal emfs, of polarity shown. If $Zb = Zd$, the two currents cancel in Zc, and the power divides equally between Zb and Zd. Delivering power to the transmitter is of no consequence. What is important is that one half the power into the hybrid is delivered to the receiver, and so hybrid transformer loss is at least three decibels.

Drawing C shows the hybrid when the transmitter of the telephone is active. Since all windings have a turns ratio of $1:1$, equal emfs are induced in the three coils. Power is delivered to the line, Za, in the transmit direction. To see what happens on the right side, consider the current in the upper circuit loop first. A counter-clockwise current induces a voltage across the coil and a voltage across Zc. Now, if the voltage induced in the coil in the lower loop equals that across Zc and is of the same polarity, no current will flow in the lower loop. (Both ends of Zc will be at the same potential.) Because no current flows in this loop, it is effectively an open circuit. With the turns ratio in the winding in the upper loop and the winding driving the load, Za, equal to one, the load seen by the source (generator plus Zb) is $Za + Zc$. Thus one half of the available power is delivered to the line impedance, Za, one half to the balance network, Zc, and none to the receiver, Zd. This will happen only if Zc is identical to Za.

(For the balanced condition, because no current is flowing in the lower loop, the open assumed above may now be closed with no change in results.) *Zc* is called a *balance network* because it should equal the line impedance and so balance out induced emfs across the receiver. In all hybrid applications, effort is devoted to obtaining the best possible match between the (two-wire) line impedance which is highly variable, and the impedance of the balance network. The name *compromise balance network* is often used to emphasize the fact that the network does not match all lines, but is a *best fit* in some sense. One method of finding a best fit is presented in Chapter 7. Hybrid transformers discussed here for illustrative purpose are decreasing in use, less expensive integrated electronic hybrid circuits are available that perform the same functions.

5.1.4 Sidetone

The situation of perfect balance of a telephone set hybrid to the line, discussed above, may occur in practice and no current will flow in the receiver due to transmitter action. A similar but more likely case is the occurrence of line impedances that come close to a match with the balance network, and only a small current will flow in the receiver. These two cases would make for a poor telephone connection. If no current flows in the receiver when a person talks into the transmitter of a telephone handset, the set sounds *dead* and the speaker may pause to investigate the situation, for he does hear himself when he speaks. Having overcome his suspicions in some way, the talker may begin a conversation with the far-end user. It is very likely that he will talk excessively loud because he does not hear himself in the receiver. The voice may even rise to a shout, to the detriment of the far-end listener. The mechanism by which the talker does hear himself is called the *sidetone path*.

The complete sidetone path consists of three parts. The first is the path through the air from the speaker's mouth to his ears. Even if one ear is covered by the receiver portion of a handset, there is some leakage to the ear because the seal of earcap to ear is not perfect. The second path is sound conduction from the talker's vocal track to his ear through connecting tissue, primarily the bone structure of the skull. The third is called the *telephone path* here. It has two parts: one is conduction through the plastic and the hollow center of the handset from transmitter area to the ear; the other is electrical, discussed in detail in the following. With a well-designed handset in usual environments, the electrical path is controlling, and proper control of its loss is essential for good telephony. As indicated above, too much loss results in the speaker raising his voice and, conversely, too little loss will cause the speaker to lower his voice, to a whisper in extreme cases. In many telephones, a separate coil, associated with the hybrid coils, is used to provide a minimum level of sidetone no matter what the line impedance may be.

Sidetone control is not simple. The problem lies in determining what impedance should be used for the balance network. The function of the balance network can be thought of as canceling the reflection (as explained in Chapter 1, Section 1.1.5) from the junction of the telephone with the line. This is not the only source of reflections, however. The junction of the loop with the equipment at the *central telephone office* (CO) causes another. Impedance irregularities in any part of the connection give rise to still more, the junctions of two-wire and four-wire trunks for example.

The two impedance factors associated with the local loop, that of the loop itself, and that presented at the CO, are lumped together under the term *loop impedance,* and are discussed in Chapter 6. Considering reflections from the two loop impedance discontinuities as though coming from one is permissible because the time delay between them is small, less than one millisecond as was illustrated in Chapter 1, Figure 1.4.

A more complex case, and one in which the time delay must be addressed, is illustrated in Figure 5.7. Three separate contributions to the total sidetone signal are shown. Path *A* in the drawing represents the one for the intentional minimum sidetone. Path *B* is equivalent to the one due to the net loop impedance of the preceding paragraph, but in this case is due to a mismatch between the station loop (the wire from the telephone to the PBX), and the impedance of the PBX station port. The spread of the impedances presented by the station loops to the PBX station ports is small compared to that seen by the PBX port facing the CO loops because there is much less variation in cable type and length. Thus the choice of an appropriate balance network for this case (B) is less difficult.

The central office loop mismatch must now be addressed by the *balance* network (*C*) in the CO port of the PBX. If the PBX uses a digital switch, a round trip delay on the order of milliseconds may be introduced in the third path, *C*. This delay, constant at all frequencies, significantly changes the effective impedance of the CO loops. For two milliseconds it is equal to a rotation of 360° at 500 Hz, 720° at 1000 Hz, *etcetera*. The reflection returned by this path will add to (and subtract from) the others shown in the drawing in a complex manner dependent upon the relative phases of the two signals at each frequency. This is discussed in more detail in the next section.

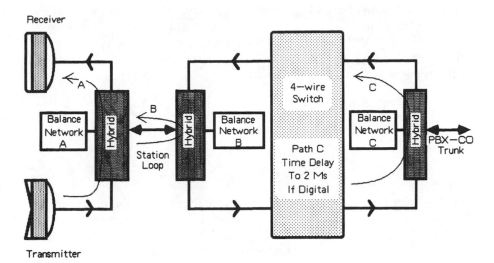

Echoes From Paths A,B, And C Add On A Voltage Basis For Net Sidetone
Echo On Path A Is The Design Controlled Sidetone

Fig. 5.7 Echo Paths in a PBX Connection

5.1.5 Digital Handsets

Digital handsets are those that use an analog-to-digital (A/D) converter between the transmitter and the output of the handset[2] and a digital-to-analog (D/A) converter between the electrical input to the handset and the receiver. Such sets are designed for use with digital PBXs and, more recently, for ISDN service. The noise due to the quantizing process must be handled as discussed in Chapter 4, Section 4.1.6 and normally is of no particular consequence. The delayed sidetone mentioned in the previous section may be of concern, however. The problem is not a direct result of the digital handset, but is intrinsic to the processing delays in digital machines. An apparent advantage of the digital telephone systems, intrinsic four-wire operation, aggravates it, so it shows up more often with digital sets. Digital PBXs may impose a round trip delay of up to 3.0 ms[3] in the echo path labeled C in Figure 5.6.

Because the handset and the machine are four-wire devices, echo path B does not exist and this is the apparent advantage mentioned. That is, better control over the sidetone level should be possible if there is one less echo to contend with. However, consider the summation of intentional sidetone introduced in the handset analog circuitry, and the delayed echo on path C at the receiver. With the signal amplitude on path A equal to A, and that on path C equal to C, the combined signal sidetone, $s(t)$, at one frequency, f, can then be written as

$$s(t) = A \cos[\omega t] + C \cos[\omega \cdot (t + \tau)], \qquad \omega = 2\pi f \tag{5.1}$$

If $A = C$, then

$$s(t) = \{2A \cos[\omega \cdot (\tau/2)]\}\cos[\omega(t + \tau/2)] \tag{5.2}$$

and the sidetone response is seen to be multiplied by a function of the frequency and the delay. If the delay is small compared to a period of the highest frequency of interest, the multiplier is near one and nearly constant with frequency, but for delays of about one or two ms, the response appears to have passed through a comb filter as illustrated in Figure 5.8.

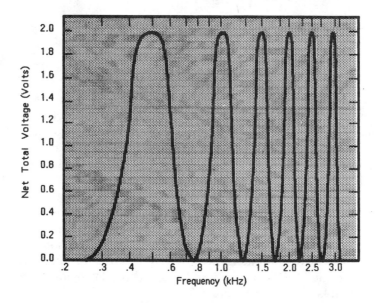

Effective Sidetone Voltage With Equal Amplitude
(1.0 Volt Peak) Sidetone And Delayed Reflection Voltages
In A Digital PBX. Round Trip Delay Equal To 2.0 ms.

Fig. 5.8 Comb Filter Effect

The subjective effects of this phenomena apparently have not been properly quantified, but are known to be quite annoying for echoes of magnitudes often encountered. It may be necessary to design the intentional sidetone level higher than optimum to mask out the annoyance; the higher level being preferred to the combined signals at a lower level.

In the case of an analog handset with the A/D conversions in the switch, echo path *B* adds a slightly phase shifted (delayed) signal to the intentional one, path *A*. The masking effects of the two on that from path *C*, seem to be much greater than the one, so the subjective effects of the comb filter action are not as pronounced.

5.1.6 Handset Design Factors

Telephone handsets are so familiar that one is not apt to consider what factors must be addressed to design a good one. Many of these are illustrated in Figure 5.9. Starting at the top left of the drawing, the surface surrounding the receiver should physically couple to the ear in a way to make as effective an acoustic seal as possible. Proceeding clockwise, it is noted that if the case or shell of the handset is made in more than one section, there is a danger of room noise leaking into the set where it can be picked up by the transmitter, partially mask the talker's voice and degrade signal-to-noise ratio. Seals between sections should be acoustically sound.

Next, the angle that the plane of the transmitter makes with the plane of the earpiece is critical. For the best pickup by the transmitter, it must be positioned with respect to the talker's lips in a well-defined way. The exact orientation to the lips for best pickup may vary among designs but small departures from the optimum have significant effects. The position is essentially determined when the receiver is placed against the ear, hence the importance of the angle between the two planes.

The pattern and sizes of the holes opening to the transmitter play a major role in determining the frequency response of the output of the telephone, and the effective area of the mouthpiece has an effect on the acoustic pressure reaching the transmitter because the obstruction in the acoustic path has a baffle effect. The pressure in front of the microphone is greater than it would be if the obstacle were not there. Handsets in use today may vary in sensitivity by as much as two decibels due to differences in this baffle effect because of differences in the size and shape of handset mouthpieces.

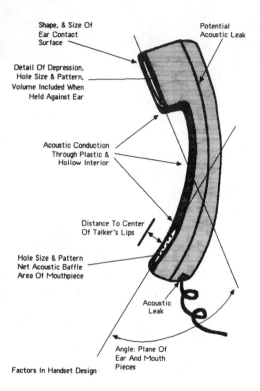

Shape, & Size Of
Ear Contact
Surface

Potential
Acoustic Leak

Detail Of Depression,
Hole Size & Pattern,
Volume Included When
Held Against Ear

Acoustic Conduction
Through Plastic &
Hollow Interior

Distance To Center
Of Talker's Lips

Hole Size & Pattern
Net Acoustic Baffle
Area Of Mouthpiece

Acoustic
Leak

Angle: Plane Of
Ear And Mouth
Pieces

Factors In Handset Design

Fig. 5.9 Factors in Handset Design

The distance between the mouthpiece and the talker's lips is important. (See [6] for methods of measuring this quantity.) The speech sound waves can be considered as emanating from a point just inside the talker's lips. The reduction in acoustic pressure as the wavefront travels outward is very rapid over the first few centimeters of distance. The decrease is approximately proportional to the inverse square of the distance. That is about 12 dB for each doubling of the distance.

Continuing the clockwise rotation (on the drawing) upward from the mouthpiece, it is pointed out here that the acoustic conduction from transmitter to earpiece is important. It should be minimized to allow maximum control of sidetone by the electrical path.

Finally, it is noted that the volume of air enclosed by the receiver earpiece, the ear, and the walls of the ear canal when the receiver is held closely to the ear must be considered. The efficiency and frequency response of the receiver are affected by this volume. Receiver test equipments are classed by (among other things) the volume of an *artificial ear* to which receivers are clamped during the test.

There are at least two nontechnical design considerations that are easily overlooked in handsets. One is the need for cheek clearance. Many people do not grasp the handset around the throat of the device, and do not have a part of the hand between the instrument and the cheek. Nevertheless, when properly positioned, the handset design should provide adequate space for cheek clearance. Many find it uncomfortable if the space is not there, and males with beards find that whiskers rubbing on the plastic can be heard very clearly as sidetone. The other factor is the ease with which the handset is cradled. When the instrument is supported by resting it on the shoulder, and held in place by tilting the head down and sideways in order to free both hands, it is said to be cradled. Good physical design makes for easy cradling.

5.1.7 Room Noise

Room noise is picked up by the transmitter and heard as sidetone. It is also transmitted to the far-end listener. If it is loud enough it interferes with the conversation. Handsets can be designed to work satisfactorily in very noisy environments, up to 90 dBA or more. These are known as specialty handsets and are discussed in the next section. For environments other than extremely noisy ones, in the range of 40 to 75 or 80 dBA, the effects of room noise (Hoth spectrum) can be translated into equivalent circuit noise and included in grade-of-service calculations. One method of doing this is to simply measure the output of a telephone with a noise meter, for various levels of room noise. Data of this type are illustrated in Figure 5.10 for a handset with a carbon and a linear transmitter. The equivalent circuit noise shown in the drawing has been translated to a value at the central office for calculations of noise transmitted to the far-end listener; and to equivalent circuit noise at the telephone receiver input, for the sidetone value. The values given in Figure 5.10 are illustrative only. They are not necessarily representative of actual transmitter performance. Use of such data is illustrated in Section 5.2 and Chapter 7.

5.1.8 Specialty Handsets

A number of handsets are designed for use in very noisy environments, or for persons with impaired hearing. They are of two general types. One type uses auxiliary amplification in the receiver or transmitter, under control of a switch or volume control. The other type uses one of several so-called noise canceling microphone designs. (See [2] and [3] for details of these.)

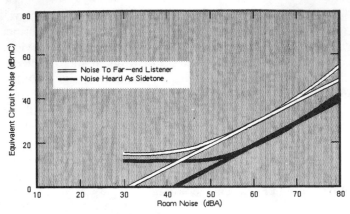

Straight Lines Are For Linear Transmitters.
Curved Lines Are For Carbon Transmitters.

Presentation Assumes Use In Identical Telephones,
With Gains Set For Equal Conversion Efficiencies.

Fig. 5.10 Example of Translation of Room Noise to Equivalent Circuit Noise

If extra amplification is placed in the transmitter, a push-to-talk switch is usually provided. In the rest position, the switch is arranged to disable the transmitter, and room noise that would have entered the side-tone path, causing masking of speech from the far end, is blocked. When the user wishes to speak, he presses the switch to enable transmission. If amplification is added to the receiver circuitry, a simple volume control may be supplied to enable adjustment of the received loudness and improve the signal-to-room-noise ratio. These features may be combined in one design. The enhanced receiver design is also useful to people with impaired hearing.

Noise canceling microphones make use of the different characteristics of the sound emanating from a talker's mouth and that of room noise. As mentioned earlier, the acoustic pressure in front of the talker's lips is decreasing roughly according to an inverse square law. The sound field in this case is referred to as the *near field*. Room noise acoustic pressure is much more uniform in nature, a characteristic of *far fields*. The decreasing pressure of the near field is used to advantage by designing a microphone exposed to the field on opposing sides; one toward the talker, and one away from the talker. The microphone is made to respond to the difference in pressure between the two openings. The pressure of the far field is nearly constant over the depth of the device and so does not activate the transmitter. These instruments are also referred to as *close talking* microphones. They are used only in very noisy locations because of the inconvenience of maintaining the correct modal distance while speaking into them.

5.2 OPTIMIZING GOS

This section considers the interactions between station design and loop and network parameters and methods to select the station parameter values to yield an optimum grade-of-service. A fixed cable size, room noise, and variable length loops are discussed.

5.2.1 Loop Effects on Transmission Ratings

For good voice terminal design, the loop and station set must be considered together to determine transmission ratings. The loop cable can be analyzed in a number of ways. The general approach is described, for example, in Chapter 20 of [1]. The basic cable parameters needed for the calculations, resistance, capacitance, and inductance per mile, are provided on pages 822 and 823 in the above reference for many types of telephone cables. A more elegant and simpler method using ABCD parameters is presented by Webb in [7]. The characteristics used in the following example are the frequency response and input impedance as a function of length, for various source and load impedances. The configuration considered is a 600-ohm station set for a PBX[4], working through 26-gauge cable to a central office termination of 900 ohms as in Figure 5.11. The drawing is similar to the recommended IEEE test configuration. Note that in the receive direction, central office toward the station, the source impedance is 900 ohms and the load 600. The feeder circuit of the central office is ignored because it has an effect of only a few tenths of a decibel on the transmit and receive ratings. However, it strongly affects sidetone, especially at loop lengths of less than 1000 feet, and should be taken into account in real designs. (See Chapter 7.) Variations in sidetone loss are not addressed in this example.

The transmission ratings used here are those of the IEEE, TOLR, *etcetera*. Cable losses were calculated for this configuration in terms of IEEE loudness loss, or just loudness loss, using Webb's method for the circuit characteristics, and (2.23) for loudness loss. The results, normalized to one kilohertz, are presented in Figures 5.12 and 5.13. Figure 5.12(a) shows the frequency response of the cable at 9000 ft, a common reference length, normalized at one kilohertz. The loudness loss due to the frequency shaping of the cable at 9000 ft is as shown on the drawing, 0.48 dB. Figure 5.12(b) shows how (IEEE) loudness insertion loss

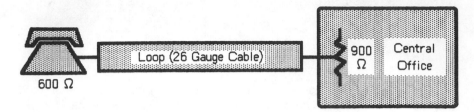

Fig. 5.11 Telephone and Loop Arrangement

changes with cable length when the terminations are 600 and 900 ohms. It is the same in both directions. The plots are terminated at 15,000 ft because that is about the length at which loading begins and beyond which the loudness loss may be considered constant.

Figure 5.13 shows the magnitude and angle of the input impedance of the loop, at one kilohertz, as a function of loop length. Figure 5.13(a) is for the case of a 900-ohm termination. Figure 5.13(b) shows the impedance when the terminating impedance is 600 ohms. Curves are plotted for three frequencies, 0.3 and 3.3 kHz as well as the 1000-Hz values. The 0.3 and 3.3 kHz are the band edges for purposes of loudness loss calculations. Note that the one-kilohertz impedances for either the 600- or 900-ohm terminations appear to be becoming equal beyond 14,000 ft. Actually they are still decreasing and would eventually approach the surge or characteristic impedance of the cable, about 900 ohms.[5] This figure serves as the first illustration of the problems associated with trying to find a balance network such as the one denoted *C* in Figure 5.7. This problem is explored further in Chapter 6, Section 6.10.

For convenient use, a least squares fit to the *insertion loudness loss* (ILL) of Figure 5.12 was made. It appears to be nearly linear, and a linear approximation is often quite adequate, but there is a slight curvature. A second order fit provides more accuracy. With length, x, in 1000 ft, the insertion loudness loss is

$$\text{ILL}(x) = -0.015 + 0.511 \cdot x + 0.00312 \cdot x^2 \text{ dB} \qquad (5.3)$$

The negative value, or gain at zero feet, is an artifact of the fit. This equation was based on the IEEE metric for loudness loss and assumes that 26-gauge cable is used. For other cable sizes, termination impedances, or metrics, an equivalent expression must be derived.

The goal of station design is to optimize the grade-of-service over all transmission conditions to be encountered. The method used to calculate GoS is that described in Chapter 4, using R_{LN} only. The values needed

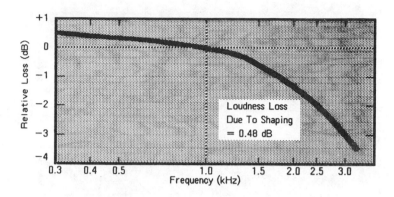

(a) Frequency Response At 9.0 Kft

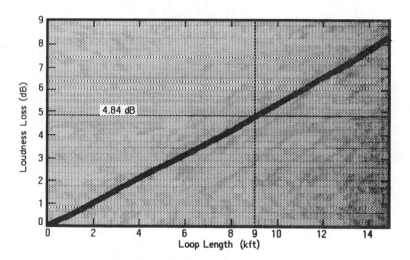

(b) Insertion Loudness Loss As A Function Of Cable Length

Fig. 5.12 Loudness Loss (LL) and Insertion Loudness Loss (ILL). ILL = LL plus One-Kilohertz Insertion Loss

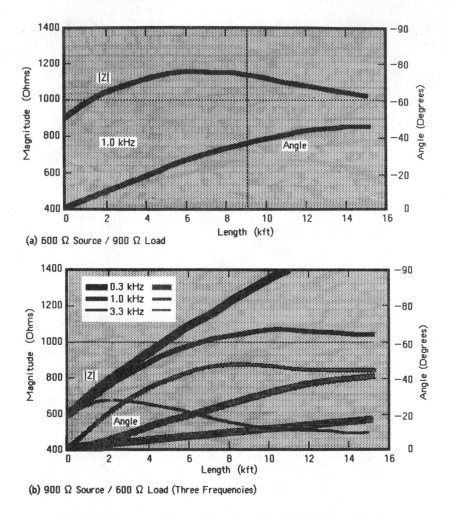

(a) 600 Ω Source / 900 Ω Load

(b) 900 Ω Source / 600 Ω Load (Three Frequencies)

Fig. 5.13 Impedance of Loop as Function of Length for Different Terminating Impedances

from the loop analysis are TOLR, ROLR, and overall station-to-station loss or OOLR, equal to TOLR + ROLR. The basic definitions for these quantities, (3.5), (3.6), and (3.8), can be expressed in logarithmic form using dBV instead of mv, and dBSPL (sound pressure level in dB above 20 μPa) instead of Pa. The necessary conversions are: dBV = dBmv − 60 and dBSPL = dBPa + 94. The results are

$$TOLR = dBSPL - dBV - 154, \qquad (5.4)$$

$$ROLR = dBV - dBSPL + 154 - 6 \qquad (5.5)$$

The value 154 results from the conversions, and the −6 from the division by two in the ROLR definition. It is left separate to emphasize that the dBV refers to the open circuit source voltage. These values are calculated using the simplified circuits (no battery feed) of Figure 5.14. Electronic stations with linear transmitters are assumed. Gain is supplied by the transmit and receive amplifiers G_t and G_r. Transducer conversion efficiencies are expressed in decibels and are determined by the choice of handset components. Some typical values, −115 dB (pressure in dBSPL to voltage in dBV) for a transmitter and +122 (dBV to dBSPL) for a receiver, are used here. The voltage E_1 in dBV is $E_1 =$ dBSPL − 115 for the transmitter, and dBSPL = dBV at E_4 + 122 for the receiver.

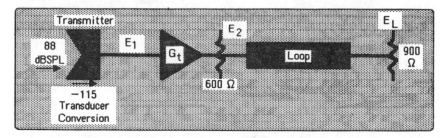

(a) Transmit Arrangement

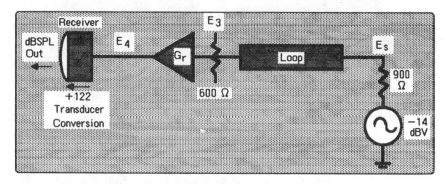

(b) Receive Arrangement

Fig. 5.14 Circuits for Determining Transmission Ratings

In the equation for TOLR from Chapter 4,

$$\text{TOLR} = -20 \log(v/p) \qquad (v \text{ in millivolts, } p \text{ in Pascals})$$

the voltage refers to E_L of the drawing because the TOLR value must apply from the transmitter to the CO. This can be found by computing the input voltage to a 900-ohm load (zero loop) and reducing it by the loop loudness loss:

$$dBV_{E_L} = dBSPL - 115 + G_t + 6 + 20\log[900/(600 + 900)] - ILL(x), \tag{5.6}$$

where the +6 term is added to convert the amplifier output voltage to its open circuit value, and the log term computes the voltage into the 900-ohm load to be consistent with the definition and use of insertion and loudness loss; that is, loss seen at the output impedance due to insertion of the cable. Combining terms, we have

$$dBV = dBSPL + G_t - 113.44 - ILL(x) \tag{5.7}$$

Substituting this and (5.3) for $LL(x)$ in the expression for TOLR yields

$$TOLR = -40.55 - G_t + 0.511 \cdot x + 0.00312 \cdot x^2 \tag{5.8}$$

Similarly, for ROLR, from Figure 5.14(b):

$$dBSPL = E_4 + 122$$
$$E_4 = -14 + 20\log[600/(600 + 900)] - ILL + G_r$$

Substituting these and (5.3) for ILL in (5.5), we have

$$ROLR = 33.95 - G_r + 0.511 \cdot x + 0.00312 \cdot x^2 \tag{5.9}$$

5.2.2 Noise Considerations

There are two distinct sources of noise heard by the user. First is that originating within the telephone and the network, referred to as circuit noise. Second, is the acoustic or room noise. Room noise at the listener's location is picked up by the transmitter and heard as sidetone in his receiver. This is in addition to the room noise heard directly through the ear not covered by the receiver, and through the leakage path to the covered ear due to the imperfect seal between receiver and ear. It is assumed that room noise can be treated in the same manner as circuit noise if it can be translated to an equivalent amount, and then added to the circuit noise on a power basis. Data such as that shown in Figure 5.10 might be used but it can be argued that such data may not relate the subjective effects of room noise as opposed to circuit noise. The method used here for including room noise is that of Chapter 4, (4.19):

$$dBrnC = N_R - 35 + 0.0078(N_R - 35)^2$$
$$+ 10 \log[1 + 10^{(1-SOLR)/10}] \qquad (5.10a)$$

where N_R is room noise in dBA, and SOLR is the IEEE sidetone loudness loss.

This expression indicates that at low values of room noise, the effective circuit noise is obtained by subtracting 35 from the number of dBA. As the noise increases above 35 dBA, there is an increasing nonlinear effect as given by the square term. The contribution attributable to the sidetone path is contained in the last term. For SOLR values of about ten or larger, the sidetone room noise is negligible. When the SOLR equals one, the sidetone and direct paths are equal and the net equivalent noise increases by three decibels. For smaller values of SOLR, the sidetone path becomes dominant. The equation is shown plotted for several values of SOLR in Figure 5.15.

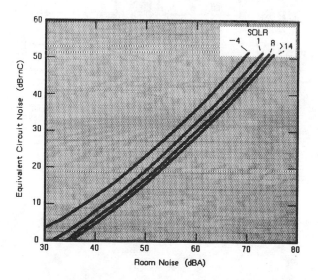

Fig. 5.15 Equivalent Circuit Noise of Room Noise Referred to a Receiving Circuit with ROLR = +46
Source: Adapted from *Red Book, Volume V,* CCITT, Geneva, 1985

The equivalent circuit noise must now be added to the actual circuit noise. A little reflection raises the question as to where to add this noise; at the transmitter, receiver, or where? To see how the room noise equivalent is treated, refer to Figure 5.16. The two connections illustrated in the drawing have the same OOLR but different ROLRs and TOLRs. The circuit noises at the CO, the room noises, and the SOLRs are identical. The equations of Chapter 4, relating transmission ratings to opinion score distributions, are only valid for the top picture, ROLR = +46 dB, because all the subjective test data concerning transmission are referred to these standard values. The effect of the room noise must be the same in both cases, however. We are concerned with ROLR, the receive loop rating, even though we are talking about the effects of room noise on the speaker, because this room noise must be added to the circuit noise being delivered to the receiver.

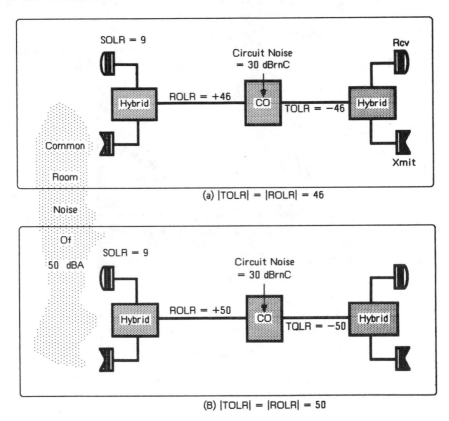

Fig. 5.16 Connections with Same OOLR but Different TOLR and ROLR

Now, the R_{LN} equation assumes that all noise sources are at the CO and drive a loop with an ROLR of +46. Equation (5.10a) gives the effective noise at the CO from room noise over this loop. The subjective effect is, therefore, that due to noise at the CO heard at the receiver after being attenuated by the loss of such a loop. The loop with the +50 ROLR in Figure 5.16, has more loss. The effective noise due to room noise, measured at the CO, is less than that measured for the circuit in Figure 5.16(a). In order for the equivalent room noise from the loop with greater loss to be the same, it must be increased by the difference in losses of the two loops. The equation above, therefore, needs a correction term equal to +(ROLR − 46). ROLR, of course, refers to that of the loop being considered. The term adjusts the room noise to a value that then measures the same at the CO as does the unadjusted noise over the loop with the ROLR equal to +46. Equation (5.10a) then becomes

$$
\begin{aligned}
\text{dBrnC} = {} & N_R - 35 + 0.0078(N_R - 35)^2 \\
& + 10 \log[1 + 10^{(1-\text{SOLR})/10}] \\
& + (\text{ROLR} - 46)
\end{aligned}
\tag{5.10b}
$$

If the reader is using the CCITT CRE metrics, the standard reference receive end has an RCRE of +1 dB, the correction term becomes (RCRE − 1). The exponent term in (5.10b) also changes. It becomes 7 − STCRE, in keeping with the conversion, STCRE = SOLR + 6 in Table 3.6.

A special case exists for digital loops or for any analog-to-digital-to-analog conversions within the station to CO loop. The resulting quantizing noise has different subjective effects and is handled in a different way than noise from analog sources. The subjective effects of quantizing noise are different for at least two reasons: The peaks of the noise are limited to be one half the value of the largest quantizing interval, and it is a correlated noise; that is, it exists only when speech is present. The expression for quantizing noise equivalent circuit noise was given, (4.20), as

$$
N'_{Qe} = \text{Vo} + 89 - \text{SNR}
\tag{5.11}
$$

with an explanation of, and methods to find, the terms on the right side. This expression was also derived to calculate a noise value referred to the CO at one end of a +46-ROLR loop. The relative positions of the noise source (point of measurement) and the CO must be taken into account to translate noise from an arbitrary point to the CO. Note that the quantity Vo, the speech level in VU, is also taken as the value as received on a +46-TOLR loop, from the far end. In this case, the noise given by (5.11) is modified by the actual TOLR of the talker's loop to find the value at the CO where it can then be power added to any other noise being considered. SNR is a subjectively equivalent signal-to-noise ratio determined by the algorithm of the encoders associated with the overall analog-digital-analog process. Reference [8] gives the computational methods needed for several types of encoding.

Having computed the effective values of the room and quantizing noise (if required) and referred them over the ROLR and TOLR of interest to the CO as $N1$, they are power summed, to the circuit noise, $N2$, to give a total noise value in dBrnC:

$$N(\text{total}) = 10 \log(10^{-N1/10} + 10^{-N2/10}) \tag{5.12}$$

Another noise that may be important is room noise at the far-end location. This is not taken into account in the methods of Chapter 4. It can be done with data such as were illustrated by the white colored curves in Figure 5.10. Such data are taken for a fixed loop length, or TOLR. Assume that the data shown were for a TOLR of -48. The equivalent circuit noise values in Figure 5.10 are for a noise at the CO serving such a loop. By reasoning in a manner similar to that used for ROLR sensitive values, this time for TOLR, the correction factor for this noise is found to be

$$\text{Corrected CO noise} = \text{noise}_{(\text{for loop of TOLR} = X)} - (X + 48),$$

where the TOLR (a negative number) is that for the loop being considered.

This is the equivalent circuit noise at the loop side of the CO serving the far-end listener and must then be translated through any intervening loss to the loop side of the CO serving the near-end subscriber. If the CO is common to both loops, it may be assumed that the loss through the office, so-called *switching loss,* is one half of a decibel. If the COs are different, then knowledge of the intervening connection loss is required. A value often used for the North American (analog) Network is 7 dB, representative of network loss on a typical toll call. The one half decibel for the local office loss must still be added. (The 7 dB includes the one half for the far-end office.) This noise is then power summed with the others as in (5.12).

The total noise value, in dBrnC is then used in the expression for R_{LN}, (4.18), along with the connection OOLR, to yield a transmission

rating. That transmission rating number determines the integration limits
A and *B* (choose equations from Table 4.1 for the data base desired) to
estimate the GoS, %GoB, %PoW or both %Gob and %PoW.

5.2.3 Variable Loop Lengths

The procedure just outlined provides an estimate of GoS for a given
connection or pair of loops. Obviously, the GoS will be different for each
loop length. A typical example of variation with loop length, for a handset
using linear transducers is shown in Figure 5.17. The shape of this curve
can be changed by shifting gain between the transmit and receive portions
of the station. Exchanging loss between TOLR and ROLR has the effect
of shifting the relative location of circuit noise with respect to the receivers.

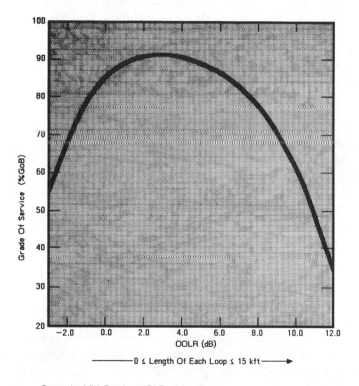

Connection With Two Loops Of Equal Length
Net OOLR Shown On Abscissa.

Fig. 5.17 Example of Grade-of-Service Variation with Loop Length

In choosing the amount of gain to be used in the station set, all loop lengths that may be encountered must be considered. One approach is to observe that the GoS curves in Figure 4.7 indicate that an overall CRE of about 8.5 (OOLR = 2.5) is preferred, no matter what the noise. One could then design the station gain to yield that number on the average loop. This approach has two problems. One is deciding what is meant by the average loop. For the United States, data such as in [9] can be used to generate a large number of averages. For example, the physical length of the loop, the geographic distance from CO, or the 1000-Hz insertion loss could be averaged. Use of any such average implies that the customers for whom the station is intended, are distributed in exactly the same way as the population of loops that were averaged. In general, this is not true. For example, business customers are typically more tightly clustered than suburban residence customers who are more tightly distributed than rural customers, *etcetera*. The target population must be defined and appropriate data used. The second problem is that using an average will give poorer service to customers on one side of the average than the other because the GoS curves are not symmetric about the line of preferred OOLR. (Refer to Figure 4.7.)

To arrive at a best compromise design, GoS must be computed for all (or many) loop lengths, and then weighted by the distribution of loops for the projected customers. When that is done, the GoS for each loop extreme must be examined, especially in terms of %PoW. The best weighted average may yield very poor service at one tail of the distribution and call for an adjustment to the first *optimum*.

The station design obtained may still not be the correct one. The values arrived at for station gains in the transmit and receive portions may be best for local calls, but calls over the network have not yet been considered. Station designers are not apt to have detailed network statistics or network simulations available to optimize designs. Average network loss and noise contributions can be used. Such data can be found in [10] and [11] for the U.S. Values of 7 dB for loss and 28.5 dBrnC for noise (averages for medium toll distances) might be used for example. The GoS distributions over the loop populations used for the local connection results must be recomputed with these increases in loss and noise. The penalties will be obvious and the task is to distribute them in some equitable fashion by readjusting the station gains. Again, special attention must be devoted to the tails of the distributions to assure that no small group is unduly penalized.

At this point, it may be said that station design would be much easier, and service to all equal, if total compensation for loop length variability were built into the station. Certainly with modern electronics, this should be possible. It is possible, but the cost may be prohibitive for a competitively priced station. That possibility should be examined in the early design stages. If the stations are to be used behind a modern PBX, key, or hybrid system, it may be possible to provide compensation in the switch. At some time in the future, digital loops with zero loss, will become ubiquitous and the loop variation problem will vanish. Because of echo considerations it is likely that there will still be a difference in loss between local and long distance calls, but this will be a relatively trivial situation with which to cope.

Other transmission factors can be mapped into a composite transmission rating as discussed in Chapter 4, and %GoB and %PoW computed. Some of these, such as echo delay and loss, are strictly network considerations. Any that are under control of the designer can and should be investigated for optimum design configurations.

Going through this procedure one time clearly demonstrates the sensitivity to small design changes that exist in the tails of GoS distributions. The often-heard excuse that there is no point in worrying about a decibel or two, because carbon transmitter efficiencies range over 15 dB (including aging effects), is found to be senseless. The distribution of the transmitter efficiencies is a variation that adds to the standard deviation of overall grade-of-service. It pushes both tails to even greater extremes. Any bias resulting from a suboptimum design, pushes a larger segment of users down one side or the other of the performance curve. As Figure 5.17 illustrates, those curves become steeper moving away from center. Even a good design leaves a percentage of cases to the left and right of preferred service. A bias may improve one side but the percent poor or worse will be disproportionately increased on the other side. Examples of some of these considerations are presented in Chapter 7.

NOTES

1. A four-wire trunk is a transmission facility that uses distinct transmission paths for the two directions of transmission.
2. Some implementations use an analog handset and do the A/D and D/A conversions in the station base. These are properly referred to as *digital stations* instead of digital handsets. For purposes of this discussion there is no difference.
3. 3.0 ms is the maximum delay time advised by the CCITT in the Yellow Book Rec. Q.517.
4. PBXs are usually designed to have 600-ohm impedances because that is the value most often associated with private-line networks.
5. Surge, or characteristic impedance is the ratio of voltage to current for a wave traveling on an infinitely long transmission line.

REFERENCES

1. International Telephone and Telegraph Corporation, *Reference Data for Radio Engineers,* Fourth Edition, American Book-Stratford Press Inc., March 1957.
2. Sessler, G.M., and West, J.E., *J. Acoust. Soc. Am.,* Vol. 46, 1969, pp. 1081–1086.
3. Matsuzawa, K., *J. Phys. Soc. Japan,* Vol. 13, 1958, pp. 1533–1543.
4. Sessler, G.M., and West, J.E., *J. Acoust. Soc. Am.,* Vol. 40, 1966, pp. 1433–1440.
5. CCITT, "Effect of Transmission Impairments," *Yellow Book, Volume V,* Annex D to Rec. P.11, ITU, Geneva, 1980.
6. CCITT, "Definition of the Speaking Position for Measuring Loudness Ratings of Handset Telephones," *Red Book, Volume V,* Annex A to Rec. P.76, ITU, Geneva, 1985.
7. Webb, P.K., "Computation of the Characteristics of Telephone Connections," *British Post Office Research Department Report,* No. 630, 1977.
8. CCITT, "Transmission Rating Models," *Red Book, Volume V,* Supplement No. 3, ITU, Geneva, 1985.
9. Manhire, L.M., "Physical and Transmission Characteristics of Customer Loop Plant," *Bell System Technical Journal,* Vol. 57, No. 3, May 1960, p. 431ff.
10. AT&T, "1969–70 Switched Telecommunication Network Connection Survey," *Bell System Technical Reference,* AT&T Pub., April 1971.
11. Carey, M.B., *et al.,* "1982/83 End Office Connection Study: Analog Voice and Voiceband Data Transmission Performance Characterization of the Public Switched Network," *AT&T Bell Laboratories Technical Journal,* Vol. 63, No. 9, November 1984.

Chapter 6
Networks

This chapter introduces network terminology and structures and points out factors in network design and performance that affect the quality of transmission. As mentioned in Chapter 2, the first electrical communication systems were telegraph facilities and digital in nature. The invention of the telephone spurred the interest in analog transmission and led to the creation of very large networks comprising analog facilities and switches. The encoding of analog signals for transmission over digital facilities was introduced in the early 1960s. The encoding scheme selected, eight-level *pulse code modulation* (PCM), and a sampling rate of 8 kHz, requires about four times the bandwidth of analog systems, nominally 32 kHz, *versus* 4 kHz, per voice channel. The impetus for this reversal of technologies was the promise of cost reductions in switching and multiplexing equipment far exceeding the cost of the larger bandwidth. The economic savings, plus the integration of communication and computing with resultant efficiencies in operations and maintenance, have been so great that almost no analog facilities are being installed today. Nevertheless, replacing existing analog facilities takes time, and networks will contain a mix of facility types for many years. Therefore, it is appropriate to consider both analog and digital networks in our discussion.

In digital networks, the transport of analog signals between endpoints becomes a completely different problem than it is in analog systems. The quality of the reproduced signal is almost entirely defined by the choice of quantizing and encoding schemes. Once that decision is made, the only transmission detail to be looked after, outside of actual facility design, is that of matching the amplitudes of the signals to be encoded to the voltage window of the quantizer. In principle, the encoding can be done at the station and the network designer and transmission engineer are relieved of quality considerations of the type that have been emphasized so far in this material. Network concerns now center on the preservation of a bit stream between two points. The impairments of interest are principally those discussed in Sections 1.3 and 1.4 and, of course, signal-to-noise ratio. The emphasis in network quality is shifting to reliability aspects such as those discussed in Sections 4.2.3 and 4.2.4, and outage or availability measures. Standards for these kinds of measures are currently in the formative stage because of the lack of acceptability criteria and experience with them.

We begin the discussion with analog networks, describing the growth, and introducing terminology, from simple private line point-to-point connections to the international switched network. We present some detail of basic network design by describing two types of loss plans for echo control. A separate section is devoted to some important transmission considerations peculiar to the local distribution system or exchange area, and some transmission characteristics of existing exchange area plant are pointed out.

6.1 PRIVATE NETWORKS

Figure 6.1 illustrates the growth of simple private line networks in three stages. Figure 6.1(a) shows a point-to-point network, a single trunk between two locations. It is shown to point out that the station equipment can be anything from a single telephone to a large PBX. In the latter case, the single trunk is probably a trunk group of some size, but it is dedicated to connecting only those two points. This is the basic service for a customer with only two locations.

The PBXs connected by the solid black lines in Figure 6.1(b) exemplify the first extension of point-to-point, a multipoint. If the ends of a multipoint are connected to form a closed loop (the shaded line), it is then called a round robin. Some form of route selection is required in these networks and a minimum of switching intelligence is needed, *i.e.,* if a PBX receives a call that does not terminate at that location, it must have intelligence to simply pass it along to the next. These arrangements are relatively inexpensive but suffer from lack of alternate routing capability and potentially serious failures. For example, in the case of a multipoint, a failure in one link isolates the two segments formed. The total number of locations is also limited because, as shown in Section 6.8, loss and echo place a limit on the number of trunks that can be placed in tandem. Quality of service suffers noticeably when more than eight or nine are connected, but there are networks where some calls may traverse as many as 15 or so.

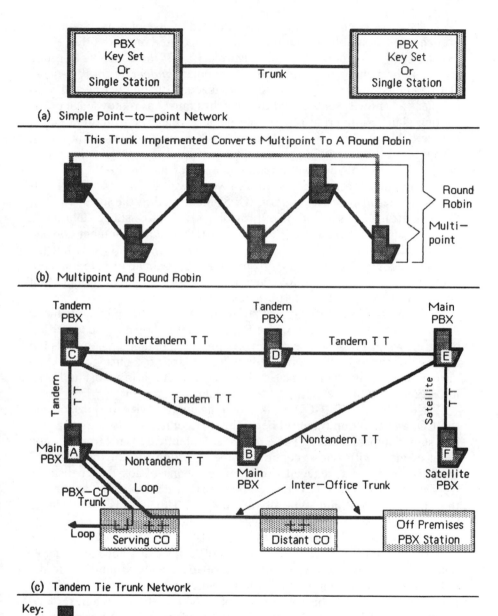

(a) Simple Point—to—point Network

This Trunk Implemented Converts Multipoint To A Round Robin

Round Robin

Multi-point

(b) Multipoint And Round Robin

Tandem PBX

Tandem PBX

Main PBX

Intertandem T T

Tandem T T

Tandem T T

Nontandem T T

Main PBX

Nontandem T T

Main PBX

Satellite T T

Satellite PBX

PBX—CO Trunk

Loop

Loop

Inter—Office Trunk

Serving CO

Distant CO

Off Premises PBX Station

(c) Tandem Tie Trunk Network

Key:

PBX   Switch

Fig. 6.1 Some Private Line Network Configurations

The next level of complexity is achieved by use of branch points. A branch point is simply a PBX that connects to three or more other PBXs instead of two, as those shown in Figure 6.1(b). If the concepts of round robin and branching are combined, the popular *tandem tie trunk network* (TTTN), illustrated in Figure 6.1(c), is realized. True numbering and routing plans are now essential, and the switches must have knowledge of the routing plan. Numbering plan is the name given to a structured set of numbers that provide identification of a PBX as well as the station served within that PBX. A routing plan is the algorithm for order of selection of possible routes between call origination and destination PBXs.

A popular feature seen with TTTNs as well as other types of networks, is the *off-premises station* (OPS) illustrated in the lower right of Figure 6.1(c). This is a method of providing service to a single station at a location remote from a PBX at minimal cost. The station number appears to the switch in the PBX as a local *on-premises station* (ONS), but the station is geographically remote from the PBX. A single private line connects the PBX to the remote location, passing directly through (not switched) the CO that would normally serve that station.

General off-net (off the private network on to the public switched network) access is provided with PBX-CO trunks shown in the lower left of the drawing. These are nearly always provided in tandem networks but can also occur in a multipoint, although they are not considered a standard network feature in that case. That is, separate tariffs (cost and pricing arrangements) may apply to these adjuncts.

Note that the PBX-CO trunk providing connectivity to loops local to the CO, and the loop (facility) beginning the connection to the off-premises station appear similar. They are, in fact, identical from a transmission point of view with one exception. If the insertion loss of the PBX-CO trunk is expected to exceed 4 dB, some means of loss compensation (repeater) must be used. The one engineering rule applicable to them is that there is a loss limit, typically about 4 dB. This limit is equivalent to the loss in about 1.5 miles of 26-gauge cable, more for larger gauges, up to about 11 miles for 19-gauge cable with H-88[1] loading. *Loading* is the expression used to describe the insertion of series inductance in the cable pair to form a low-pass filter with a cutoff frequency around 3.5 to 4.0 kHz, thus providing a more uniform frequency response within the band. (Figure 6.2 illustrates this for the case of 35,000 ft, about 6.6 miles, of 19-gauge cable.) A loss limit is applicable to the off-premises station as well, but is generally larger, by two or more decibels, than that for the PBX-CO trunk. Also, compensation, if required, can be placed anywhere in the path between the PBX and the remote station.

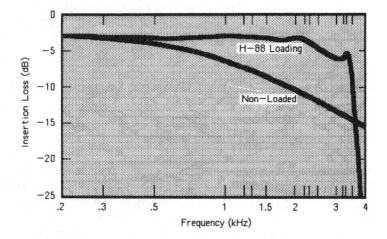

Fig. 6.2 Insertion Loss, 35,000 ft, 19-gauge Cable

Nomenclature applicable to TTTNs is illustrated in Figure 6.1(c). The major entities are the main PBXs. Any or all customer locations may be designated as *mains*. All the trunks are known in general as *tie trunks* (TTs). If a connection is completed between any two PBXs over one trunk, as shown between *A* and *B*, that trunk is serving as a *nontandem tie trunk* for that connection. If two trunks are required to complete the call (*A* to *C* to *B*), the point where they connect serves as a tandem PBX for that call, and each trunk serves as a *tandem tie trunk* (TTT). If three or more trunks (*A,C,D,E*) are needed, all but the two end ones serve as *intertandem tie trunks* (ITT). Trunks and PBXs thus may change designations depending upon the connection being served.

Some rules apply for control of loss on connections comprising more than one trunk, so-called *built up connections*. The rules are variable depending upon network complexity and individual trunk length but are of this nature: All trunks may be designed to VNL plus two plus two or VNL + 4 rules (VNL stands for a loss value that is determined by the length of the trunk and VNL design is described in Section 6.8). Each *plus two* represents a switchable pad (attenuator that can be switched in or out of a connection) that may be provided within a PBX. The PBXs serving the connection *A–B* in Figure 6.1(c) would switch these pads in when that connection is established; for the connection *A–C–B*, mains *A* and *B* might switch both pads out and Main *C* switch both in, the object being to complete all connections with a total loss of VNL + 4. The pad switching depends upon the function of the particular trunk and PBX in each connection. The numbering plan must be designed in such a way that each switch can determine its function and provide appropriate pad switching as the connection is established.

Any main PBX may have CO access and, as indicated, any tandem may be a main. The only distinctive designation is that of satellite PBX. A *satellite* is any PBX that cannot access a CO directly and has trunk access to only one other PBX. One of these is shown at the right center of Figure 6.1(b). Users at a satellite location must make use of a *satellite tie trunk* (STT) to its adjacent PBX to place a call off-net or to any network station.

The transition from point-to-point to tandem tie trunk network illustrated in Figure 6.1 presents an attractive set of service offerings. There is something there for many different sizes of organizations, small to large. In addition, one may begin with a point-to-point and grow rather painlessly to a large tandem configuration. This potential for growth is why transmission considerations become interesting. For any network of these types, there is no way to anticipate total size, so there can be no overall network engineering rules. Whatever a customer orders initially may be considered to be the entire network. If additions are made later, the new size is the entire network. This can lead to degraded performance and call for an engineering evaluation, adjustments or complete redesign of older portions of a network that has experienced substantial growth of this type as shown below.

Assume that the overall noise objective for a long trunk or a long connection is 40 dBrnC maximum. If a simple point-to-point circuit is installed between two distant locations, the noise on that circuit is then permitted to be 40 dBrnC. If the locations are far apart, say across the country, the noise may, in fact, be that high.[2] In time, company growth calls for a new location, so another point-to-point circuit is added to make a simple multipoint. In principle, there is no rule that says that this new trunk cannot also have 40 dBrnC of noise if it is also very long, even though the first was allocated the maximum for a complete connection. The process of addition of trunks, each operating at or near the maximum value of noise for its length, can continue indefinitely from a tariff point of view. Even if the initial network is a complete TTTN, later additions may be constructed by simply adding trunks for connectivity without considering end-to-end performance. An aggravating factor (in the U.S.) is that such networks, if geographically large, must be supplied under tariffs which vary from state to state. It may, therefore, be more economical to lay out parts of the network in ways that are not conducive to good overall transmission but cost less. For these reasons, transmission quality on tandem tie trunk networks may be highly variable.

6.2 TWO-WIRE AND FOUR-WIRE TRUNKS

Before considering more structured topologies used for various network configurations, a brief word is needed about the two major types of trunks that can be used and can make significant differences in their cost and transmission performance. These are called *two-wire* and *four-wire trunks,* reflecting the number of wires needed to implement them on cable facilities. Two-wire trunks use the same medium, with a single transmission channel, for communication in both directions. The correct direction for the flow of information on the channel is determined by devices placed at either end called hybrids. (See Section 5.1.3.) Such a trunk is economic to construct but the hybrids cannot function perfectly; the result is reflections of signals back and forth between the two ends. These reflections degrade transmission quality either directly, by their interference with the desired transmission, or by the increased loss required in a trunk of this type to attenuate the reflections to a tolerable level. Four-wire trunks use a separate channel for transmission in each direction. They, therefore, are roughly twice as expensive to implement. Four-wire trunks, when connected via four-wire switches, do not require hybrids and so do not incur reflections. Even if a network is designed using only four-wire trunks, a conversion from four-wire to two-wire operation is required at the end offices or PBXs, because telephone loops or the station loops are two-wire entities.[3] Loss needed to attenuate reflections arising only from the ultimate four-wire to two-wire conversions may be distributed and shared among many trunks. Individual trunks may, therefore, be designed with less loss than if they each gave rise to reflections and intrinsically offer better transmission. When the topology of a network is described, it is not uncommon to prefix the type name with one of the descriptors, *two-wire* or *four-wire.*

6.3 HUBS

In the networks discussed so far, the routing algorithms are fairly simple and a minimal control intelligence is required. Connections are completed essentially by hopping from one near switch to the next until the desired location is reached. This has two negative aspects: First is an extended time to complete the connection (call set-up time) because each switch has to decide the disposition of the call and complete appropriate signaling to the next. Second, as the number of trunks in tandem increases, transmission quality diminishes.

One method of reducing these problems, providing more flexibility and easier transmission engineering, is use of the *hub,* shown in Figure

6.3(a). A hub may have any number of branches (>2) and be connected to other hubs *via* a single trunk or a multipoint arrangement. Large networks can be constructed by interconnecting these star-like configurations, and routing algorithms devised that impose limits on the number of trunks and switches any connection may use.

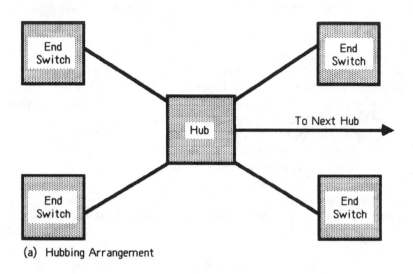

(a) Hubbing Arrangement

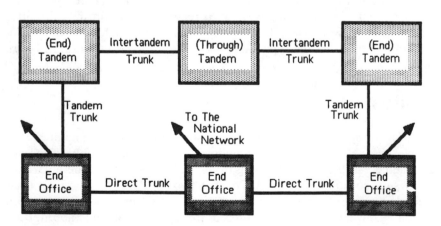

(b) Metropolitan Tandem Arrangement

Fig. 6.3 Two Switching Topologies

6.4 METROPOLITAN TANDEM NETWORKS

Figure 6.3(b) shows a special type of network, *metropolitan tandem,* that is an adjunct to the complex hierarchical one discussed in the next section. It is presented here to maintain the sequence of complexity. In very densely populated areas, a large number of COs are needed and the local traffic demands high connectivity between all of them. A fully connected topology[4] is expensive, so the solution is the metropolitan tandem arrangement. The COs involved are connected in an arrangement resembling a tandem tie trunk network. (See Figure 6.3.) The COs, labeled *end offices* (EO) are connected *via* direct trunks, and through one or more *tandem offices* that serve in a capacity similar to tandem PBXs in TTTNs. The only traffic permitted through these tandem offices is that which remains within the small network, access to the larger hierarchy for regular switched network calls is directly from each end office. This rule, together with the fact that no more than four trunks are needed to complete any connection, permits a transmission design of the tandem trunks that is more economical than required in the lower levels of the complete hierarchy. This adds to the total savings in the configuration.

6.5 HIERARCHICAL STRUCTURE

The North American network, as it existed until about 1982, is used as an illustration of hierarchical structures. It is diagrammed in Figure 6.4. Individual subscribers (customers) are connected locally by the end office, or class-5 office. These offices, in turn, are grouped by connections to toll or class-4 offices, and so on, up the hierarchy to the regional or class-1 offices. In the mid-1970s there were six regional offices, five in the U.S. and one in Canada, with full connectivity in the U.S. A call between two distant end offices, EO 1 and EO 2, would generally be routed up the hierarchy, from EO 1, on an *up link,* then across to the *down link* to EO 2. The crossover from the up-to-down links is made at the first opportunity, current traffic density and so, trunk occupancy, being the determining factor. The trunks between the class-1 regional centers are sometimes referred to as *Finals* because they represent the last chance for an available connection between pairs of offices that could normally be served by going from the left-to-right sides of the diagram by using trunks at a lower level. The lower-level high-usage trunk groups are supplied as required to meet traffic demands between any pair of offices. These are implemented when the traffic reaches an intensity that makes it more economical to install such a group than to continue to route that traffic to a higher level before transferring to the down link.

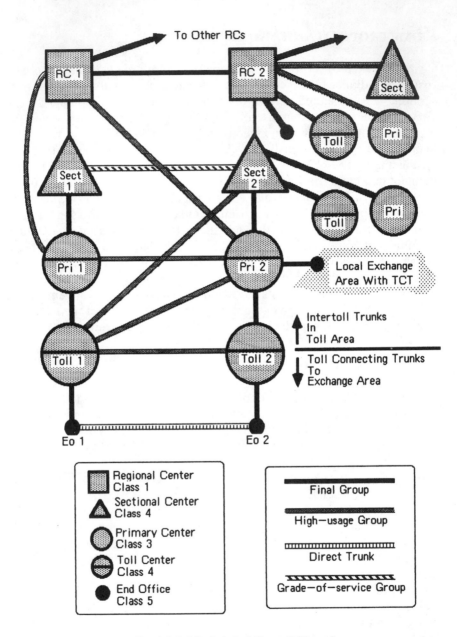

Fig. 6.4 Public Switched Network Hierarchy

The direct trunk, shown between the two end offices in Figure 6.4, is the most uncommon because of the use of metropolitan tandem networks described above. They appear in cases of a very high community of interest between two local exchanges that are geographically distant. For example, traffic between an exchange in the financial district in New York and its counterpart in Los Angeles might justify a direct trunk (group).

Grade-of-service groups serve a similar purpose but at some other level in the hierarchy. They are designed for a very low probability of blocking (blocking occurs when all trunks are in use), and may have no alternate choices. In this case, they are the last resort for a connection between two endpoints. (That situation is not shown explicitly in the generalized view of Figure 6.4.)

Toll connecting trunks (TCTs) also serve end offices that happen to home on offices in the higher levels of the hierarchy. The switching machine at such a location serves the dual role of toll center and whatever other function is called for by its placement in the hierarchy. The office designated primary center 2 in Figure 6.4 is shown in this dual role.

The routing or switching algorithm calls for the least number of trunks in any connection. As a call is being established, trunks are connected to proceed up the hierarchy. At each node, the first choice trunk group is the one which would result in the most direct path to the destination. If that group is busy, the next most direct choice is examined, *etcetera*. Only if all *alternate* routes are busy is a call forced to proceed all the way up one side and all the way down the other side for a total of nine trunks in tandem. It is estimated that on average, the probability of this occurrence is less than 0.001. The excess capacity implied by that number is needed during peak traffic periods.

Note here that, for reasons that will become apparent in Section 6.8 on loss plans, toll trunks are implemented as four-wire designs even if realized on cable, and toll connecting trunks are treated as two-wire designs even if realized on carrier facilities because the end-office switch must be of a two-wire type to connect to the local loop. Some switches above the class-5 office may be two wire as well, but the required hybrids can be well balanced with modest effort and these two-wire to four-wire junctions do not give rise to significant reflections. The reason for this is that impedances of the trunks to which they connect can be relatively well controlled compared to the situation for the class-5 office which connects to the local or exchange area. Section 6.10 describes the exchange area and its unique problems in some detail.

6.6 DYNAMIC NONHIERARCHICAL ROUTING

In 1984 introduction of a new switching plan was begun in the U.S. portion of the North American network. It is called *dynamic non-hierarchical routing* (DNHR). It uses the flexibility of electronic switches under program control to enable changing from one preplanned routing arrangement to another during normal operations, up to ten times per day. Traffic throughout the network is monitored continuously at a central location and the optimum plan selected and implemented as the pattern changes. The optimum plans are based on the gradual shift of peak traffic times across the country due to the four time zones.

DNHR is structured around a large number of high capacity machines (92 or more by 1988) forming a network that replaces the two highest levels of the original hierarchy. These machines are heavily interconnected, though not fully. No more than two trunks are permitted in a path through this new high level switching arrangement. The three lower levels of the hierarchy function as before, but can now be considered a multiplicity of independent subnetworks. The DNHR network interconnects these in a more efficient manner than the upper levels of the hierarchy did. The high-level DNHR network functions as a movable one that can be assigned, in part, to small geographic areas as needed, and still serve the inter-regional connectivity role of the original two top levels. The ability to change preferred routing plans, coupled with the greater sharing capabilities of the DNHR network, spreads traffic more uniformly over the total network, increasing trunk efficiency by about 15%.

6.7 LARGE PRIVATE SWITCHED ARRANGEMENTS

The last topology discussed here is an adaptation of the national hierarchical structure to private networks; it is illustrated in Figure 6.5. Another set of nomenclature applies for these networks. *End office* of the national hierarchy is replaced by the *main PBX,* and *toll connecting trunks* become access lines. Access to the public network is over *off-net access lines* (ONALs) instead of PBX-CO trunks as in tandem tie trunk networks, and the satellite PBX of TTTNs becomes a tributary PBX. The switching offices or nodes are labeled simply SS1, SS2, or SS3. Few, if any networks are so large that all three levels are needed for satisfactory service, but the generic design rules can accommodate them.

There are at least three distinctive classes of networks that follow this topology. The first is referred to as CCSA for *common control switching arrangement.* Later ones, taking advantage of features available in electronic switches, are called ETN for *electronic tandem network,* and

EPSCS for *enhanced private switched communications service*. EPSCS networks are designed using four-wire trunks and switches. The resulting elimination of hybrids at switches permits lower end-to-end losses and a generally improved quality of transmission.

There are a number of other special arrangements for networks in the broad class of private networks, as well as variations on those mentioned here. Section 4 of [1] provides a great deal of detail about a number of them.

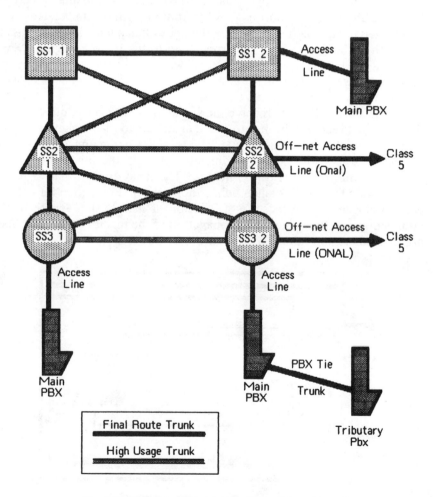

Fig. 6.5 Hierarchical Structure for Large Private Networks

6.8 NETWORK LOSS PLANS

As discussed in Section 4.1.6 and illustrated in Figure 4.9, echo path loss must be increased to maintain a consistent grade-of-service for echo performance as connection length increases. This section provides a brief overview of some ways of providing adequate echo protection through loss control.

Figure 6.6 is a simple diagram of a connection between two stations, A and B, emphasizing the two major reflection points, the junctions of its two-wire and four-wire portions. When station A transmits, a talker echo is returned *via* the path indicated by the heavy black line. The total loss in this path is the connection loss from A up to the hybrid serving B, and back again, plus the loss incurred crossing the hybrid device (transhybrid loss) near B. Some transmission across the device will occur because a perfect balance is never realized.

The station at B will receive the signal more than once, directly over the desired direct path from A to B and again due to the listener echo path indicated by the heavy shaded line. The total listener echo path loss is the talker echo path loss plus the connection loss in the forward direction again plus the transhybrid loss near A. The delay times, relative to the transmission from A, for talker echo, and to the first reception for listener echo, associated with these, is a function of the physical length of the connection and the velocity of propagation of the facilities used.[5] Echoes are of concern in both voice and data transmission.

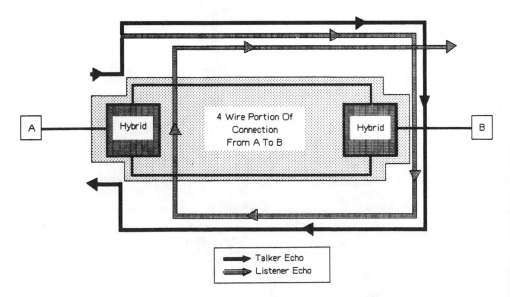

Fig. 6.6 Paths for Talker and Listener Echo on a Connection

In the case of data transmission, the effects of talker echo may be eliminated by opening the receiver portion of the station during, and for a short time after, completion of transmission. Such an arrangement would be useful in a half-duplex mode of communication, that is, when only one direction of transmission is permitted at a time. The delay in enabling reception after a transmission poses a bound on turn-around time in an interactive communication. Full-duplex transmission, communication simultaneously in both directions, may be accomplished by partitioning the available bandwidth and using different, nonoverlapping, frequency spectra for the two directions. This involves a speed or bit rate penalty but avoids the echo problem. Higher speeds may be achieved with full use of the available spectrum by incorporating (self) echo cancelers in the modems. Listener echo is usually of no consequence in data transmission if the circuit is adequate for speech.

For speech transmission, the subjective effects of echoes are a function of their delay and magnitude. Longer delays require greater echo attenuation. The permitted echo magnitude as a function of delay has been determined through subjective tests. The measure of magnitude is expressed in terms of the loss in the echo path. The loss, in turn, is expressed as a weighted measure of the frequency response of the echo path. One such measure is the *weighted echo path loss* (WEPL), given by (1.15):

$$\text{WEPL} = -20 \log\left[(1/3200) \int_{200}^{3400} 10^{-\text{EPL}(f)/20} \, df \right] \text{dB} \qquad (6.1)$$

where EPL(f) is the absolute frequency response, in decibels, of the echo path. The *weighting* is seen to be simply the average voltage loss of the echo path over the band 200 to 3400 Hz, expressed in decibels.

Subjective opinion of talker and listener echoes, as reflected in transmission ratings for echo, R_E and R_{LE}, are plotted in Figures 6.7(a) and 6.7(b) as a function of delay. Echo magnitudes are expressed in terms of their path loss in WEPL. It is apparent that talker echo, Figure 6.7(a) is controlling. That is, for a given delay, talker echo requires considerably more loss, for the same transmission rating, than does listener echo.

The curves indicate that, for small delays, zero loss is acceptable. In these cases, loss needed for singing margin, to prevent oscillations through the echo path loop, or maintenance of stability on the circuit, is controlling. Allowing for component variations and drift, a minimum value of four to six decibels is usually assumed adequate for singing margin protection. For long delays, Figure 6.7(a) shows that the desired loss becomes quite large. It cannot be realized without excessive transmission loss, so echo suppressors or cancelers become necessary. In a

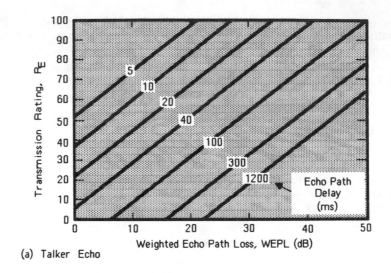

(a) Talker Echo

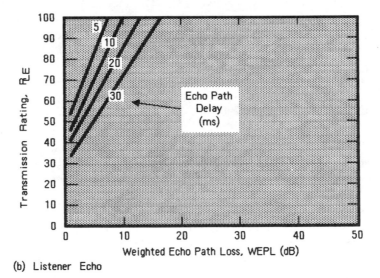

(b) Listener Echo

Fig. 6.7 Transmission Ratings for Echo with Various Losses and Delays
Source: Adapted from *Red Book, Volume V*, CCITT, Geneva, 1985

switched connection environment, a trunk may serve in a short connection on one call and a long connection on the next. The choice of loss to be assigned to trunks becomes complex. The factors to be considered are:

- Stability
- Echo tolerance
- Total connection loss
- Total connection length

For the analog North American network, the solution of this multifaceted problem was reduced to a procedure known as *via net loss* (VNL) design. Its derivation is summarized in the following paragraphs.

Referring to Figure 6.6, and assuming equal loss, L, in both directions of transmission (between hybrids), the talker echo path loss is

$$EPL = 2 \cdot L + RTL \tag{6.2}$$

where RTL is the return loss at the hybrid near B, which in a connection is the far-end toll office, the junction of the four-wire toll network and the far-end two-wire toll connecting trunk. For satisfactory echo performance, EPL, as given by (6.2), should be large enough to adequately suppress the echo for every length of connection.

The amount of echo that may be permitted in a connection of arbitrary length is expressed in terms of the EPL required for satisfactory operation and is called echo tolerance, a subjectively determined quantity. Echo tolerance is a statistical quantity because tolerance varies with people.

As mentioned earlier, echo tolerance decreases with increasing delay. That is, the longer the delay, the greater is the required echo path loss. For each value of delay, there is a distribution of opinion as to the required loss. The standard deviation, s_t, of the distributions is nearly constant at about 2.5 dB, independent of the average value. Average values are plotted as a function of delay in Figure 6.8. This curve, a direct evaluation of echo, does not go to zero loss at zero delay but to about one decibel. While one decibel of loss at zero delay is acceptable from an average echo annoyance view, it is still not adequate for singing margin. Four decibels is considered to be the minimum for that. Thus, the curve of Figure 6.8 must be modified at the left side to intersect the ordinate at 4 dB.

With such an adjustment, the curve of Figure 6.8 would describe the required EPL, to be met by (6.2), as a function of echo delay for an average listener. It must be shifted vertically by an amount proportional to the standard deviation of echo tolerance to satisfy most people instead of only about half of them. Adjustments other than this are necessary, however, because the quantities L and RTL in the equation are also statistical. The variation of these terms must be examined. We consider RTL first.

The hybrids in Figure 6.6 represent those at the junction of the toll and toll-connecting trunks. The return loss at these junctions is strongly influenced by the distribution of the impedances of the loops to which the toll-connecting trunks are switched at the class-5 office. Measurements (see [2]) have shown that the average return loss of loops, measured against a standard reference impedance of 900 ohms in series with a 2.16-microfarad capacitor, is about 11 dB with a standard deviation, s_{RL}, of about 3.5 dB. (The standard deviation was estimated from the curves given in [2].) Reference [3] provides data on the return loss of toll-connecting trunks terminated in the same reference impedance. Those data show a mean return loss of about 20 dB and a standard deviation of about 5.5. The actual return loss presented by the junction during a real connection will be the result of the trunk impedance, with variations reflected in the data of [3], working into the loop impedances, with variations reflected in [2], instead of as measured to the reference impedance. The real return loss will not be the sum of those measured, so no estimate of the working mean and standard deviation can be made from these numbers. The mean echo return loss will, however, be something less than the 11 dB, and the standard deviation greater than the 5.5.

The L term in (6.2) is the loss of all the intertoll trunks in the connection. Its variability is a function of the variation on individual trunks and the number of trunks in the connection. The variability of trunk losses, s_L, is taken as 2 dB.[6]

Equation (6.2) must satisfy the constraints imposed by the echo tolerance curve of Figure 6.8, and allow for the variabilities just mentioned. Solving that equation for L yields

$$L = (\text{EPL} - \text{RTL})/2 \tag{6.3}$$

where EPL is the echo tolerance curve, a function of delay or connection length, and the connection loss L should be adequate to provide satisfactory performance in 99% of all connections. That is, L should be large enough to include 99% ($2.33 \cdot s_{\text{total}}$) of all cases encountered.

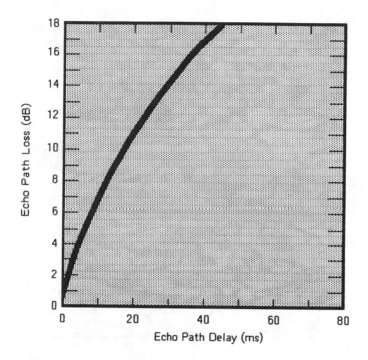

Fig. 6.8 Average Talker Echo Tolerance
Source: Telecommunications Transmission Engineering, Volume 1, Principles, Second Edition, American Telegraph and Telephone Company, 1980. Reprinted with permission of AT&T—All Rights Reserved

The total variability, s_{total}, may be found from the three components above, tolerance, RTL, and trunk loss variabilities as

$$s_{\text{total}} = (s_t^2 + s_{\text{RL}}^2 + N \cdot s_L^2)^{1/2} \tag{6.4}$$

where N is the number of trunks in a connection. (s_L is the value for only one trunk.)

The expression for L, including the variability, becomes

$$L = (\text{EPL} - \text{RTL} + 2.33 s_{\text{total}})/2 \tag{6.5}$$

This equation, based on the tolerance curve, is plotted in Figure 6.9 for several values of N. Inspection shows that L should increase by about 0.4 dB for each trunk in the connection. If the 4-dB minimum for singing margin is added to this 0.4 value, the linear approximation for the one trunk connection may be drawn as illustrated in Figure 6.8. The linear equation for the case of one four-wire trunk is, as shown,

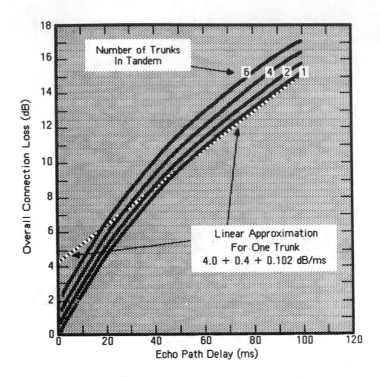

Fig. 6.9 One-Way Loss for Satisfactory Echo on 99% Of Connections
Source: From *Telecommunications Transmission Engineering, Volume 1, Principles,* Second Edition, American Telegraph and Telephone Company, 1980. Reprinted with permission of AT&T—All Rights Reserved

$$L_{\text{one trunk}} = 4.0 + 0.4 + 0.102 \text{ ms} \quad \text{(dB)} \tag{6.6}$$

where 0.102 is the slope of the upper portions of the set of dark curves. The 4.0 dB is assigned to the toll-connecting trunks, 2.0 dB each, providing an echo path loss of 4 dB by virtue of the echo traversing the near end trunk twice. (See Figure 6.10.) The equation for loss of a single toll trunk then becomes

$$L_{\text{trunk}} = 0.4 + 0.102 \text{ ms} \quad \text{(dB)} \tag{6.7}$$

with the time in milliseconds referring to the round-trip delay. This delay is simply two times the length of the trunk in miles, divided by the velocity of propagation. The velocity of propagation is dependent upon the type of facility and a further simplification for a working equation has been established by tabulating velocity factors, known as VNL Factors (VNLF), for each type of facility. (See [6, p. 599] for example):

$$\text{VNLF} = (2 \cdot 0.102)/(\text{velocity of propagation}) \text{ dB/mi} \qquad (6.8)$$

The trunk loss for VNL design is then given by

$$\text{VNL} = \text{VNLF} \cdot (\text{length in miles}) + 0.4 \text{ dB} \quad (\text{VNL trunk loss})$$
$$(6.9)$$

One may ask about the fact that two-wire switches are frequently used in the toll environment. These require hybrids, and should not suppression of echoes from these be included in the trunk loss? Such hybrids connect intertoll trunks to intertoll trunks. The impedance of these, including the effects of office cabling, *etcetera* can be, and is, well controlled. The return losses associated with these configurations (called *through return loss*) are maintained to a median value of about 27 dB. If, when measured, a toll-switching office has a distribution of through return loss with a median less than 27 dB, trunks to and from that office must have a small additional loss, called a *B* factor, added to the VNL design value. *B* factors range from 0, for a median office return loss of 27 dB, to 1.5, for an office with a through return loss median of 14 dB. See [7, p. 56] for a list of *B* factors.

The North American network is presently evolving from an all-analog-to-an-all-digital network. When the transition is complete, a fixed loss plan will be in effect. That plan calls for zero, four, or six decibels of loss to be inserted in a connection when it is established. The choice depends on the distance between the end offices. The values are compromises judged to be suitable for the range of connection lengths within the three groups.

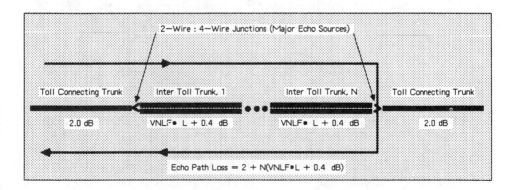

Fig. 6.10 Partitioning of the VNL Requirement

Other countries are most often considerably smaller, geographically, than North America, and fixed loss plans are more common even for analog networks. A single value for trunk loss can be chosen as the optimum value, based on talker echo tolerance, for the longest possible connection, and assigned to all connections without undue transmission loss penalty. An advantage of a fixed loss plan is that contrast, changes in received volume from call to call, is reduced because the variation of loss between connections is minimized.

6.9 INTERNATIONAL CONNECTIONS

This section introduces the international network structure, its terminology, and the magnitudes to be expected for some of the impairments.

6.9.1 International Analog Networks

The generalized international connection is considered comprising three parts as illustrated in Figure 6.11(a). The national systems on either end are interconnected *via* the international chain. (Facilities for the international chain, such as submarine cable and satellite systems, are usually owned and administered through cooperative ventures of national administrations.) The CCITT recommends that the international chain and its connecting circuits be constructed entirely of four-wire facilities as shown in Figure 6.11(b). These are then collectively referred to as the four-wire chain.

The national systems are under jurisdiction of the individual administrations and their structures vary, but generally resemble the hierarchy arrangement of the analog North American network, as in Figure 6.4. A structure of this general type is illustrated in Figure 6.12 to show the nomenclature and symbolism used internationally. Digital connections to the international network are illustrated, but analog connections are also used, of course.

The switching locations illustrated in Figure 6.12 are labeled with the generic terms, exchange or transit centers. The four-level hierarchical nomenclature, corresponding to toll office through regional center of Figure 6.4, is simply: *quaternary center* (QC), *tertiary center* (TC), *secondary center* (SC), and *primary center* (PC). The local exchange is the equivalent of the class-5 office in the U.S. The four-level structure above the two local exchanges in the lower corners of the drawing is intentionally

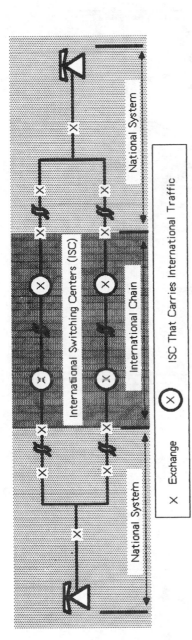

(a) The Three Parts Of An International Connection

| × Exchange | ⊗ ISC That Carries International Traffic |

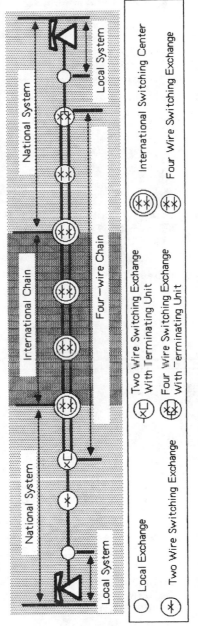

(b) Sample International Connection Showing Nomenclature

Fig. 6.11 Structure of International Connections
Source: Adapted from Red Book, Volume III—Fascicle III.1, CCITT, Geneva, 1985

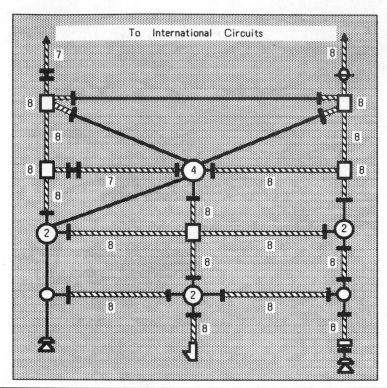

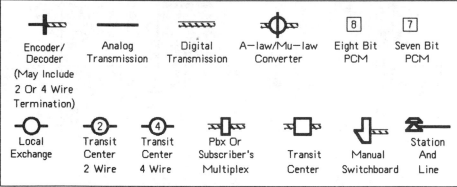

Fig. 6.12 CCITT Nomenclature for a National Network
Source: Red Book, Volume III—Fascicle III.1, CCITT, Geneva, 1985. Reprinted with permission

ambiguous to show that hierarchy position is not always self-evident. It is clear that the three two-wire transit centers are QCs (class 4) because they connect directly to local exchanges. Above that, alternate route trunking interconnects the switches so heavily that the structure is blurred. The four-wire transit center in the middle of the drawing could be an SC (class 2), the upper digital centers PCs (class 1), and the remaining centers could all be TCs (class 3). Several other assignments of rank might be envisaged.

In the design of an international transmission plan, the CCITT had to allow for interconnection of any two global locations and to consider the existing plans and facilities of individual administrations. The following trunking rules were established: The number of tandem international circuits[7] is limited to four, and the number of national circuits limited to six in each national system. These rules allow up to 16 circuits in tandem on an international connection. In practice, the maximum number of permitted circuits is almost never realized. The CCITT sponsored a survey to gather data on numbers of circuits used in connections in 1973. The results are published in [8]. More than 50% of connections used five or less; almost 99% used eight or less; and the relative occurrence of 11 circuits in tandem was found to be 0.02%. The frequency of occurrence of 14 circuits in tandem was estimated at about $5 \cdot 10^{-8}$ percent.

Transmission levels on international connections are referred to points called virtual analog switching points. These define the (transmission) junction between the national and international networks and may or may not actually occur at a switch, which is the reason for the word virtual. They are defined as the point in the four-wire path of a national network that is closest to the switch providing international access, and that has a transmit dBr of -3.5 and a receive dBr of -4.0. These values were chosen to be as compatible as possible with the existing networks of many administrations and yet lend themselves to good circuit design practices. The losses in the transmit and receive portions of the national networks are controlled by specifying send and receive REs (SCRE and RCRE) to and from these points.

Transmission levels on digital facilities have no intrinsic meaning, but the concept must be preserved for compatibility with interconnecting analog facilities. All digital trunks are assumed to be operating at a zero level. If the virtual analog switching points occur in a digital facility connected through digital switches, both receive and transmit dBr values are taken as zero on the facility. A level measurement would be made by means of a codec to convert to an analog signal, and a test pad would be inserted for the analog level measurement.

Circuits in the four-wire chain, see Figure 6.11, are designed for 0.5 dB loss if analog, and zero-decibel loss if they are digital. Stability of these trunks is assured by requiring a minimum six-decibel return loss at the first adjacent four-wire to two-wire conversion, wherever it may occur. For purposes of specifying mean objective levels, the international portion of the four-wire chain is assumed to contain one analog circuit (trunk) and so has the 0.5-dB loss reflected in the dBr values, transmit minus receive equals 0.5. National transmission level plans must provide international access compatible with the levels at the virtual analog switching points. When the international plan was adopted it was recognized that time and costs would be incurred by administrations in required upgrading or rearranging of national networks for compliance. There are, therefore, two sets of international loss objectives: interim and long term.

The interim short-range objective is 13 to 25.5 dB, overall CRE (OCRE) (see [9]). Long-term objectives for complete international connections call for a traffic weighted average OCRE of between 13 and 16 dB. Recall that the preferred OCRE when only noise and loss were considered, was shown to be 8.5 dB in the discussion of R_{LN} in Section 4.1.4. The connections here are thousands of miles in length and echo considerations prohibit designing to such a small absolute loss (about 2.5 dB).

Interim values for SCRE (send) are 11.5 to 19.0 dB and 2.5 to 7.5 dB for RCRE (receive). Long-term SCRE and RCRE objectives are 11.5 to 13.0 dB, and 2.5 to 4.0 dB. These are summarized and listed with conversions to (NOSFER) REs and IEEE metrics in Figure 6.13, together with a sketch showing where they apply.

Complete discussions of the international connection and its components, for all impairments, are given in the Red Book, Volume III, Fascicle III.1. In particular, Rec. G.103, G.111, and G.121, should be consulted.

Design objectives for noise on international connections are given in Rec. G.123. For the portions of the national systems of Figure 6.11 up to the four-wire chain, also called national extensions, noise objectives are defined in two length categories. Those up to 1500 km in length may introduce noise equivalent to 4 pW0p/km (6.02 dBrnC0/km), but those of greater length should meet the more stringent objective of 2 pW0p/km (3.01 dBrnC0/km). Note that these rules apply to the complete national extension, without regard to its circuit (trunk) composition. The administration may allocate the requirement among circuits in whatever manner it deems appropriate. The rule simply places a maximum on the noise introduced to the international connection. No length categories are defined for circuits within the four-wire chain, but it is assumed that the maximum length of any one circuit is 7500 km, so the 2 pW0p/km would apply in an allocation for a specific length. The CCITT recognizes the time variability

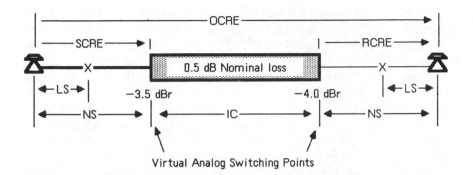

Virtual Analog Switching Points

Mean Transmission Ratings Per CCITT Rec. G.111 and G.121				
	Interim		Long Range	
	Min	Max	Min	Max
SCRE	11.50	19.00	11.50	13.00
SRE	9.02	14.61	9.02	10.17
TOLR	−41.00	−33.50	−41.00	−39.50
RCRE	2.50	7.50	2.50	4.00
RRE	1.74	5.87	1.74	3.00
ROLR	48.1	53.10	48.1	49.60
OCRE	13.00	25.50	13.00	16.00
ORE	10.17	19.17	10.17	12.42
OOLR	7.00	19.50	7.00	10.00

Corrected REs as Given in References
REs Calculated Per G.111, Annex 1
IEEE LRs Per Table 3.6

Fig. 6.13 Transmission Levels: Analog International Connection
Source: From *Red Book, Volume III—Fascicle III.1*, CCITT, Geneva, 1985

of noise and, for planning purposes at least, average noise power on a circuit in the four-wire chain is considered to be 1 pW0p/km. The worst circuit in the chain may exhibit a noise power of 3 pW0p/km (4.77 dBrnC0/km), and during the most adverse conditions of the busy hour, noise power on a circuit may rise to 4 pW0p/km (6.02 dBrnC0/km). This last requirement recognizes that in a well-designed broadband (high capacity) facility, noise build-up due to low-level intermodulation products just begins to dominate noise from other sources under peak load conditions.

Figure 6.14, adapted from [10], is a representation of a hypothetical international connection with average and maximum noise values, and typical loss values for each component. Only one leg of the transmission path is shown in the four-wire sections. The direction of transmission for the portion shown is from the transmitter at top to the receiver at the bottom. Noise values (underlined) for the circuits are given as measured at the receive side of the switch at each exchange, and represent the noise due to the circuit connecting the adjacent switches. The noise contribution for each switch is shown adjacent to the switch. Minimum and maximum noise values are shown in pWp0. The average values are given in dBrnC also, in the third column. Noise build-up through the connection is given in dBrnC in the right-most column, at each switch point. Noise is accumulated by taking the total noise at a switch, reducing it by the circuit loss to the next switch, and then adding it to the noise contribution of the circuit traversed and to the noise contribution of the next switch, using power addition. The noise on each loop, transmitter and receiver portions, is assumed to be measured at the *local exchanges* (LE). The sum adjacent to the transmitting LE is the noise on the transmit loop, 20 dBrnC plus the noise of the LE, another 20 dBrnC, for a total of 23. The noise summation is terminated at the receiving LE but includes the noise contribution (20 dBrnC) of the receive loop. It is the sum of the 28.3 dBrnC at the lower *primary center* (PC), reduced by the 5.5-dB loss of the next circuit, and added to the 20 from the office and the 20 from the receive loop. Two losses are shown between the lower ISC and the PC, where the receive-end hybrid (terminating set) is installed. The one labeled 0.5 dB is the trunk loss, the 4.0 dB is a combination of terminating set loss, a nominal 3.5 dB, plus 0.5 dB for receive dBr adjustment. Level (dBr) adjustment pads depend upon the make-up of the international links. Details are provided in [10].

6.9.2 International Digital Networks

The equivalent connection for an all-digital network is illustrated in Figure 6.15. Two cases are considered: The first, on the right, is a digital network with analog subscriber access. On the left, digital subscriber access is shown. Connections using digital trunks and digital switches have zero loss, and the noise is essentially that due to the quantizing operation at the LE (local exchange) or in the station so no loss or noise values are included in the network diagram.

Diagrams such as these are referred to as *hypothetical reference connections* (HRX). The one shown, 2000 km in length, is an HRX for a moderate length connection. The HRX for a long connection is 27500 km,

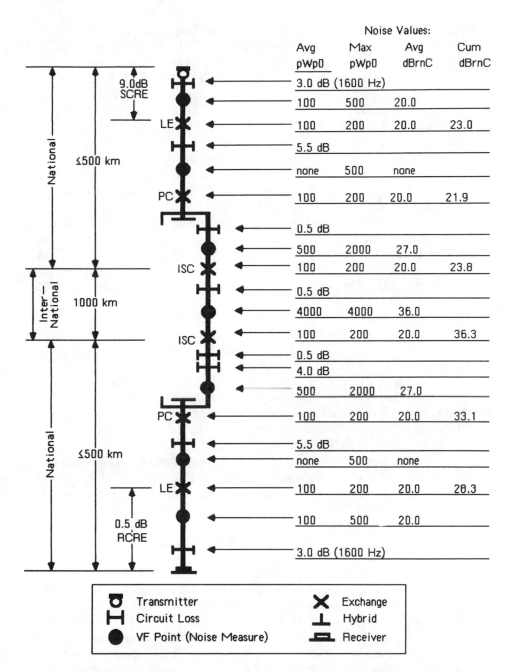

Fig. 6.14 Hypothetical International Analog Connection of Moderate Length
Source: Red Book, Volume III—Fascicle III.1, CCITT, Geneva, 1985. Reprinted with permission

comprising 12 digital links plus the two end links as compared to the five digital links plus two end links of Figure 6.15. These diagrams are useful for planning purposes when considering digital impairments that may accrue through a connection; *i.e.,* digital errors, jitter, and slips.

In case one, the hybrid loss, incurred for conversion from four-wire to two-wire operation, is shown as a pad adjacent to the hybrid. An additional pad is shown next to this to maintain generality. Because transmission level has no meaning in an all-digital connection, such a pad might be used for echo control on moderately long connections, if echo cancelers were not employed. The pad implementation might be digital, as implied in the drawing, or analog following the codec required to return to analog transmission.

Objectives for digital network impairments were discussed in general terms in Section 1.3. For ISDN applications, the CCITT has established some connection performance objectives for a 64-kb/s channel based on bit error rate (BER) measures. The objectives apply to the 27500-km HRX. Procedures for proportioning them to shorter connections are outlined in [11].

In recognition of the problems associated with the interpretation of error rates, as discussed in Section 4.2.4, a rather elaborate set of definitions, measurements, and criteria have been defined. The connection evaluation procedure is as follows:

A long measurement interval, T_L, is used as a base period over which averages may be taken. In [11], a period of one month is suggested for T_L. Observations are made for each one second (S) interval in T_L, and the number of errors in each S is recorded. The total number of seconds in T_L is S_{total}. The record is examined in detail.

An occurrence of ten consecutive S intervals that contain more than 64 errors each begins an unavailable period. The period ends at the start of any ten consecutive S intervals with less than 64 errors each. The unavailable periods are removed and the ends of the record gaps created are joined to form a new continuous record.

The sum of all the removed S intervals is designated (circuit) unavailable seconds, $S_{unavail}$.

Total available seconds is then defined by $S_{avail} = S_{total} - S_{unavail}$.

Within all S_{avail} seconds, count the number that contain at least one error. This total is designated S_{error}. Note that the remaining seconds are the *error-free seconds* (EFS) of Section 4.2.4.

Within all S_{avail} seconds, count the number that contain more than 64 errors. These are called severely errored seconds, and their total is designated $S > 64$. These intervals, at the 64-kb/s rate, are seconds in which the bit error rate exceeded $1 \cdot 10^{-3}$.

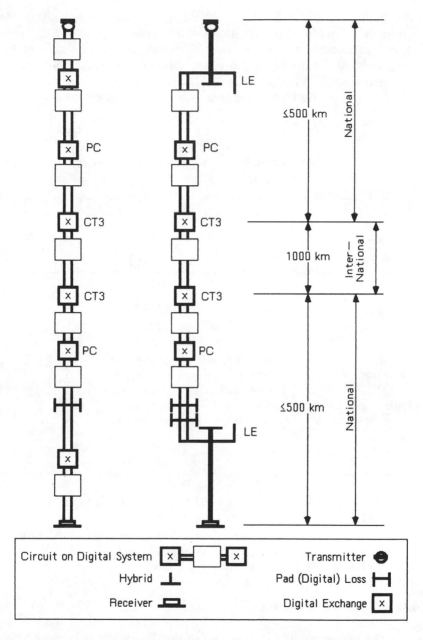

Fig. 6.15 Hypothetical International Digital Connection of Moderate Length
Source: Red Book, Volume III—Fascicle III.1, CCITT, Geneva, 1985. Reprinted with permission

Remove $S > 64$ from the record and group the remaining S by consecutive counts of 60, forming one-minute (M) packets. Count the number of packets containing more than four errors to obtain $M > 4$. These intervals, at the 64-kb/s rate, are minutes in which the bit error rate exceeded $1 \cdot 10^{-6}$. Finally, define $M_{avail} = S_{avail}/60$.

Using these measures, the objectives can be summarized as in Table 6.1.

Table 6.1
ISDN Error Rate Performance Objectives

Criterion	Objective	Calculation
Degraded minutes	Fewer than 10% of one-minute intervals to have a bit error rate worse than $1 \cdot 10^{-6}$.	$\dfrac{M > 4}{M_{avail}} < 10\%$
Severely errored seconds	Fewer than 0.2% of one-second intervals to have a bit error rate worse than $1 \cdot 10^{-3}$.	$\dfrac{S > 64}{S_{avail}} < 0.2\%$
Errored seconds	Fewer than 8% of one-second intervals to have any errors. (92% EFS)	$\dfrac{S_{error}}{S_{avail}} < 8.0\%$

The evaluation procedure appears complex at first but a little review shows that it is one which could be automated rather easily. As experience is gained with these measures and criteria, the measurement interval T_L will, no doubt, be specified instead of suggested, and it is not unlikely that other measures or procedures will evolve.

6.10 EXCHANGE AREA

The exchange area includes all of the equipment necessary to connect subscribers to the class-5 office, the class-5 office itself, and short local direct or intraexchange trunks between class-5 offices. Two major transmission characteristics distinguish the exchange area from the toll area of the national network. First is the nominal termination impedance of 900 ohms as opposed to 600. Second is the relative lack of homogeneity in the exchange area. This second characteristic gives rise to transmission concerns unique to these local areas. The reasons for the nonhomogeneous nature and the resulting transmission considerations are the subjects of this section.

Exchange area plant or equipment is relatively expensive compared to toll equipment because it is not shared among many users, as in the case of a toll trunk for example. Engineers controlling the exchange area continually search for more economic arrangements. New, less expensive technologies or methods are quickly put to use in new construction; so variety is found everywhere. Even without changes of this type, economics dictate variety in exchange plant layout.

A simple example illustrating the economic use of cable size or cross section (the number of pairs in a sheath) and the wire gauge in the cable is shown in Figure 6.16. The total transmission loss in the cable to subscribers at all distances from the class-5 office must be limited and fine gauge, high loss, multipair cable is the least expensive. This type of cable is used to serve locations near the CO, where distances are short and the population density high, by choice of CO location. Smaller cable sizes and larger gauges are employed as distance from the CO increases. The various gauge cables should be run all the way from the CO to the customer's premise; mixed gauges are not allowed. Drop wire, shown in Figure 6.16, is simply the wire used to run from the telephone pole to the subscriber's house. It is different because it must be physically tough to exist without benefit of a sheath. The distribution principles illustrated in Figure 6.16 explain the variety of cable make-up found in subscriber loops. Now we consider the effects on transmission.

The sections of cable shown extending beyond the points of connection to a residence are often purposefully left intact to accommodate future changes and rearrangements of the plant. Electrically they are open-circuited transmission line stubs of various lengths and are called *bridged taps*. The wavelength of a one-kilohertz signal in exchange area cables ranges from about 10 to 20 miles. Bridged taps are usually too short, normally less than a few miles, to exhibit pronounced stub resonance or tuned transmission line effects at voice band frequencies. They do add excess capacitance to the loops, however. (Cable capacitance may range from about 0.06 to 1.0 microfarad per mile.) A loop design plan known as *dedicated plant* calls for the elimination of bridged tap. This design improves transmission at the expense of flexibility but is often used. The population of subscriber loops is a mix of those with and without bridged taps of various lengths. This mix, by itself, causes loops of the same gauge cable and working length (physical length of cable from

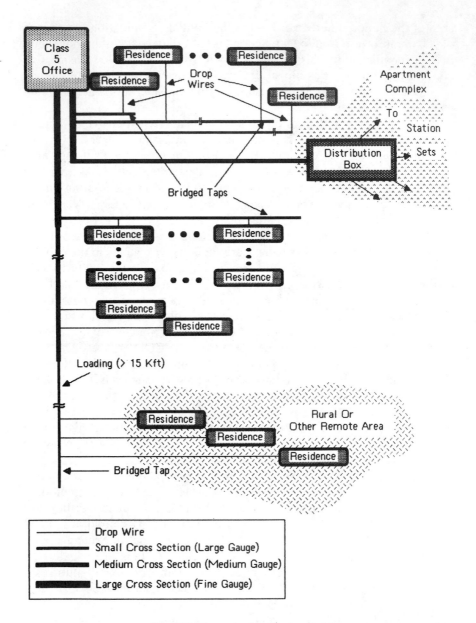

Fig. 6.16 Elements of an Exchange Area

the CO to the subscriber location) to show a variety of impedances as seen from the CO.

Different gauge cables have different characteristic impedances. Significant demographic changes within an exchange area may result in different cable sizes serving locales of about the same distance from a CO. This also contributes to the impedance variations found in loops of the same length.

The cables serving residences near the office are not long enough for the characteristic impedance of the pairs to become dominant so their terminations are significant to their impedance as viewed from the CO. Figure 6.17 illustrates this. The magnitude and angle of the one-kilohertz impedance of 26-gauge cable terminated in 600 and 900 ohms is shown as a function of length in kilofeet. At very short lengths, the impedance is close to that of the terminations. The magnitude increases with length to about 6,000 or 12,000 ft, depending on the termination, and then slowly decreases to the surge, or characteristic impedance of the cable which is about 918 ohms. The angle begins at zero, increases gradually in the negative direction, reflecting the capacitance of the cable, and approaches the characteristic angle of about −43°.

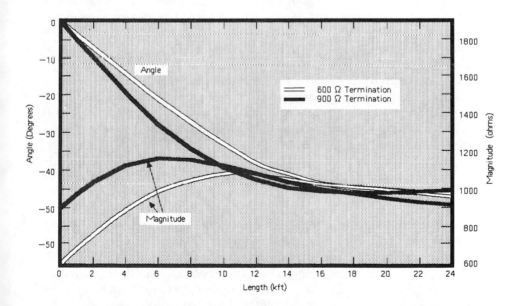

Fig. 6.17 Impedance, 26-Gauge, 0.083 Capacitance Cable, with 600 Ω and 900 Ω Terminations

All of these factors combine to make loop impedances dispersed. The wide range of impedances encountered in the exchange area is of importance in the design of compromise equalizers for use in hybrids in systems that connect to the exchange plant. Two examples are four-wire PBXs and four-wire digital class-5 offices. The balance networks for hybrids in telephone sets must operate in this impedance environment, of course, and, as mentioned in Section 5.14, that is what makes control of sidetone difficult.

Scatter diagrams of these impedances, as determined in a survey conducted in 1959 are given in [12][8] along with details of the kinds of loop terminations that were used. (That detail can be important in the design of balance networks for special applications.) The essence of those scatter diagrams is illustrated in Figure 6.18. The relative densities of portions of the scatter plots in [12] are approximated by the different shaded areas in Figure 6.18. The plots in the reference were arbitrarily broken into three density areas (by eye estimate) to construct the figure. Nonloaded and loaded loops are shown separately, Figures 6.18(a) and 6.18(b), because of the distinctively different patterns of their scatter diagrams. The total variability in the case of nonloaded loops is smaller. The greater variability in loaded loops may be due to different types of loading, the effects of bridged taps which may or may not have the loading extended to their ends, and perhaps the greater opportunity for more bridged taps to exist. Some detail for the reasons the scatter plots exhibit the shapes they do is provided in Section 7.1.

The scatter plots chosen for representation here are for impedance values at 3 kHz only. That frequency was picked because the scatter tends to increase with frequency, and so balancing problems are most severe at the higher frequencies. The plots for 3 kHz illustrate a worst case situation.

The diagrams shown represent impedances as seen from the subscriber's location looking toward the central office. The type of termination at the CO is a major factor in these impedances. Four cases are often considered: Connections to two-wire and four-wire toll connecting trunks, termination in a 900-ohm resistor in series with a 2.0-mfd capacitor,[9] and connection to another loop terminated in a telephone. The scatter diagrams of Figure 6.18 represent loops terminating in a 22-gauge four-wire toll-connecting trunk. The mean impedance for this termination as a function of frequency is shown in Figure 6.19(a) for both loaded and nonloaded loops. The values shown at 3 kHz may not appear to be the topological centers of the corresponding scatter plots. The means are self-weighted averages of the points in the scatter diagrams and it is not apparent from those where the means would fall.

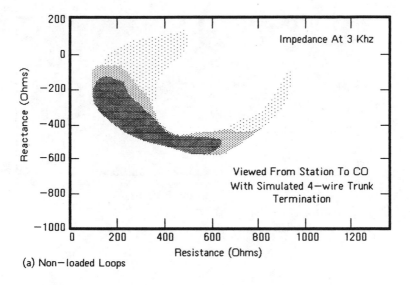

(a) Non—loaded Loops

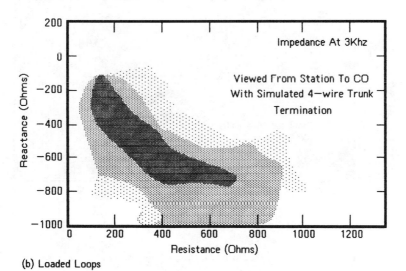

(b) Loaded Loops

Fig. 6.18 Approximate Envelopes of Scatter Diagrams of Loop Impedances (Shading Indicates Relative Density of Scatter)
Source: Adapted from P.A. Gresh, "Physical and Transmission Characteristics of Customer Loop Plant," *Bell System Tech. J.*, Vol. 48, No. 10, December 1969 from the Bell System Technical Journal. Copyright 1969 AT&T

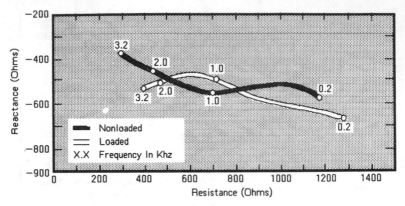

(a) Loaded And Nonloaded With Simulated Four−wire
Trunk Termination At CO

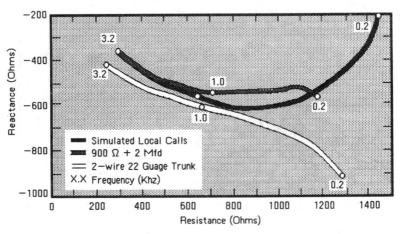

(b) Nonloaded Loops With Various CO Terminations

Fig. 6.19 Mean Input Impedance of Loops, as Viewed from Station toward
CO, as Function of Frequency
Source: P.A. Gresh, "Physical and Transmission Characteristics of Cus-
tomer Loop Plant," *Bell System Tech. J.,* Vol. 48, No. 10, December 1969.
Reprinted with permission from the Bell System Technical Journal. Copy-
right 1969 AT&T

The change in mean impedance with frequency, for the two-wire
trunk termination case, is shown as the white line in Figure 6.19(b). The
900 ohm/2.0 mfd case is illustrated by the gray line, and termination on
another loop is shown as the black line in the figure. Though not repro-

duced in any of the figures here, those in [12] show the scatter diagram at 3 kHz for this last case, loop-to-loop connection, to have imaginary impedance components from +350 to −1700 ohms, and real components from about 40 to 2200 ohms. Thus impedance changes of more than 50 : 1 can be expected in the exchange area.

NOTES

1. Loading designations always have at least one letter and two digits. The letter(s) stands for the specified spacing between load coils, the digits are the value, in millihenries, of the inductance per coil.
2. C-MESSAGE noise generally increases with distance and, for any (analog) transmission system, is taken as doubling (add three decibels) for every double distance.
3. The equivalent of four-wire station loops are coming into use as part of the preparation for all digital networks and ISDN services. See Section 5.1.5 for a discussion of digital telephones.
4. A fully-connected topology is one in which every node is connected to every other node and requires $N(N − 1)/2$ trunks, where N is the number of nodes.
5. Velocity of propagation ranges from about 3% to 99% of the speed of light, with loaded cable at the low end and satellite facilities at the upper, the latter being determined by the delay in the satellite and the ground stations. (See Table 1.2.)
6. Two decibels for this standard deviation is an estimate made at the time of the derivation of VNL design rules. By recent standards it is quite conservative because that is closer to the standard deviation found on complete connections. See [4] and [5], for example.
7. The CCITT uses the term circuit in the same sense that trunk is used in the U.S.
8. These data are still valid, except perhaps in small detail. Even though different loop design plans have been used since 1959, and new station sets introduced, nothing has happened that should significantly change the variations illustrated.
9. This termination is often used as an average CO input impedance.

REFERENCES

1. AT&T, *Telecommunications Transmission Engineering, Volume 3, Networks and Services,* Second Edition, Bell System Center for Technical Education, 1977.
2. Gresh, P.A., "Physical and Transmission Characteristics of Customer Loop Plant," *Bell System Technical Journal,* Vol. 48, No. 10, December 1969.
3. Ingle, J.F., *et al.,* "1983 Exchange Access Study: Analog Voice-Frequency Transmission Performance Characterization of the Exchange Access Plant," *IEEE Transactions on Communications,* Vol. COM-35, No. 1, January 1987.
4. AT&T, "1969–70 Switched Telecommunication Network Connection Survey," *Bell System Technical Reference,* AT&T Pub., April 1971.
5. Carey, M.B., *et al.,* "1982/83 End Office Connection Study: Analog Voice and Voiceband Data Transmission Performance Characterization of the Public Switched Network," *AT&T Bell Laboratories Technical Journal,* Vol. 63, No. 9, November 1984.
6. AT&T, *Telecommunications Transmission Engineering, Volume 1, Principles,* Second Edition, Bell System Center for Technical Education, 1977.
7. Freeman, R.L., *Telecommunication Transmission Handbook,* Second Edition, John Wiley and Sons, New York, 1981.
8. CCITT, "General Characteristics for International Telephone Connections and International Telephone Circuits," *Red Book, Volume III,* Rec. G.101, ITU, Geneva, 1985.
9. CCITT, "Corrected Reference Equivalents (CREs) and Loudness Ratings (LRs) in an International Connection," *Red Book, Volume III,* Rec. G.111, ITU, Geneva, 1985.
10. CCITT, "Hypothetical Reference Connections," *Red Book, Volume III,* Rec. G.103, ITU, Geneva, 1985.
11. CCITT, "Error Performance of an International Digital Connection Forming Part of an Integrated Services Digital Network," *Red Book, Volume III,* Rec. G.821, ITU, Geneva, 1985.
12. Alexander, A.A., Gryb, R.M., and Nast, D.W., "Capabilities of the Telephone Network for Data Transmission," *Bell System Technical Journal,* Vol. 39, No. 3, May 1960, p. 431*ff.*

Chapter 7
Applications

In this chapter we illustrate the use of some of the tools developed in the preceding material. We begin by considering the design of a voice terminal, which we call a station, incorporating a linear transmitting element for general application. As a yardstick for performance, comparisons will be made with an *average* station using a carbon transmitter, referred to as the *carbon set*. The IEEE metrics, appropriate to North America, will be used. Results may be converted to approximate values in other metrics by the use of Table 3.6.

7.1 THE ENVIRONMENT

To begin, a description of the electrical and acoustical environments in which the set is expected to operate is required. For the electrical environment we draw upon the descriptions of subscriber loop characteristics provided in [1]. The data used will be presented as we go along. To keep this example relatively simple, we restrict our efforts to loops equal to 18,000 ft or less in length. This restriction allows us to ignore loaded loops and at the same time (from Figure 2 of [1]) include about 88% of subscribers.

From the data shown in Figure 9 of [1], the distribution of cable gauge by loop length, it is estimated that 94% of all loops of 18,000 ft or less use 22-, 24-, or 26-gauge cable. Further, about 72% of these are constructed with either 22- or 24-gauge and the use of these two is nearly equal. A simple model of a local (intraexchange) connection may therefore be taken as a 22-gauge loop switched through a central office to a 24-gauge loop. The lengths of the loops will be varied to see how performance changes. For such a model we need to examine:

1. The insertion loss of each loop, the central office, and the end-to-end connection.

2. The impedance seen by each station.
3. The return losses at each station.
4. The direct current in the loops.

Each of these parameters is a function of loop length and dependent upon the characteristics of the stations. Toll connections can be studied by inserting network impairments at the CO to far-end loop interface.

We will assume that all station impedances are 600 ohms when viewed as sources or sinks. This is a gross simplification in the case of the carbon set but reasonable for so-called electronic sets incorporating linear transducers, referred to as *linear sets*. For return-loss calculations, a compromise balance network will be introduced and used in both the carbon and linear sets. The model, with sufficient station detail, is shown in Figure 7.1. The central office circuitry is modeled after Feed Circuit II of the IEEE standard for measuring telephone connections, Std 269-1971 (see [2]).

A check on the reasonableness of the model can be made by examining the expected distribution of impedances seen looking into each of these loops from the station end and comparing it with the data for non-loaded loop impedances in [1]. For this purpose, the loops are assumed terminated in a simulated four-wire trunk, a 900-ohm resistor in series with a 2.01-mfd capacitor, and the results compared with Figure 35 of the reference. It shows the distribution of input impedances of loops so terminated. The input impedance may be calculated using the procedures described by Webb in [3], and summarized in the Appendix to this chapter.

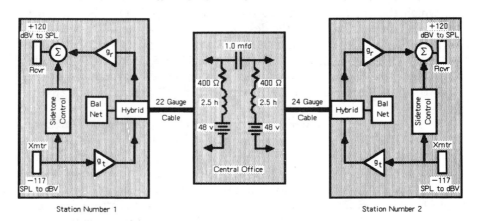

Fig. 7.1 Connection Model

Results for 22-, 24- and 26-gauge cable are shown in Figure 7.2 together with a representation of the scatter diagram of actual loop impedances presented in [1]. The essence of the scatter plots of [1] is captured by simply showing areas with three levels of shading as was done in Section 6.10. A high density of points in the scatter is indicated by the darkest areas, intermediate density by the middle level, and sparse scattering by the light. The boundaries are determined by eye estimates. The three long curves show the impedance at 3.15 kHz as a function of loop length.[1] The scatter plot values are at 3.0 kHz but differences due to a shift of 150 Hz are negligible for this purpose. It is seen that the three cable sizes essentially cover the left central through the right extreme of the scatter plot.

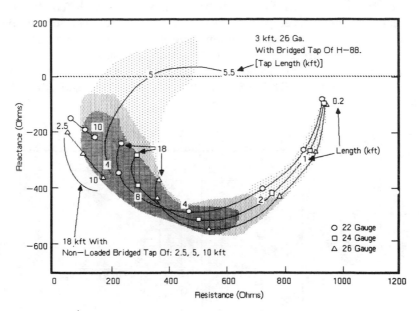

Fig. 7.2 Scatter Plot Representation, Loop Impedances at 3 kHz with Calculated Impedances Superimposed
Source: Adapted from P.A. Gresh, "Physical and Transmission Characteristics of Customer Loop Plant," *Bell System Tech. J.*, Vol. 48, No. 10, December 1969

At the far left of the figure, the effects of bridged tap are illustrated. Eighteen thousand feet of cable with bridged tap lengths of 2500 to 10,000 ft have significantly different impedances than the 18,000 ft with no bridged tap. The width and density of the scatter diagram at the left side indicate a frequent occurrence of bridged tap and, in fact, [1] shows that most loops do have some, and the mean length is about 2500 ft. The general dispersion in the scatter plot is no doubt attributable to their effects. Less than one percent of the scatter plot is contained in the lightly shaded area extending into the positive reactance quadrant of the impedance plane. Normal loop construction should not exhibit inductive properties. The occasional inadvertent appearance of loaded bridged tap can account for this, however, as illustrated. It is concluded that the loops chosen in the model should be adequate to describe the range of impedances to be encountered in normal situations. It would be advisable to check performance in some of the extreme conditions, however. This will be discussed later.

It is expected that performance, as measured by grade-of-service, may be strongly affected by different loop lengths and compromise may be necessary. An estimate of the distribution of loop lengths can be used to obtain a weighted average GoS. This will then reflect user demographics and may assist in choosing among alternatives. Such an estimate is available from Figure 2 of [1], the distribution of loop length. A derived histogram, for loops up to 18,000 ft is shown in Figure 7.3. The cell-height values are entered near the top of each bar. These will be used as weights later.

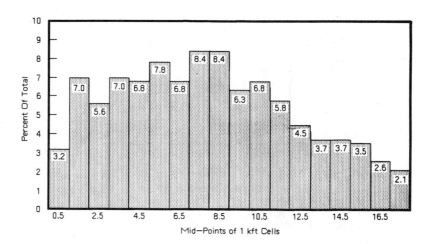

Fig. 7.3 Histogram of Subscriber Loop Lengths to 18,000 ft
Source: From: P.A. Gresh, "Physical and Transmission Characteristics of Customer Loop Plant," *Bell System Tech. J.*, Vol. 48, No. 10, December 1969

A significant part of the operating environment is the ambient room noise level. The noise level, in dBA, will be used as a parameter in comparisons with the carbon set and in optimization of design variables. The range of noise levels of interest may be taken as 35 to 80 dBA. Some appreciation of these numbers can be gained from the data below, adapted from Figure 4-8 of [4]. Sound levels, in dBA, grouped according to general consensus subjective descriptions, for private offices and conference rooms (left column below), and secretarial offices with business machines (right column) are approximately:

Level (dBA)	Description	Level (dBA)
40 to 50	Quiet	48 to 61
50 to 61	Moderately noisy	61 to 73
61 to 71	Noisy	73 to 85
71 to 82	Very noisy	85 to 99
above 82	Intolerably noisy	above 99

On this same scale, a *normal voice at nine feet* is rated at about 54 dBA. The qualifiers on these ratings of private offices and secretarial offices are important because expectation enters the subjective descriptors agreed to. It is also noted on the graph that telephone use is rated as difficult for levels between 73 and 87 dBA, and unsatisfactory for levels greater than 87. As will be seen, the descriptors for use of the telephone at these noise levels are most kind.

The preceding paragraphs describe the environment in gross terms. A detailed description, including other factors, is provided in Table 7.1 as a parameter list. Table 7.1 comprises a minimum set of parameters that should be considered in station design.

Table 7.1
Parameters

1. Loop length	100 to 18,100 ft (zero to 18)
2. Room noise	35 to 80 dBA
3. Circuit noise	18 to 35 dBrnC
4. Toll network loss	7 dB (typical)
5. Talker echo delay	10 to 30 ms
6. Far end return loss	15 to 60 dB
7. Toll network low frequency cut-off	310 to 400 Hz
8. Toll network high frequency cut-off	2.8 to 3.2 kHz
9. Modal distance	0.2 to 1.5 in (0.918 std)

7.2 VARIABLES AND OPTIONS

Design variables to be considered are listed in Table 7.2. The items are all identified in Figure 7.1.

Table 7.2
Variables

1. Linear transmitter efficiency:	Fixed at -117 dB
2. Receiver efficiency:	Fixed at 120 dB
3. Carbon transmitter efficiency:	See below
4. Receiver efficiency carbon set:	See below
5. Net station gain in the transmit direction:	g_t
6. Net station gain in the receive direction:	g_r
7. Loop length transmit gain adjust:	See below
8. Loop length receive gain adjust	IEEE ROLR, 0 to 14 dB
10. Sidetone frequency shaping:	IEEE SR, 0.0 to 6.0
11. Balance network optional:	Yes or No
12. Transmit noise protection optional expander:	Yes or No
13. Frequency response of transmitter and handset	See below
14. Frequency response of receiver and handset	See below
15. Frequency response associated with g_t	See below
16. Frequency response associated with g_r	See below

The frequency response of items 13 through 16 will be taken as flat here, another simplification. For real cases, the loudness loss of each shape can be computed by (2.23) and added to the total TOLR or ROLR as appropriate. Transmit and receive frequency shaping can have significant effects on performance but their effects have not been analyzed in detail. For some recommendations of preferred shapes see [5]. Gross effects can be included as modifications of channel slope for example in (7.32) and (7.33) in Section 7.5.

Net station gain in the transmit direction must include the total path from the output of the transmitting element to the loop interface. Net station gain in the receive direction includes a corresponding path from the loop to the receive element.

The efficiencies of the transmitter and receiver of the assumed standard reference carbon set are dependent upon loop current (no g_t or g_r appear in the carbon set). For this example the efficiencies are described by the equations:

$$\text{TOLR} = -37.36 - 6.75 \log(i) \text{ dB} \tag{7.1}$$

$$\text{ROLR} = 36.08 + 5.64 \log(i) \text{ dB} \tag{7.2}$$

where i = loop current in mA.

These relations provide a net station OOLR of about -2.8 + (Loudness Insertion Loss) dB on long loops (20 mA) and about -3.4 dB on a near zero length loop (80 mA). With 22-gauge cable these translate to a net change in OOLR of about 8 dB for loops ranging from zero to 18,000 ft. This represents reasonable or even better than typical performance for a carbon station. On finer gauge cable the variation may be as great as 15 dB or more for real telephones.

Linear stations may incorporate loop length compensation for the transmitter or receiver. Such a device may be described as inserting a loss in the transmit or receive path inversely proportional to loop current. The relation describing the loss is

$$\text{Loss} = (i - 20)L_{\text{max}}/80 \text{ dB} \tag{7.3}$$

where $8/6L_{\text{max}}$ is the loss inserted for a zero length ($i = 80$) loop.

The optional balance network can be designed using a number of procedures; one method is outlined in the Appendix to this chapter. The design of the one chosen for this example is discussed in the next section.

As discussed in Section 5.1.7 and illustrated in Figure 5.10, linear transmitters or microphones typically are much more sensitive than carbon devices to low and modest levels of room noise. Figure 7.4 shows the relations between room noise in dBA and equivalent noise in dBrnC used

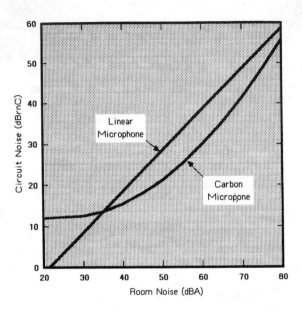

Fig. 7.4 Circuit Noise at CO, Produced by Room Noise (TOLR = −46)

here. It translates room noise to noise at a CO at the end of a loop with a TOLR of −46. The curves are plotted from the equations used for this model. The equation for the carbon microphone, with y = room noise in dBA, $20 \leq y \leq 80$ is

$$\text{dBrnC} = 19.33 - 0.627(y) + 0.013414(y)^2 + 0.0000051283(y)^3 \tag{7.4}$$

and for the linear microphone:

$$\text{dBrnC} = -21.361 + y \tag{7.5}$$

These equations were chosen to have both microphones yield the same circuit noise for a room noise of 35 dBA. They are not atypical, however. The calculated values must be adjusted according to the TOLR of the station-loop configuration being considered. (A more negative value of TOLR gives rise to more equivalent circuit noise.) With the actual TOLR designated TOLR_a, this is accomplished by

$$\text{dBrnC}_{\text{adjusted}} = \text{dBrnC} + (-46 - \text{TOLR}_a) \tag{7.6}$$

This is the noise due to near-end room noise that will be heard by the far-end listener, so the TOLR$_a$ used is that for the near-end station. In the model, noise heard in the receiver that originates at the local CO is the power sum of the circuit noise and the room noise from the other end. The equivalent noise calculated by (7.6) appears at the CO at the side of the switch on which the near-end loop terminates. It must be reduced by the switch loss, taken as 0.5 dB, to translate it to the side of the switch terminating the far-end station before the power sum is calculated. In the case of a toll connection it is further reduced by the toll network loss before the sum is calculated.

For use in the GoS equations, the total noise at the far-end side of the switch must be adjusted to a value corresponding to an equivalent value of a loop with an ROLR of 46. For example, if the actual far-end loop has an ROLR of 49, it will attenuate the noise more than would the 46-ROLR loop, so it must be reduced. The correction equation, with $x =$ (the value from (7.6) adjusted for switch and toll network loss summed with circuit noise), is

$$\text{corrected COnoise} = x - (\text{ROLR} - 46) \tag{7.7}$$

Because we are not considering any quantizing noise, this result is the quantity N for use in the R_{LN} equation (4.18).

7.3 CHOOSING A BALANCE NETWORK

Even though loaded loops are being ignored, for a balance network design we will consider them as an illustration of the limitations on return loss imposed by the mix of loaded and nonloaded loops. The two scatter plot representations presented in the discussion of the exchange area, Figure 6.18, are combined in Figure 7.5, with the low density portion of loaded loops omitted.

Proceeding by the method described in the Appendix, a point, labeled C_8, is chosen and a circle satisfactorily encompassing most of the scatter is drawn. The point and radius are arbitrary and may be refined later if desired. In this case, the circle is that with the radius R_8 in the figure. The letter R used, in this section only, represents a circle radius, not the transmission rating. The vector from the origin to C_8 is constructed and extended to intersect the far side of the circle. The intersections with the circle are labeled Z_a and Z_b to delineate points of measurement. All scatter plot impedances that lie on the circle have the

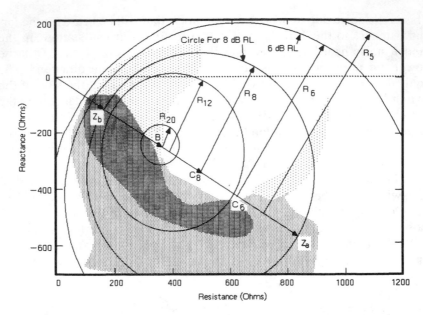

Fig. 7.5 Construction of the Balance Network with the 3 kHz Impedance Scatter Plot

same return loss, to be determined, with a balance network, as yet unknown.

The length of R_8, distance Z_a to C_8, and the magnitude of C_8, origin to C_8, are measured and found to be about 415 and 600. The ratio $|C_8|/R_8$ is the parameter A defined in the Appendix and the magnitude of the reflection coefficient $|\rho|$ is, by (A7.24):

$$|\rho| = A - (A^2 - 1)^{1/2} \tag{7.8}$$

which yields $|\rho| = 0.40$, and with the relation (A7.14):

$$\text{RL} = -20 \log|\rho| \tag{7.9}$$

we find the return loss for impedances on the circle to be 7.92 or about 8 dB, which is the reason for the subscript 8.

The center C_8 is found by measuring to be located approximately at the point $(500, -345)$. The location of the balance network impedance, Z_1 of the Appendix, at 3 kHz is found from (A7.22):

$$|Z_1| = R_8 \left[\frac{1 - |\rho|^2}{2|\rho|} \right] \tag{7.10}$$

so $|Z_1|$ is equal to 434.5 or about 434. This is marked off on the vector C and the coordinates of the balance network measured as about $(360, -250)$, so the impedance of the network at 3 kHz is $= 360 - j250$ ohms.

A similar procedure using the 1-kHz scatter plots in [1] (not shown here), yields the network impedance at 1 kHz. It is found to be $680 - j520$ ohms. A network can be synthesized from these frequency and impedance values by a number of procedures such as those in [12].

A simple network that satisfies these impedance requirements closely is a parallel RC pair of 1100 ohms and 0.2 mfd in series with 300 ohms. The actual impedances of this network are $360.5 - j250.7$ ohms at 3.0 kHz and, at 1.0 kHz, $677.9 - j522.4$ ohms.

To explore the expected return loss at points on the fringes of the scatter diagram, it is only required to calculate new C and R values for arbitrary choices of $|\rho|$ (or return loss) using (A7.21) and (A7.22) with the choice of Z_1 that was just made. For smaller values of return loss or larger values of $|\rho|$, the circle center migrates away from the origin, eventually approaching infinity. The circle radius R also increases, but slightly faster. Results for a number of values of return loss, less than and greater than 8 dB, are indicated in Figure 7.5. It is seen that the entire scatter diagram will be included within the circle for a return loss of 5 dB. This is deemed acceptable for our purposes.

The return losses expected from these calculations are those that would apply for a toll call because the impedances in the scatter plots used were calculated against a simulated four-wire trunk, 900 ohms in series with 2.01 mfd. Return loss is generally more critical for toll calls because of significant echo delay that can be encountered. Different values are to be expected for our local connection model.

7.4 MODEL DETAIL

To use the equations of Sections 4.1.5 and 4.1.6 to compute transmission rating values, R, and then find grade-of-service, a number of input variables need to be calculated. The equations use connection loss, OOLR and return loss, RL, at the station-loop interface. Also, for the noise calculations, as mentioned in Section 7.2, TOLR and ROLR values must be known. To evaluate GoS for both ends of the connection, these quantities must be determined for both directions of transmission: station 1 to station 2, and station 2 to station 1. (See Station identifier numbers, 1 and 2, in Figure 7.1.)

Referring to Figure 7.1, quantities applicable to the GoS of station 1, or determined by the station-loop configuration of station 1, will be subscripted with 1. For example, the OOLR applicable to 1 is the loss from station 2 to station 1 so, with Lmd defined as the modal distance loss correction to account for variation in position of the handset:

$$OOLR_1 = TOLR_2 + ROLR_1 + \text{switch loss} + \text{toll loss} + Lmd_2$$

which describes the acoustic to acoustic loss when the person at 2 is speaking.

For use in the analysis, the following quantities are defined:

Definitions

XmitEff: the conversion efficiency of the transmitter (-117 dB linear)
RcvEff: the conversion efficiency of the receiver (120 dB linear)
XmitG: gain of the transmit amplifier, g_t, in decibels
RcvG: gain of the receive amplifier, g_r, in decibels
LIL: loudness insertion loss of the loop
LIL$_s$:   loudness insertion loss of the switch
SOLR: sidetone loudness loss attributed to the sidetone control
SOLRnet: effective sidetone loss when the return loss is considered

---

With these definitions the following relations can be established:

$$TOLR_1 = XmitEff_1 - XmitG_1 + LIL_1 \qquad (7.11)$$

$$ROLR_1 = RcvEff_1 - RcvG_1 + LIL_1 \qquad (7.12)$$

$$OOLR_1 = TOLR_2 + ROLR_1 + \text{switch loss} + \text{toll loss} + Lmd_2$$
$$(7.13)$$

Let

$$s_l = 10^{-SOLR/10} \qquad (7.14)$$

$$r_l = 10^{-RL/10} \qquad (7.15)$$

then, assuming power addition of sidetone and reflected voltages:

$$\text{SOLRnet}_1 = -10 \log(s_{l1} + r_{l1}) \tag{7.16}$$

$$\text{TOLR}_2 = \text{XmitEff}_2 - \text{XmitG}_2 + \text{LIL}_2 \tag{7.17}$$

$$\text{ROLR}_2 = \text{RcvEff}_2 - \text{RcvG}_2 + \text{LIL}_2 \tag{7.18}$$

$$\text{OOLR}_2 = \text{TOLR}_1 + \text{ROLR}_2 + \text{switch loss} + \text{toll loss} + \text{Lmd}_1 \tag{7.19}$$

$$\text{SOLRnet}_2 = -10 \log(s_{l2} + r_{l2}) \tag{7.20}$$

The assumption of power addition for the sidetone signal with the reflected signal in (7.16) and (7.20) is reasonable but not accurate. A correct value can be determined by vector addition of the two signals and then calculating an effective loudness loss. (The phase of the reflected signal is determined by ρ and the phase characteristics of g_t and g_r.) A very pessimistic value can be estimated by assuming voltage addition in (7.16) and (7.20) and using 20 instead of 10 as the exponent divisor in (7.14) and (7.15). Such a calculation might be used to find a lower bound for the net sidetone loss.

For insertion loss and return loss calculations it is convenient to use the ABCD parameters of the connection elements. For the loop cables these can be found by Webb's method given in the Appendix to this chapter. What is needed are equations for input and output impedances and for cable insertion loss. Beginning with the fundamental cable constants, let

  Rc = resistance per unit length

  Cc = capacitance per unit length

  Lc = inductance per unit length

  Gc = conductance per unit length

and

  $Zc = Rc + j\omega Lc$, series impedance per unit length

  $Yc = Gc + j\omega Cc$, shunt admittance per unit length

The cable propagation constant, γ, and the characteristic impedance, Zo, are given by

$$\gamma = (YcZc)^{1/2} \tag{7.21}$$

$$Zo = (Zc/Yc)^{1/2} \tag{7.22}$$

For any length of cable, x, the ABCD parameters are

$$A = \cosh(\gamma x) \tag{7.23a}$$

$$B = Zo \cdot \sinh(\gamma x) \tag{7.23b}$$

$$C = \sinh(\gamma x)/Zo \tag{7.23c}$$

$$D = \cosh(\gamma x) \tag{7.23d}$$

At voice frequencies both the inductance and conductance of the cable are often ignored; we include the inductance here because it is found to have small effect on the return loss calculations, usually less than two percent at the highest frequency, 3.4 kHz, used in the WEPL calculation. (WEPL, *weighted echo path loss,* is defined in Section 1.2.5.) The cable constants used are

24 Gauge

Rc = 274 Ω/mi

Cc = 0.083 mfd/mi

Lc = 0.9539 mh/mi

22 Gauge

Rc = 173 Ω/mi

Cc = 0.0825 mfd/mi

Lc = 0.8744 mh/mi

The ABCD parameters of the other elements in Figure 7.1 can be computed by known techniques.[2]

The input or load impedance, Z_L, at any point in the system is given through the ABCD parameters of the system from that point to the load as

$$Z_L = A/C \tag{7.24}$$

The output or source impedance, Z_s, at a point is given in terms of the ABCD parameters of the system from source to the point as

$$Z_s = B/A \tag{7.25}$$

and the return loss, RL, is

$$RL = 20 \log(|Z_L + Z_s|/|Z_L - Z_s|) \qquad (7.26)$$

The *inserted connection loss,* ICL, of any section of the total transmission system working between a source impedance, Z_s, and a load impedance, Z_L, is as shown in the Appendix:

$$ICL = 20 \log(|AZ_L + B + CZ_LZ_s + DZ_s|/|Z_L + Z_s|) \qquad (7.27)$$

7.5 SUMMARY OF TRANSMISSION RATING EQUATIONS

The preceding sections presented the model we will use and values, or equations to find values, for its comprising parameters and variables. Before any results are discussed, a review of the parameters and equations for the transmission rating, R, is appropriate. Use is made of all the relations developed in Section 4.1 except those relating to quantizing noise. First the parameters needed are listed alphabetically and then the equations, in the order in which the calculations proceed.

The parameters are:

f_l Lower frequency channel 10-dB cut-off (hertz).

f_u Upper frequency channel 10-dB cut-off (hertz).

LEPD   Listener echo path delay (milliseconds).

LEPL   Listener echo path loss, expressed as WEPL (decibels).

N' Composite noise at the CO, calculated as the power sum of the result of (7.6) and assumed circuit noise (dBrnC). The prime (') is used here to indicate that this noise sum will have noise from other sources added. It is not yet the total.

N_R Room noise in dBA.

S_l Channel frequency response slope below 1 kHz (decibels per octave).

SOLR   The net SOLR computed by (7.16) or (7.20) as appropriate.

SR   Sidetone frequency response factor, $0 \le SR \le 6$.

S_u Channel frequency response slope above 1 kHz (decibels per octave).

TEPD   Talker echo path delay (milliseconds).

TEPL   Talker echo path loss, expressed as WEPL (decibels).

The calculation of R values always begin with R_{LN}. Having found OOLR and N' as above, the circuit noise plus far-end room noise contribution, N', must be modified by the contribution of near-end room noise.

Equation (4.19) is used to find an equivalent circuit noise, N_{Re}, for near-end room noise at the station end of a receive loop with an ROLR of 46. With room noise, R_N in dBA, N_{Re} is

$$N_{Re} = R_N - 35 + 0.0078(R_N - 35)^2$$
$$+ 10 \log[1 + 10^{(1-SOLR)/10}] \text{ (dBrnC)} \tag{7.28}$$

The result of (7.28) is then summed by power addition with N' to yield N, the total effective circuit noise:

$$N = 10 \log(10^{N'/10} + 10^{N_{Re}/10})$$

The OOLR and N are used in (4.18) to find R_{LN}:

$$R_{LN} = 147.76 - 2.257[(OOLR - 2.2)^2 + 1]^{1/2}$$
$$- 1.907N + 0.02037 \cdot N(OOLR) \tag{7.29}$$

The bandwidth rating, R_{LNBW}, is computed next, but to do this we need the various K factors:

$$K_1 = 1 - 0.00148(f_l - 310) \tag{7.30}$$
$$K_2 = 1 - 0.000429(f_u - 3200) \tag{7.31}$$
$$K_3 = 1 + 0.0372(S_l - 2) + 0.00215(S_l - 2)^2 \tag{7.32}$$
$$K_4 = 1 + 0.0119(S_u - 3) - 0.000532(S_u - 3)^2$$
$$+ (S_u - 3)(S_l - 2) \tag{7.33}$$
$$K_{BW} = K_1 K_2 K_3 K_4 \tag{7.34}$$

These are used in (4.27) to yield

$$R_{LNBW} = 22.8 + K_{BW}(R_{LN} - 22.8) \tag{7.35}$$

The effects of sidetone and sidetone frequency response are now taken into account through the sidetone weighting factor, K_{ST}, computed from (4.26):

$$K_{ST} = 1.021 - 0.002(SOLR - 10)^2 + 0.001(SOLR - 10)(SR - 2)^2 \tag{7.36}$$

K_{ST} is used to modify the loss-noise-bandwidth rating, R_{LNBW}, to find (equation (4.28)) the loss-noise-bandwidth-sidetone rating R_{LNBWST}:

$$R_{LNBWST} = K_{ST}(R_{LNBW}) \tag{7.37}$$

The reader may have noticed that the parentheses around subscript members, *i.e.*, (BW)(ST) are not being used here as they were in Chapter 4. In this exercise we are using all parameters. The parentheses were shown in Chapter 4 to indicate that certain parameters may be left out if they are not of interest in a particular analysis. Leaving out the optional parameters is equivalent to setting them equal to 1.0. Note that the structure of (7.35) and (7.37) is such that with K_{BW} or K_{ST} set to 1, the ratings are not changed.

Echo delay and loss effects are calculated next. Listener echo is handled first. It is used to modify R_{LNBWST} as given by (7.37); then a talker echo rating is computed and combined with the listener echo result for the composite transmission rating R_{total}. The listener echo rating R_{LE} is given by (4.29):

$$R_{LE} = 9.3(LEPL + 7)(LEPD - 0.4)^{-0.229} \tag{7.38}$$

It is used in (4.30) to get the loss-noise-listener echo composite, R_{LNLE}. The tedious notation is relaxed a little now, the optional parameters, BW and ST, are dropped in designating the result given by (4.30). Thus $R_{LNLEBWST}$ is shortened to

$$R_{LNLE} = \frac{R_{LNBWST} + R_{LE}}{2} - \left[\left(\frac{R_{LNBWST} - R_{LE}}{2}\right)^2 + 169\right]^{1/2} \tag{7.39}$$

A talker echo rating factor, R_E, is calculated now from the talker echo path loss and delay, in (4.31) with $D = TEPD$, and $E = TEPL$:

$$R_E = 106.4 - 53.45 \log\left[\frac{1 + D}{[1 + (D/480)^2]^{1/2}}\right] + 2.277(E) \tag{7.40}$$

Talker echo effects are strongly influenced by sidetone and this is accounted for through use of (4.32) and the echo-sidetone rating, R_{EST}:

$$R_{EST} = R_E + 2.6(7 - SOLR) - 1.5(4.5 - SR)^2 + 3.38 \tag{7.41}$$

Finally, R_{EST} is combined with R_{LNLE} with an equation nearly identical to (7.39):

$$R_{total} = \frac{R_{LNLE} + R_{EST}}{2} - \left[\left(\frac{R_{LNLE} - R_{EST}}{2} \right)^2 + 100 \right]^{1/2} \qquad (7.42)$$

These transmission rating equations, together with the preliminary ones given in Sections 7.2 and 7.4, and the relations for %GoB and %PoW, (4.17a) and (4.17b) are programmed to compute grade-of-service based on the Murray Hill and CCITT data bases, defined in Table 4.1. The program is used to compare performance of a basic linear station design with a carbon station and show how the variables of Table 7.2 can be manipulated to optimize the design.

7.6 INITIAL DESIGN

For a first design we adjust the transmit and receive amplifiers and the sidetone control network of the linear station to provide a TOLR, ROLR, and SOLR about equal to that of our representative carbon set. These values for the carbon set have been chosen to represent a typical telephone used in North America. Typical is difficult to define because of the large variations found in practice. This variation is reflected, for example, in EIA Standard RS-470, [5], which establishes mandatory bounds for TOLR and ROLR. It states that TOLR must lie within the range -53 to -36 and that ROLR must be between 40 and 55. These are measured with 9,000 ft of 26-gauge cable used in the test configuration specified in [2]. More restricted ranges for these quantities are also given in RS-470 and labeled as desirable; mean values are also stated under this heading. The desirable recommendations are

$$-51 \le TOLR \le -41, \quad mean = -46$$
$$+43 \le ROLR \le +53, \quad mean = +48$$

There is no sidetone loss value, SOLR, stated in the standard, it merely notes that SOLR can influence performance. We chose the recommended mean values of TOLR and ROLR to describe our carbon set. SOLR is taken as 9 dB, a value that some may consider too small, but is not atypical.

Following some trial runs between which small adjustments were made to the equations for the carbon set to force it to have the desired

TOLR and ROLR, all other variables were set to conditions we refer to as *default*. Room noise is set to 35 dBA, echo delay to 0.5 ms, far-end echo return loss to 60 dB, *etcetera*. The default conditions are chosen so that the parameters they represent will have a negligible effect on performance. For the carbon set they are listed in Table 7.3.

Table 7.3
Default Conditions; Carbon Configuration

	Station 1	Station 2
Transducer Type.................:	Carbon	Carbon
Loop Length ....(1000 ft):	0	0
Xmit Efficiency ....(dB):	0	0
Rcve Efficiency ....(dB):	0	0
Xmitter Gain ....(dB):	0	0
Receiver Gain ....(dB):	0	0
Impedance ....(Type):	Bal-Net	Bal-Net
Modal Distance..............(in):	0.918	0.918
Sidetone: SOLR ....(dB):	9.00	9.00
Room Noise................(dBA):	35.00	35.00
Xmit Loop i Cmp.........(dBmax):	0	0
Rvc Loop i Cmp.........(dBmax):	0	0
Expander Used ....:	NO	NO
Sidetone Enhance ....(0–6):	0	0
Circuit Noise.............(dBrnC):	18.00	
Toll Ckt Loss ....(dB):	0	
Echo Delay.................(ms):	0.50	
Far-End RL ....(dB):	60.00	
Low-End Slope..........(dB/Oct):	2.00	
High-End Slope ....(dB/Oct):	3.00	
Low-Frequency Cut-off........(Hz):	310.00	
High-Frequency Cut-off ....(Hz):	3200.00	

The upper portion of the table describes the conditions established at station 1 and at station 2. The lower portion describes the network conditions. Several station parameters are set to zero because they are not applicable to the carbon set. In particular, transmitter and receiver efficiencies are dependent upon loop current (7.1) and (7.2), and are evaluated within the program when loop cable type and length are specified.

The row labeled impedance merely indicates whether a balance network is in use. As mentioned earlier, for source and sink purposes, 600 ohms is used throughout all calculations, the balance network affects return loss only.

The rows labeled *Xmit Loop i Cmp* and *Rcv Loop i Cmp* state the values for Lmax of (7.3), loop current compensation. When set to zero, no compensation is possible. Use of the optional expander is indicated with a yes or no. The row labeled *Sidetone Enhance* states the numeric for the descriptor of sidetone frequency shaping and provides the value of SR for (7.36). The circuit noise of 18 dBrnc represents a nominal exchange area noise level.

Table 7.4 is the computer output describing the performance of the configuration of Table 7.3 for loop lengths from 100 ft to 18,100 ft. The value of 100 is used to inhibit the evaluation of a true length of zero. Such a length is unrealistic and the effective SOLR often changes significantly and rapidly as the length of cable between the set and the CO impedance approach zero. A calculation at zero length merely confuses the issues of interest.

The table columns associated with loop length give the GoS measures for the two stations for both the Murray Hill and CCITT data bases. The two rows immediately following give the average GoS values over the 18,000 ft of loop length variation, weighted by the user loop length distribution of Figure 7.3 (Wtg. AV:) and the arithmetic average (Avg:). Grade-of-service is so good for the default conditions that only very minor differences appear between the two.

The lower half of Table 7.4 is a printout of some results and some intermediate calculations for a loop of exactly 9000 ft, the recommended length for standard tests. The row names are mostly self-explanatory.

The TOLR and ROLR in the table are only near to -46 and $+48$, the desired mean values. This is because the cable gauges are 22 and 24 as opposed to the test configuration use of 26. These cable insertion losses are slightly different than for 26 gauge, but of more significance is the direct current resistance difference giving rise to different amounts of current and so affecting the efficiency of the carbon set. (26-gauge cable has a direct current resistance of 440 ohms per mile.)

Note that the fifth row, labeled *Ckt + FE Room Noise,* which we will refer to simply as *COnoise* is the sum of circuit noise and effective Far-End room noise only. Near-end room noise is not included in this value. Near-end room noise is left out at this point to provide a measure of the effect of far-end room noise. The 18 dBrnC of circuit noise, adjusted from an ROLR of 48.96 (station 1) to 46, would be 15.04 dBrnC. The default room noise of 35 dBA raised it only tenths of a decibel to 15.95.

The next row is labeled *Total Received Noise.* It is the COnoise of the fifth row converted to sound pressure in dBSPL, delivered from the receiver into an artificial ear, added to the room noise entering through the sidetone path.[3] The *Total Received Noise* values are not of concern in this example but may be used to compute signal-to-noise ratios for estimating articulation index in very severe environments. Articulation index estimates intelligibility on the basis of signal-to-noise ratio; see [7] for the method of computation.

The last ten rows comprise intermediate results and are listed for the station 2 side of the model only. *Total Noise* is the power sum of all noise, including near-end room noise through the sidetone path, at the input to the near-end receiver. (*Near end* means station 2 in this case.)

Tables 7.5 and 7.6 show the default settings for the first choice linear design and results using these conditions. The results are quite similar to those for the carbon set. Somewhat degraded performance is noticed for short loop lengths, 7.8% MH PoW for the linear as opposed to 1.1% for the carbon at 100 ft. The droop in performance for short lengths shows up in the average values as a few percent difference when compared to the carbon averages. Once again the weighted and nonweighted averages are essentially the same.

Our model of the typical carbon set has not incorporated any receive or transmit frequency shaping and so is unrealistic. The comparison would not be fair except that the linear set also lacks these features. After a sidetone enhancement factor is added to the linear model, we will also add it to the carbon set.

Table 7.4
Default Conditions; Carbon

Length kft	Station 1 MH %GoB	MH %PoW	CCITT %GoB	CCITT %PoW	Station 2 MH %GoB	MH %PoW	CCITT %GoB	CCITT %PoW
0.1	94.4	1.1	97.7	0.1	94.4	1.1	97.7	0.1
1.1	97.0	0.5	99.0	0.0	96.7	0.6	98.9	0.0
2.1	98.2	0.3	99.5	0.0	97.9	0.3	99.4	0.0
3.1	98.9	0.1	99.8	0.0	98.7	0.2	99.7	0.0
4.1	99.3	0.1	99.9	0.0	99.2	0.1	99.8	0.0
5.1	99.6	0.0	99.9	0.0	99.5	0.1	99.9	0.0
6.1	99.7	0.0	100.0	0.0	99.7	0.0	100.0	0.0
7.1	99.8	0.0	100.0	0.0	99.8	0.0	100.0	0.0
8.1	99.8	0.0	100.0	0.0	99.8	0.0	100.0	0.0
9.1	99.7	0.0	100.0	0.0	99.8	0.0	100.0	0.0
10.1	99.7	0.0	100.0	0.0	99.8	0.0	100.0	0.0
11.1	99.6	0.0	99.9	0.0	99.8	0.0	100.0	0.0
12.1	99.5	0.1	99.9	0.0	99.7	0.0	100.0	0.0
13.1	99.4	0.1	99.9	0.0	99.7	0.0	100.0	0.0
14.1	99.3	0.1	99.9	0.0	99.6	0.0	100.0	0.0
15.1	99.1	0.1	99.8	0.0	99.6	0.0	99.9	0.0
16.1	99.0	0.1	99.8	0.0	99.5	0.1	99.9	0.0
17.1	98.8	0.2	99.7	0.0	99.4	0.1	99.9	0.0
18.1	98.5	0.2	99.6	0.0	99.4	0.1	99.9	0.0
Wtg. Av:	99.18	0.11	99.80	0.00	99.24	0.11	99.81	0.00
Avg:	99.04	0.14	99.76	0.00	99.17	0.12	99.78	0.00

Table 7.4 (cont'd)

Details of Calculations at 9000 ft

	Station 1	Station 2
OOLR...............................(dB)	4.49	3.54
TOLR...............................(dB)	−45.99	−44.46
ROLR...............................(dB)	48.96	49.53
SOLReffective (MD corrected)..........(dB)	8.26	8.78
Ckt + FE Room Noise...(dBrnC, ROLR: 46)	15.95	15.71
Total Received Noise ....(dBSPL)	29.12	28.21
Modal Distance Correction.............(dB)	0	0
Expander Loss ....(dB)	0	0
Loop Resistance ....(ohms)	294.89	467.05
Loop Current.......................(mA)	53.64	44.98
Transmit Curr. Compensation..........(dB)	0	0
Receiver Curr. Compensation ....(dB)	0	0
%GoB mh.............................	99.72	99.82
%PoW mh.............................	0.03	0.02
%GoB ccitt	99.96	99.98
%PoW ccitt............................	0.00	0.00

Loop Length......................................(kft):	9.00
Total Effective Noise (Station 2, CO).............(dBrnC):	15.84
Kbw ....Station 2:	1.0000
Kst ....Station 2:	1.0132
Rln ....Station 2:	114.9070
Rlnbw.......................................Station 2:	114.9070
Rlnbwst ....Station 2:	116.4206
Re ....Station 2:	213.5134
Rle.......................................Station 2:	2057.0339
Rlnle.......................................Station 2:	116.3336
Rtotal ....Station 2:	115.3152

Table 7.5
Default Conditions; Linear Configuration

	Station 1	Station 2
Transducer Type................:	Linear	Linear
Loop Length ....(1000 ft):	0	0
Xmit Efficiency ....(dB):	−117.00	−117.00
Rcve Efficiency ....(dB):	120.00	120.00
Xmitter Gain ....(dB):	12.02	12.02
Receiver Gain ....(dB):	−8.45	−8.45
Impedance ....(Type):	Bal-Net	Bal-Net
Modal Distance............(inch):	0.918	0.918
Sidetone: SOLR.............(dB):	9.00	9.00
Room Noise...............(dBA):	35.00	35.00
Xmit Loop i Cmp.........(dBmax):	0	0
Rvc Loop i Cmp..........(dBmax):	0	0
Expander Used ....:	NO	NO
Sidetone Enhance ....(0–6):	0	0
Circuit Noise.............(dBrnC):	18.00	
Toll Ckt Loss ....(dB):	0	
Echo Delay.................(ms):	0.50	
Far-End RL ....(dB):	60.00	
Low-End Slope...........(dB/Oct):	2.00	
High-End Slope ....(dB/Oct):	3.00	
Low-Frequency Cut-off........(Hz):	310.00	
High-Frequency Cut-off ....(Hz):	3200.00	

Before making further comparisons between the two sets, it is worthwhile to show the effect of the balance network. Table 7.7 describes the performance of the linear set with the hybrid balance network replaced with a 600-ohm resistor. Comparing Tables 7.7 and 7.6, we see that average performance on loops of 10,000 to 18,000 ft is degraded only slightly, about 5 percentage points on the GoB scale. Performance on short loops is improved, as would be expected with the identical set impedances electrically close. The significant difference exists in the effective SOLR, *SOLR(effective)* row. At 9000 ft the sidetone loss is reduced from 7.5 and 8.5 dB for stations 1 and 2, to 1.5 and 2.5 dB. The effect on performance is not significant because of the benign default conditions of insignificant, room noise, echo, and delay. As will be shown later, small changes in SOLR have large effects in more realistic situations.

A more extensive comparison of performance for the two sets is provided by making computer runs with the model for a range of room noise values. Such runs were made for values at station 1 ranging from 45 to 75 dBA while at station 2 values of 50 to 80 dBA were used. Tables 7.8 and 7.9 show the output listings for one case. Rather than pore over tables, however, the results are summarized in Figure 7.6 for station 2. (Station 1 results are similar.) The carbon set is significantly better. A point of interest in Tables 7.8 and 7.9, however, is the COnoise (ckt + FE) entries compared to those of Tables 7.4 and 7.6. The circuit noise is still only 18 dBrnC but the COnoise has increased significantly, from about 16 and 17 dBrnC (station 2) for the two types of sets with room noise of 35 dBA, to 22 and 32, carbon and linear, for 55 dBA, increases of 6.5 and 15 dB. The difference in the increases is due to the nonlinear *versus* linear noise sensitivity characteristics illustrated in Figure 7.4. Similar changes appear in the rows giving the *total noise*. In this case the increases are about 14 and 22 dB because the near end-room noise has been added.

Also note in Tables 7.8 and 7.9 that interesting differences appear between the arithmetic and weighted GoS averages over loop length. They are still not large, however, being only a few percent or less. Even such small differences should be cause to examine performance at various lengths. For both types of sets, the short loop performance is relatively inferior but, for the linear set, performance at short lengths is extremely poor compared to that of the carbon set, roughly 15% good or better *versus* 50%.

Table 7.6
Default Conditions; Linear

| Length 1000 ft | Station 1 | | | | Station 2 | | | |
| | MH | | CCITT | | MH | | CCITT | |
	%GoB	%PoW	%GoB	%PoW	%GoB	%PoW	%GoB	%PoW
0.1	76.5	7.8	83.7	1.2	76.5	7.8	83.8	1.2
1.1	86.2	3.7	92.2	0.4	86.1	3.8	92.1	0.4
2.1	90.8	2.1	95.5	0.2	91.0	2.1	95.6	0.1
3.1	93.8	1.3	97.4	0.1	94.1	1.2	97.6	0.1
4.1	95.8	0.8	98.5	0.0	96.2	0.7	98.7	0.0
5.1	97.2	0.5	99.1	0.0	97.6	0.4	99.3	0.0
6.1	98.2	0.3	99.5	0.0	98.5	0.2	99.6	0.0
7.1	98.8	0.2	99.7	0.0	99.1	0.1	99.8	0.0
8.1	99.3	0.1	99.9	0.0	99.5	0.1	99.9	0.0
9.1	99.5	0.1	99.9	0.0	99.7	0.0	100.0	0.0
10.1	99.6	0.0	100.0	0.0	99.8	0.0	100.0	0.0
11.1	99.7	0.0	100.0	0.0	99.8	0.0	100.0	0.0
12.1	99.6	0.0	99.9	0.0	99.8	0.0	100.0	0.0
13.1	99.6	0.0	99.9	0.0	99.8	0.0	100.0	0.0
14.1	99.5	0.0	99.9	0.0	99.8	0.0	100.0	0.0
15.1	99.5	0.1	99.9	0.0	99.8	0.0	100.0	0.0
16.1	99.4	0.1	99.9	0.0	99.7	0.0	100.0	0.0
17.1	99.3	0.1	99.9	0.0	99.7	0.0	100.0	0.0
18.1	99.2	0.1	99.8	0.0	99.7	0.0	100.0	0.0
Wtg. Av:	96.95	0.68	98.62	0.05	97.19	0.64	98.70	0.05
Avg:	96.86	0.73	98.50	0.06	97.09	0.70	98.57	0.06

Table 7.6 (cont'd)

	Station 1	Station 2
OOLR...............................(dB)	1.08	1.08
TOLR...............................(dB)	−47.46	−46.45
ROLR...............................(dB)	47.53	48.53
SOLReffective (MD corrected)..........(dB)	7.47	8.52
Ckt + FE Room Noise...(dBrnC, ROLR: 46)	17.81	17.10
Total Received Noise ....(dBSPL)	32.30	30.21
Modal Distance Correction.............(dB)	0	0
Expander Loss ....(dB)	0	0
Loop Resistance ....(ohms)	294.89	467.05
Loop Current.......................(mA)	53.64	44.98
Transmit Curr. Compensation..........(dB)	0	0
Receiver Curr. Compensation ....(dB)	0	0
%GoB mh.............................	99.50	99.68
%PoW mh.............................	0.05	0.03
%GoB ccitt............................	99.92	99.96
%PoW ccitt............................	0.00	0.00

Loop Length......................................(kft):	9.00	
Total Effective Noise (Station 2, CO).............(dBrnC):	17.20	
Kbw ....Station 2:	1.0000	
Kst ....Station 2:	1.0107	
Rln ....Station 2:	111.9461	
Rlnbw.......................................Station 2:	111.9461	
Rlnbwst ....Station 2:	113.1460	
Re ....Station 2:	207.3921	
Rle...Station 2:	2018.1477	
Rlnle...Station 2:	113.0573	
Rtotal ....Station 2:	112.0089	

Table 7.7
Linear: Default, No Balance Network

| Length kft | Station 1 | | | | Station 2 | | | |
| | MH | | CCITT | | MH | | CCITT | |
	%GoB	%PoW	%GoB	%PoW	%GoB	%PoW	%GoB	%PoW
0.1	84.8	4.2	91.0	0.5	84.8	4.2	91.0	0.5
1.1	86.9	3.5	92.7	0.3	87.2	3.4	92.9	0.3
2.1	87.1	3.4	92.8	0.3	87.8	3.2	93.3	0.3
3.1	86.0	3.8	92.0	0.4	87.5	3.3	93.1	0.3
4.1	85.0	4.2	91.2	0.4	87.5	3.2	93.2	0.3
5.1	85.4	4.0	91.5	0.4	88.8	2.8	94.1	0.2
6.1	87.3	3.3	93.0	0.3	91.1	2.1	95.7	0.1
7.1	89.9	2.4	94.9	0.2	93.6	1.3	97.2	0.1
8.1	92.6	1.6	96.7	0.1	95.7	0.8	98.4	0.0
9.1	94.7	1.0	97.9	0.0	97.2	0.5	99.1	0.0
10.1	96.0	0.7	98.6	0.0	98.0	0.3	99.5	0.0
11.1	96.4	0.6	98.8	0.0	98.3	0.2	99.6	0.0
12.1	96.2	0.7	98.7	0.0	98.2	0.3	99.5	0.0
13.1	95.7	0.8	98.4	0.0	98.0	0.3	99.5	0.0
14.1	95.1	0.9	98.1	0.0	97.7	0.4	99.3	0.0
15.1	94.4	1.1	97.7	0.1	97.3	0.4	99.2	0.0
16.1	93.9	1.2	97.5	0.1	97.0	0.5	99.0	0.0
17.1	94.2	1.2	97.6	0.1	97.0	0.5	99.1	0.0
18.1	95.3	0.9	98.2	0.0	97.6	0.4	99.3	0.0
Wtg. Av:	90.92	2.24	95.32	0.19	93.31	1.56	96.75	0.12
Avg:	91.45	2.07	95.67	0.17	93.78	1.44	97.01	0.11

Table 7.7 (cont'd)

	Station 1	Station 2
OOLR.............................(dB)	1.08	1.08
TOLR.............................(dB)	−47.46	−46.45
ROLR.............................(dB)	47.53	48.53
SOLReffective (MD corrected) ....(dB)	1.50	2.49
Ckt + FE Room Noise...(dBrnC, ROLR: 46)	17.81	17.10
Total Received Noise ....(dBSPL)	35.28	33.78
Modal Distance Correction ....(dB)	0	0
Expander Loss ....(dB)	0	0
Loop Resistance ....(ohms)	294.89	467.05
Loop Current........................(mA)	53.64	44.98
Transmit Curr. Compensation ....(dB)	0	0
Receiver Curr. Compensation ....(dB)	0	0
%GoB mh..............................	94.57	97.09
%PoW mh..............................	1.08	0.48
%GoB ccitt	97.82	99.08
%PoW ccitt............................	0.05	0.01

Loop Length..(kft):	9.00
Total Effective Noise (Station 2, CO) ....(dBrnC):	17.24
Kbw ....Station 2:	1.0000
Kst ....Station 2:	0.8782
Rln ....Station 2:	111.8646
Rlnbw.....................................Station 2:	111.8646
Rlnbwst ....Station 2:	98.2446
Re ....Station 2:	223.0685
Rle..Station 2:	2018.1477
Rlnle ....Station 2:	98.1566
Rtotal ....Station 2:	97.3611

Table 7.8
Carbon: Room Noise 1,2 = 55,60 dBA

Length kft	Station 1				Station 2			
	MH		CCITT		MH		CCITT	
	%GoB	%PoW	%GoB	%PoW	%GoB	%PoW	%GoB	%PoW
0.1	53.3	21.9	59.3	6.7	56.6	19.4	63.1	5.5
1.1	64.8	14.1	72.1	3.2	65.3	13.8	72.6	3.1
2.1	73.5	9.3	80.9	1.6	71.7	10.2	79.1	1.9
3.1	80.6	6.0	87.5	0.8	77.0	7.6	84.2	1.2
4.1	86.2	3.7	92.1	0.4	81.5	5.6	88.2	0.7
5.1	90.3	2.3	95.2	0.2	85.3	4.1	91.4	0.4
6.1	93.1	1.5	96.9	0.1	88.4	3.0	93.8	0.3
7.1	94.2	1.2	97.6	0.1	90.4	2.3	95.2	0.2
8.1	94.3	1.1	97.7	0.1	91.0	2.1	95.6	0.1
9.1	94.0	1.2	97.5	0.1	90.4	2.3	95.2	0.2
10.1	93.5	1.4	97.2	0.1	89.4	2.6	94.5	0.2
11.1	92.9	1.5	96.8	0.1	88.1	3.0	93.6	0.3
12.1	92.1	1.7	96.4	0.1	86.7	3.6	92.5	0.3
13.1	91.3	2.0	95.8	0.1	85.0	4.2	91.2	0.4
14.1	90.3	2.3	95.2	0.2	83.3	4.9	89.8	0.6
15.1	89.3	2.6	94.4	0.2	81.4	5.6	88.2	0.7
16.1	88.1	3.0	93.6	0.3	79.4	6.5	86.4	0.9
17.1	86.8	3.5	92.6	0.3	77.4	7.4	84.6	1.1
18.1	85.3	4.1	91.4	0.4	75.3	8.4	82.7	1.4
Wtg. Av:	87.75	3.62	92.60	0.54	83.42	5.13	89.41	0.75
Avg:	86.86	4.00	91.88	0.63	82.07	5.70	88.26	0.89

Table 7.8 (cont'd)

	Station 1	Station 2
OOLR.............................(dB)	4.49	3.54
TOLR.............................(dB)	−45.99	−44.46
ROLR.............................(dB)	48.96	49.53
SOLReffective (MD corrected) ....(dB)	8.26	8.78
Ckt + FE Room Noise...(dBrnC, ROLR: 46)	25.53	22.20
Total Received Noise ....(dBSPL)	39.98	42.58
Modal Distance Correction.............(dB)	0	0
Expander Loss ....(dB)	0	0
Loop Resistance ....(ohms)	294.89	467.05
Loop Current.......................(mA)	53.64	44.98
Transmit Curr. Compensation..........(dB)	0	0
Receiver Curr. Compensation ....(dB)	0	0
%GoB mh.............................	94.05	90.51
%PoW mh.............................	1.21	2.25
%GoB ccitt	97.52	95.29
%PoW ccitt.............................	0.06	0.16

Loop Length......................................(kft):		9.00
Total Effective Noise (Station 2, CO).............(dBrnC):		31.14
Kbw ....Station 2:		1.0000
Kst ....Station 2:		1.0132
Rln ....Station 2:		86.8398
Rlnbw....................................Station 2:		86.8398
Rlnbwst ....Station 2:		87.9838
Re ....Station 2:		213.5134
Rle.......................................Station 2:		2057.0339
Rlnle......................................Station 2:		87.8980
Rtotal ....Station 2:		87.1069

Table 7.9
Linear: Default, Room Noise 1,2 = 55,60

Length 1000 ft	Station 1 MH		Station 1 CCITT		Station 2 MH		Station 2 CCITT	
	%GoB	%PoW	%GoB	%PoW	%GoB	%PoW	%GoB	%PoW
0.1	4.3	84.7	3.1	72.8	4.2	85.0	2.9	73.4
1.1	7.8	76.5	6.4	60.1	7.2	77.8	5.8	62.0
2.1	12.1	68.3	10.9	48.7	10.6	71.1	9.3	52.3
3.1	17.3	59.7	16.7	38.1	14.4	64.4	13.4	43.7
4.1	23.7	50.8	24.2	28.5	18.8	57.6	18.4	35.7
5.1	31.1	42.0	33.0	20.4	23.7	50.8	24.2	28.5
6.1	39.2	33.7	42.7	13.9	29.3	44.1	30.8	22.2
7.1	47.5	26.4	52.5	9.2	35.2	37.6	38.0	16.9
8.1	55.4	20.3	61.7	5.9	41.3	31.7	45.3	12.6
9.1	62.4	15.6	69.4	3.8	47.1	26.7	52.1	9.3
10.1	67.4	12.6	74.7	2.7	51.4	23.3	57.1	7.4
11.1	68.9	11.7	76.3	2.4	52.1	22.7	57.9	7.1
12.1	68.5	12.0	75.8	2.5	50.6	23.9	56.2	7.7
13.1	67.4	12.6	74.7	2.7	48.4	25.6	53.7	8.7
14.1	66.1	13.4	73.4	2.9	46.1	27.6	50.9	9.9
15.1	64.7	14.2	71.9	3.2	43.7	29.6	48.1	11.1
16.1	63.2	15.1	70.3	3.6	41.4	31.7	45.3	12.5
17.1	61.4	16.3	68.3	4.1	39.1	33.8	42.6	14.0
18.1	59.2	17.7	65.9	4.7	37.0	35.8	40.1	15.5
Wtg. Av:	45.37	32.34	49.57	17.00	33.63	42.01	36.20	23.19
Avg:	47.27	31.00	51.77	16.47	34.35	41.34	37.09	22.79

Table 7.9 (cont'd)

	Station 1	Station 2
OOLR..............................(dB)	1.08	1.08
TOLR..............................(dB)	−47.46	−46.45
ROLR..............................(dB)	47.53	48.53
SOLReffective (MD corrected)..........(dB)	7.47	8.52
Ckt + FE Room Noise...(dBrnC, ROLR: 46)	37.10	32.16
Total Received Noise..............(dBSPL)	51.83	52.04
Modal Distance Correction.............(dB)	0	0
Expander Loss......................(dB)	0	0
Loop Resistance...................(ohms)	294.89	467.05
Loop Current.......................(mA)	53.64	44.98
Transmit Curr. Compensation..........(dB)	0	0
Receiver Curr. Compensation..........(dB)	0	0
%GoB mh..............................	61.76	46.59
%PoW mh..............................	16.02	27.13
%GoB ccitt.............................	68.74	51.51
%PoW ccitt.............................	3.96	9.60

Loop Length..(kft):	9.00	
Total Effective Noise (Station 2, CO).............(dBrnC):	43.33	
Kbw......................................Station 2:	1.0000	
Kst...Station 2:	1.0107	
Rln...Station 2:	62.6732	
Rlnbw.....................................Station 2:	62.6732	
Rlnbwst...................................Station 2:	63.3449	
Re...Station 2:	207.3921	
Rle...Station 2:	2018.1477	
Rlnle.......................................Station 2:	63.2585	
Rtotal.....................................Station 2:	62.5681	

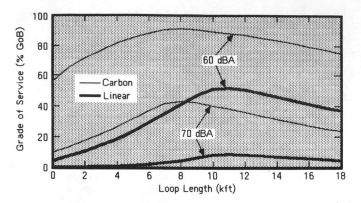

(a) Percent Good or Better With Parameter Room Noise

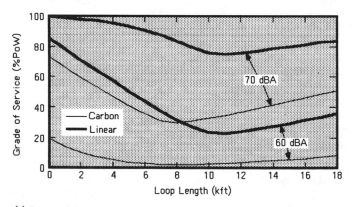

(b) Percent Poor or Worse With Parameter Room Noise

Fig. 7.6 Comparative Performance, Assumed Carbon *versus* Simple Linear

A good visual presentation of the differences in performance of the two sets is made by using average performance and room noise in area plots. Figure 7.7 shows this type of comparison for %GoB and %PoW. Quantitative differences at very high noise levels may not be accurate using this model, but it is known to the author to be quite effective for rank orderings at least, even at levels to 90 dBA. The concern over accuracy is indicated in Figure 7.7. The marked difference in performance

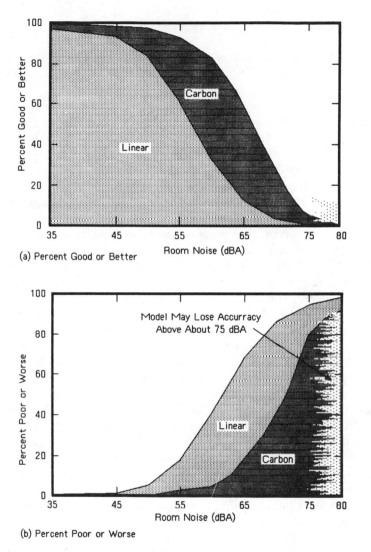

(a) Percent Good or Better

(b) Percent Poor or Worse

Fig. 7.7 Weighted Average GoS *versus* Room Noise for Similar Station Designs

of the two sets is due primarily to the difference in room noise sensitivities, Figure 7.4, and the fact that subjective response to noise is nonlinear with noise expressed in decibels. Annoyance increases very rapidly at high noise levels. This will be evidenced again in the next section.

7.7 NOISE PERFORMANCE IMPROVEMENT

An obvious approach to improving the performance of linear transmitter stations is to modify the linear characteristic shown in Figure 7.4.
There are at least two methods of doing this.

One is to use a pressure differential or gradient microphone construction for the transmitter (see [8]). These microphones are substan-
tially less sensitive to noise fields that are nearly uniform, as is most room
noise.[4] A disadvantage for general use is that untrained users move them
about excessively during conversation. The microphones' discrimination
of pressure differentials is so great that a listener is annoyed by the frequent large changes in received volume. In extreme noise environments,
greater than about 80 dBA, they can be very effective, however.

A second approach is to approximate the nonlinear characteristic of
the carbon transmitter. This can be done with an expander, a level sensitive amplifier. Expanders for this application are similar to those discussed in Section 1.3.7 as parts of companders. They have slow attack
and release times, on the order of 10 ms, and so behave more like automatic gain control (AGC) mechanisms rather than nonlinear devices.
These can be designed to have low gain for signals below some threshold
and high gain for signals above the threshold. In practice, the implied step
function change in gain is not desirable, again because of changes in
modal distance and angle of the handset mouthpiece in normal use. An *S*
shaped characteristic is best. The break points and slope may require
subjective testing for optimization. For this example, a three-step gain
function, defined below, is implemented. Expanders add to the cost of the
set and this must be weighed against the undesirable properties of carbon
transmitters: loop current sensitive gain, and poor manufacturing tolerances as discussed in Chapter 5. We will return to this consideration later.

For a general purpose expander one would not like to discriminate
against the quiet talker in a normal acoustic environment. This consideration places a limit on the lowest level at which gain begins to increase
and an upper limit on the level of room noise that can be affected. The
implementation used here provides zero loss to the transmitter output for
levels greater than 72 dBA. For levels between 70 and 72, 8 dB of loss is
introduced. Levels between 66 and 68 dBA are attenuated by 11 dB, and
levels below 68 cause 13 dB of loss to be inserted. The expander attenuation goes from full on, 13 dB of suppression, to full off, zero loss, over a 4-
dB range of room noise. (This is rather steep.) The result of this addition
to the linear design is shown in Figure 7.8. Up to the cut-off of the
expander action at 72 dBA, the new design provides better service than

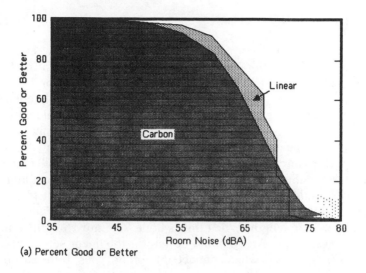

(a) Percent Good or Better

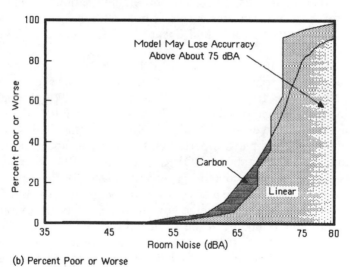

(b) Percent Poor or Worse

Fig. 7.8 Weighted Average GoS *versus* Room Noise with Expander Added
to Linear Set

the standard carbon. Note that the expander action loss steps from high-
to-low noise values are: 8, 3, and 2 dB but the improvement is in jumps of
about 15, 18, and 12 percent, illustrating the rapidly decreasing sensitivity
to room noise as it decreases through the low 70s dBA range.

7.8 LEVEL OPTIMIZATION

At the close of Section 7.6 the small differences between the arithmetic and weighted averages were noted. They pointed to the fact that performance for short loop lengths was excessively poor even at the modestly loud room noise level of 55 dBA. This inequable condition is shown in Figure 7.9(a), a plot of the data in the %GoB column of Table 7.9. In the preceding section an expander was added to improve overall noise performance as shown in Figure 7.8. Not apparent in 7.8 is the fact that the performance on short loops, though improved, is still much worse than that for medium or long ones. The OOLR as a function of loop length needs to be modified. The transmitter and receiver gains must be adjusted to manipulate the TOLR and ROLR of the stations to achieve a better balance with loop length. When this is done, one must watch compatibility with current sets. An optimum design for the linear sets used in pairs could prove disastrous when connected to other types. If the carbon set is the other type of interest, there are three combinations in which some control can be exercised: linear sets at both stations 1 and 2; carbon at station 1, linear at station 2; and linear at station 1, and carbon at station 2.

Choosing the best gain settings for the receive and transmit amplifiers can be done by trial and error or with the aid of plots such as shown in Figure 7.9. GoS is a function of OOLR, and Figures 7.9(a) through 7.9(c) show GoS plotted against OOLR (at the reference length, 9000 ft) for the three cases. The curves are plotted only for the station OOLRs in which one of the amplifiers plays a part. For the linear-to-linear case, Figure 7.9(a) applies to both stations: $OOLR_1 = TOLR_2 + ROLR_1$, and $TOLR_2$ contains g_t while $ROLR_1$ contains g_r, *etcetera*. Data for these plots can be generated by simply stepping the gain of the amplifiers to cover the desired range. The range is determined by plotting GoS for the extreme loop lengths, zero and 18,000 ft. When sufficient data points are collected to make the curves cross, that is enough. The crossover designates the optimum OOLR because then each loop length extreme is penalized equally. We assume here that this is the desired goal. User demographics could dictate some other choice.

We now remove the appropriate amplifier gains given in Table 7.5 from the proper TOLRs and ROLRs in Table 7.6 and substitute the variables $-G_t$ and $-G_r$ in their place. Remove here means to add the gain or subtract the loss. For example, $TOLR_1 = -47.46 - 12.02 - G_t = -35.44 - G_t$. If the crossover point in Figure 7.9(a) is taken as $OOLR_1 = OOLR_2 = 6.6$, the following equations can be written:

$$OOLR_1 = -34.43 - G_t + 39.08 - G_r = 6.6 \qquad (7.43)$$

$$OOLR_2 = -35.44 - G_t + 40.08 - G_r = 6.6 \qquad (7.44)$$

These yield the essentially equal results:

$$G_t + G_r = -1.95 \text{ and } -1.96 \qquad (7.45)$$

describing the generalized optimum settings for the linear-to-linear case. G_t and G_r cannot have unlimited variation as implied by (7.45). In practice, the gains are limited by network overload requirements or, in some cases perhaps, by crosstalk considerations.

In similar fashion, using Tables 7.3 and 7.4 for the carbon TOLR and ROLR values, we arrive at equations useful for the other two configurations (we use OOLR = 5.2 from Figure 7.9(b) and 6.6 from Figure 7.9(c)):

$$7.9(b): OOLR_2 = TOLR_{1\text{-carbon}} + ROLR_{2\text{-linear}}$$

or

$$OOLR_2 = 5.2 = -45.99 + (48.53 - 8.45 - G_r), \qquad G_r = -11.11$$
$$(7.46)$$

and

$$7.9(c): OOLR_1 = TOLR_{2\text{-linear}} + ROLR_{1\text{-carbon}}$$

or

$$OOLR_1 = 6.6 = -(46.45 - 12.02 + G_t) + 48.96, \qquad G_t = 7.93$$
$$(7.47)$$

Equations (7.46) and (7.47), as expected, yield a number for the sum of G_t and G_r, -2.61, that is inconsistent with (7.45). Because the slope of the curves in the all-linear case, Figure 7.9(a), are relatively shallow and performance relatively good, we examine the two lower figures in Figure 7.9 in search of a compromise. If we move right of the crossover in Figure 7.9(b) and left in Figure 7.9(c), new OOLRs can be chosen that will nearly balance the shift from optimum in each case. As it happens here, this is a trivial exercise. Simply splitting the difference between the optimum points, as indicated by the white dots, does not penalize either case much more than the other. We would then have $OOLR_{2\text{-linear}} = OOLR_{1\text{-carbon}} = 5.9$ dB.

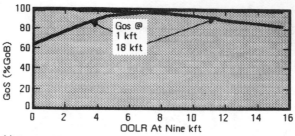

(a) Grade of Service For Linear Stations in Linear–Linear Pair

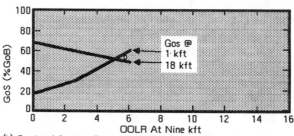

(b) Grade of Service For Linear Station in Carbon–Linear Pair

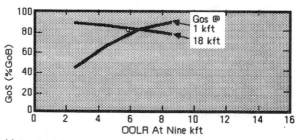

(c) Grade of Service For Carbon Station in Carbon–Linear Pair

Fig. 7.9 GoS at Extreme Loop Lengths as Function of OOLR at 9000 ft

If the value of 5.9 dB were used for the OOLRs instead of 6.6, some small improvement could be realized overall for the cases of 7.9(b) and 7.9(c). However, we choose to remain with the 6.6 value for the linear-to-linear case. The gains become $G_t = 8.2$ and $G_r = -9.6$.

In the next section we will optimize the sidetone control section of the linear set by examining performance on toll calls, but let us take a look at how performance has changed with the addition of the expander and the gain adjustments. Grade-of-service as a function of loop length for the initial design and with the successive changes, is shown in Figure 7.10 for a connection with room noise at station 1 set to 55 dBA and to 60 dBA at station 2. It is self-explanatory. One result of the next section is included in Figure 7.9, sidetone adjustment. The curve is nearly indistinguishable. For the conditions presented, sidetone appears to have no effect, except perhaps to degrade performance if it were not at optimum.

7.9 SIDETONE CONTROL

The subjective effects of sidetone loss and frequency shaping are most pronounced when significant amounts of echo and echo delay are encountered. Preferred shaping and loss are both functions of echo magnitude, echo delay, circuit noise, and room noise.

Talker echo transmission rating, R_E, and listener echo transmission rating, R_{LE}, were shown as functions of delay and echo path loss in Figures 6.7(a) and 6.7(b). Average tolerance to echo was illustrated in Figure 6.8. The curves displayed in those figures are functions of echo loss and delay only, not composite transmission ratings including noise and other parameters as variables. When these are included, sidetone factors become extremely important, as illustrated in Figure 4.10.

Choosing an optimum SOLR and sidetone enhancement factor, SE, is complicated by the many variables. To reduce the complexity, we assume that the network echo loss and delay behave according to the VNL design rules as discussed in Section 6.8. The rules are designed to satisfy the average tolerance function of Figure 6.8, modified by the standard deviation of listeners. The modification simply increases the required echo path loss of Figure 6.8 by 2.33 (99%) standard deviations (2.5 dB), or 5.8 dB. This means that the one-way connection loss must increase by half that amount. Values for round-trip echo path loss for any delay can be found from the curve of Figure 6.8 shifted by that amount. For example, for delays of 0, 15, and 30 ms, we estimate required loss values of 7.5, 14.8 and 20.5 dB.

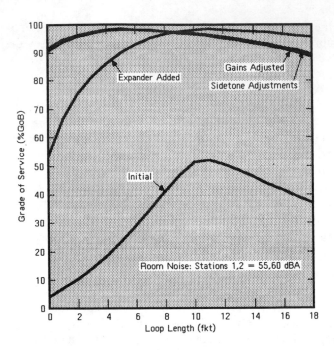

Fig. 7.10 Effects of Progressive Design Changes

A little experimentation with the model shows that grade-of-service, as a function of delay and loss, is most sensitive to the sidetone frequency shaping, SE. It is also found that the optimum enhancement is nearly constant as delay and loss vary, if they obey the VNL rule. Figure 7.11 illustrates the enhancement factor influence with room noise and SOLR as parameters. (The GoS calculations for the figure include both talker and listener echo effects.) Because enhancement effects were found to be essentially independent of them, arbitrary choices of echo delay and talker echo path loss (TEPL) were made for the calculations leading to the results in Figure 7.11. They include the secondary effects of listener echo and *Listener Echo Path Loss* (LEPL). LEPL is about 1.5 times TEPL because of the one extra traversal of the connection. By examining Figures 7.11(a) through 7.11(c), we decide:

1. Preferred enhancement is essentially independent of noise.
2. The curves flatten as noise increases.
3. Optimum SOLR appears to be about eight.
4. With an SOLR of eight, an enhancement value of 4.6 appears satisfactory.

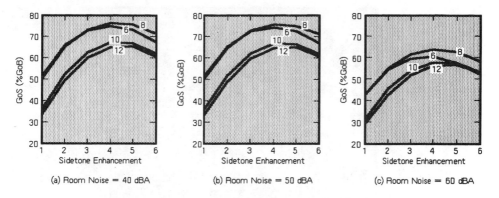

Fig. 7.11 Average GoS *versus* Sidetone Enhancement: Talker Echo Delay =
30 ms, TEPL = 20.5 dB; Parameter: SOLR (dB)

The patterns and trend indicated in Figures 7.11(a) through 7.11(c) continue with increasing room noise until the expander upper threshold, 73 dBA, is exceeded. At that point, the %GoB grade-of-service collapses to a nearly flat curve with an average value of about one or two percent.

The curves of 7.11 do not tell the whole story, however. With SOLR values near ten, as loop length changes the impedance of the cable, the effective SOLR passes through ten, effecting a rapid change in K_{ST}, the sidetone weighting factor, (7.36). It exhibits a strong quadratic behavior due to the second term. This multiplies the nearly linear behavior of R_{LN}, (7.29) through (7.37). The result on grade-of-service is shown in Figure 7.12. Viewing the figure confirms that a value of SOLR near eight is best. This exercise is an example of how one of the several nonlinear functions in the model can dominate and modify an otherwise smooth performance trend. A flag is waved, warning that all variations should be explored in detail before any design is considered final.

This result applies for toll connections, the case in which echo is of concern. A similar evaluation of SOLR for local calls under adverse conditions of noise and loss will show that the preferred SOLR value is close to 11. In view of the fact that the performance of the linear design as shown in Figure 7.10 promises to be very good, we decide to take any SOLR penalty in the local call situation. Acknowledging that a value larger than eight would be better for local calls, Figure 7.12 indicates that we might move slightly higher with only a minor effect on toll calls; however, for this example, we will favor local calls and choose 10 dB as the design SOLR.

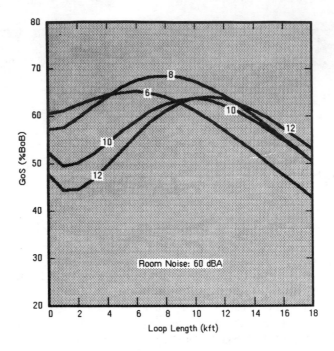

Fig. 7.12 Grade-of-Service as a Function of Loop Length; Parameter: SOLR

The resulting performance for local calls is illustrated in Figure 7.13. The upper curve, (a), shows the performance of station 2 (an improved linear set) when connected to another improved linear set. The room noise at the two locations is 55 dBA at station 1 and 60 dBA at station 2. For comparison, the performance of two carbon sets in the same environments is also shown, curve (b). For this comparison, the carbon sets were also assigned an SE value of 4.6.

To assure that the improved set is compatible with carbon sets, performance was calculated for connections of linear to carbon (station 1 to station 2), and carbon to linear. The results are shown as curves (c) and (d). Note that in almost every situation, performance with a linear set is superior to the carbon-to-carbon case of curve (b). The linear-to-carbon case (c), is poorer for loop lengths in excess of 7000 ft.

7.10 MODAL DISTANCE EFFECTS

Most of us have experienced difficulty hearing or being heard when the room noise at our or the far-end location was very loud. The common

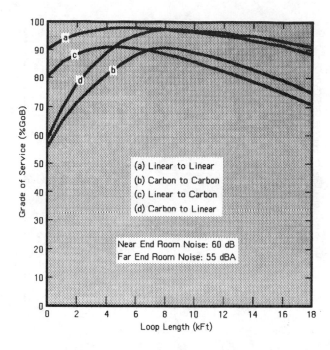

Fig. 7.13 Final Linear Design in Various Combinations Compared to Assumed Carbon Set

practice is to move the mouthpiece of the handset closer and raise our voice. The results of these actions can be analyzed by varying the modal distance in the model and the results are not altogether as expected.

A rather poor toll call was simulated by using the network survey data in [9] to pick impairment values around the 80- to 90-percent points on the CDFs of characteristics of long toll connections. Stations 1 and 2 were assigned improved linear sets and room noise of 65 dBA. Weighted loop length average GoS was calculated for modal distances from 0.1 to 1.5 inches. Recall that the standard distance is 0.915 inches. Results are shown in Figure 7.14. Contrary to expectation, when the set is moved in very close, performance is degraded significantly compared to that at the standard distance. The reason is that the received volume is simply too great and the poorer signal-to-noise ratio for larger distances is preferred. The reader may be familiar with the problems of listening to someone *shouting* into the phone and so appreciate this result. For this connection, an optimum modal distance occurs, rather sharply, at about 0.7 inches. Other combinations of impairments and noise levels would, of course, have their own best distances.

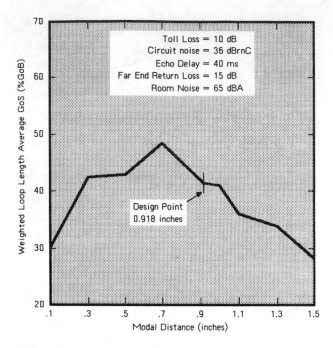

Fig. 7.14 Far-End GoS with Near-End Modal Distance Changes for a Be-low-Average Quality Connection

7.11 EVALUATION AND COMMENT

The significant steps in the design process described in the several preceding sections were:

1. Choice of a balance network.
2. Introduction of an expander.
3. Transmit and receive gain alignment.
4. Choice of sidetone shaping.
5. Choice of sidetone loss.

For each of these, other selections could have been made. For the balance network, the center of the circle in Figure 7.5 might be shifted down and to the right, placing more emphasis on loaded loop impedances, and ignoring the apparent incorrect loop implementations of the upper quadrant.

The expander steps could be modified in any number of ways to shape the difference between the areas in Figure 7.8. As mentioned earlier, however, any expander design should undergo extensive subjective testing. Besides the noise level *versus* loss characteristic, attack and release times play an important role in their subjective evaluation.

For gain and sidetone optimization, one may wish to place more emphasis on toll calls than was done here. For example, choose the SOLR to be 8 dB as indicated in Figure 7.12. Then, if 7 dB were added to the OOLRs of Figure 7.13 to examine average toll performance, some relative degradation of performance would be expected on long loops.

A consideration not mentioned previously is the anticipation of nearly ubiquitous digital networks and relatively small value fixed-loss network transmission plans that will ensue. With digital loops as well, their length will no longer be a consideration. It is likely that with fewer parameters, a rather different design, providing even better performance, could be realized.

Digital loops also allow for digital telephones, which can be arranged as true four-wire devices. End-to-end four-wire transmission is realizable and, in principle, the echo problem vanishes. Electrically, that is true. However, an acoustic feedback path must always exist between the handset receiver and transmitter. The loss of this path is highly variable depending upon the handset design and the spatial relation between it and objects in its immediate environment. A relatively high loss ensues with the handset in its expected in-use position. A low-loss value is realized if placed on a hard surface with the transmitter and receiver facing the surface. The acoustic loss between receiver and transmitter has been measured to be on the order of 30 to 35 dB for a handset of common design suspended freely in a typical room. It is not clear that networks will ever be free of a need for echo control.

In closing this section, let us compare the final design with the performance of our assumed carbon set with the area plots of Figures 7.7 and 7.8. For local calls, and with the sidetone enhancement of carbon set made equal to 4.5, as for the linear, the results are shown in Figure 7.15. For room noise levels up to 72 dBA, a point at which specialty handsets become of interest, the linear set equals or out-performs the carbon by up to around 38% in terms of good or better grade-of-service. The cost of providing that significant difference, above a basic optimized design, is that of the expander. Figure 7.10 shows that it is the major improvement factor. Recall that the 38% is an estimate of the percent of users who will notice the difference. From the other point of view, the difference in the percent poor or worse category, Figure 7.15(b), is up to about 22. The cost *versus* quality question then translates into the cost of an expander *versus* that of displeasing an additional 22% of the customers.

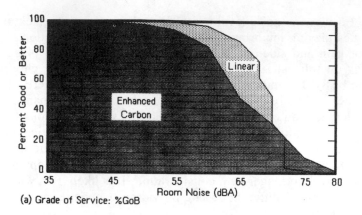

(a) Grade of Service: %GoB

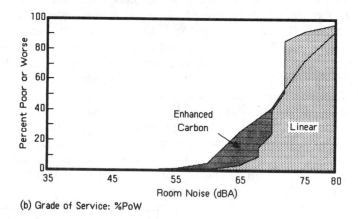

(b) Grade of Service: %PoW

Fig. 7.15 Comparative Performance: Final Linear Design *versus* Enhanced Carbon Set

7.12 DATA TRANSMISSION CONSIDERATIONS

As noted in the Preface and discussed briefly in Section 4.2.5, no criteria for acceptable data transmission have yet been determined. There is no nice model such as used above to aid in decisions concerning trade-offs or what to implement; the default course of designing to economic bounds may continue for a long time. In addition the subjective aspects of data integrity (error rate), stressed in 4.2.5 are no doubt confounded by through-put considerations, particularly in private networks, from LANs to international configurations.

These two aspects of data quality, integrity and through-put, are commonly treated as separate entities, the first under the general heading of transmission theory, the second as a traffic problem. The two do come together, however, in the design of digital transmission systems and modems. Typically, as a fixed bandwidth channel with some mix of impairments is stressed with an increasing bit rate, a point is reached at which the error rate increases so rapidly that any observer would agree that it passed from acceptable to unacceptable within a narrow range of parameter variation. These thresholds determine the operational speeds of nearly all devices and form the practical basis for published operational error rates for them. A device in this context includes individual transmission facilities. Assemblies of devices, networks, must allow for many tandem devices and so network performance may be several orders of magnitude less. Digital networks may incorporate error correction schemes so the user is not aware of the underlying error rates but this is at the expense of through-put at the network level.

In designing devices, the operational environment is often either well defined, or limits can be stated for individual impairments in terms of the thresholds mentioned above. Predicting performance when a terminal is connected to a network is another matter. This is where the transmission line simulators, described in Section 4.2, are important. Data taken by testing modems on simulators can be used to generate models analogous to the subjective models for speech transmission. Analytic descriptions of the effects of individual impairments, and trade-offs between them can be determined.

If the intended network can be described in detail equivalent to that in [9], together with measures of correlations between impairments, a network simulation may be used in conjunction with the terminal descriptors to evaluate performance. The process is time consuming and when complete, is valid only for the particular modulation and detection methods, and bit rate(s) used in the tests. To date, it appears that no statistical description applicable to a class of techniques has evolved.

At the network level, a designer has guidelines for the maximum permissible amount of each of numerous impairments. References [10] and [11] for example contain many. Reference [10] is a speech transmission oriented document but much of the material is useful in data considerations. The designer must know the characteristics of individual facilities and switching entities in order to plan a network to meet given objectives of transmission performance. Ideally, these data should include performance under all anticipated environmental conditions. The task of the designer is then to assemble a network providing the required connectivity with transmission paths that will not exceed the transmission requirements.

For data transmission this task is not simple because many of the impairments do not add in some simple manner, such as power addition for noise. Estimating nonlinear distortion on a built-up connection, for example, is particularly troublesome. For the four-tone method of measurement, two rules-of-thumb are: second-order distortion will add on a power basis, and third-order distortion on about a (17 log N)-basis. That is, the factor ten used in power addition, or 20 in voltage addition, is replaced by 17.

The build up of impairments such as (analog) phase jitter or envelope delay distortion can only be approximated in switched networks using many types of facilities. For large networks, problems of this type may best be studied using Monte Carlo techniques in computer simulations. When reasonably reliable statistics on the performance of individual network components are available, very good results can be achieved and network costs minimized.

NOTES

1. Most calculations in this example are carried out at one or more of the ISO preferred frequencies, in keeping with a recognized approach of spanning the spectrum. When loudness loss values are used, 0.3 and 3.3 kHz are included in accordance with the IEEE equations.

2. See Chapter 6 of [13] for example for an introduction to the ABCD parameters, also known as the *transformation* or *U matrix*. In [3], Webb also gives the basic definitions.

3. Recall that in Section 2.5.1 it was stated that sound pressure of speech, measured two inches from the talker's lips, is about 88 to 92 dBSPL. With desirable OOLRs on the order of zero to six decibels, speech pressures out of the receiver are on the order of 82 to 92 dBSPL.

4. A uniform noise field is one in which the pressure is uniform at all points; that is, the field is uniformly distributed in pressure. Sounds measured close to the source (the talker's lips) do not have uniform fields but display pressure changes with distance, roughly proportional to the inverse of the distance.

Appendix 7A

In this appendix we show the model and present the results of Webb's analysis for the ABCD parameters of a transmission line, and the equation for its insertion loss. We then present a graphical method for finding a balance network that will provide the best possible match to an arbitrary set of impedances.

7A.1 ABCD PARAMETERS FOR A TRANSMISSION LINE

The basic characteristics of transmission line are:

R resistance per unit length (ohms)

L inductance per length (henrys)

C capacitance per unit length (farads)

G resistive conductance per unit length (mhos)

Let

$$Z = R + j\omega L = \text{impedance per unit length} \tag{7A.1a}$$

$$Y = G + j\omega C = \text{conductance per unit length} \tag{7A.1b}$$

Then, for the elemental length of line shown in Figure 7A.1(a), given V_2 and I_2, V_1 and I_1 can be written as

$$V_1 = V_2 + I_2 Z\delta l \tag{7A.2}$$

$$I_1 = I_2 + V_1 Y\delta l \tag{7A.3}$$

and for arbitrary length, l, the expressions for V_1 and I_1 become

$$V_1 = V_2 \cosh[(YZ)^{1/2}l] + I_2(Z/Y)^{1/2} \sinh[(YZ)^{1/2}l] \qquad (7A.4)$$

$$I_1 = V_2(Y/Z)^{1/2} \sinh[(YZ)^{1/2}l] + I_2 \cosh[(YZ)^{1/2}l] \qquad (7A.5)$$

The quantity $(YZ)^{1/2}$ is defined as the propagation constant, γ, and $(Z/Y)^{1/2}$ the characteristic impedance, Z_0.

The ABCD parameters may be defined as shown in Figure 7A.1(b), with the positive direction of I_2 as indicated. Applying the definitions to (7A.4) and (7A.5) and using γ and Z_0, we have

$$A = \left.\frac{V_1}{V_2}\right|_{I_2=0} = \cosh(\gamma l) \qquad (7A.6)$$

$$B = \left.\frac{V_1}{I_2}\right|_{V_2=0} = Z_0 \sinh(\gamma l) \qquad (7A.7)$$

$$C = \left.\frac{I_1}{V_2}\right|_{I_2=0} = \frac{1}{Z_0} \sinh(\gamma l) \qquad (7A.8)$$

$$D = \left.\frac{I_1}{I_2}\right|_{V_2=0} = \cosh(\gamma l) \qquad (7A.9)$$

7A.2 INSERTION LOSS

Insertion loss of a device, or ICL for *inserted connection loss* as used in the text, is the ratio of the power delivered to a load from a source without the device between them, to the power delivered with the device inserted. The ratio is expressed in decibels.

For a source V_1 with impedance Z_s, a network described by its ABCD parameters, and a load Z_L receiving a voltage V_2, the schematic and applicable matrix relations are shown in Figure 7A.2. The power ratio required is

$$\text{ICL} = 20 \log(V_{2wo}/V_{2w}) \qquad (7A.10)$$

where the *wo* and *w* indicate without and with the network inserted. These voltages are related to V_1 through the parameter A, $V_2 = V_1/A$. From the right-hand matrix of the first equation in Figure 7A.2:

$$V_{2wo} = V_1 \frac{Z_L}{Z_L + Z_S} \qquad (7A.11)$$

and from the second:

$$V_{2w} = V_1 \frac{Z_L}{AZ_L + CZ_LZ_s + B + DZ_s} \tag{7A.12}$$

and

$$ICL = 20 \log \frac{|AZ_L + CZ_sZ_L + B + DZ_s|}{|Z_L + Z_s|} \tag{7A.13a}$$

Inverting the network matrix and reversing the direction of transmission:

$$ICL_{rev} = 20 \log \frac{|AZ_L + CZ_sZ_L + B + DZ_s|}{|AD - BC| \, |Z_s + Z_L|} \tag{7A.13b}$$

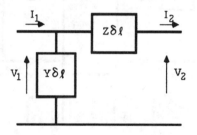

(a) Elemental Transmission Line Of Length $\delta\ell$

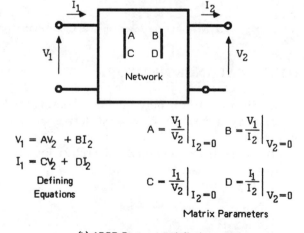

$V_1 = AV_2 + BI_2$

$I_1 = CV_2 + DI_2$

Defining
Equations

$$A = \frac{V_1}{V_2}\bigg|_{I_2=0} \qquad B = -\frac{V_1}{I_2}\bigg|_{V_2=0}$$

$$C = \frac{I_1}{V_2}\bigg|_{I_2=0} \qquad D = -\frac{I_1}{I_2}\bigg|_{V_2=0}$$

Matrix Parameters

(b) ABCD Parameter defintions

Fig. 7A.1 Elemental Line and Parameter Definitions

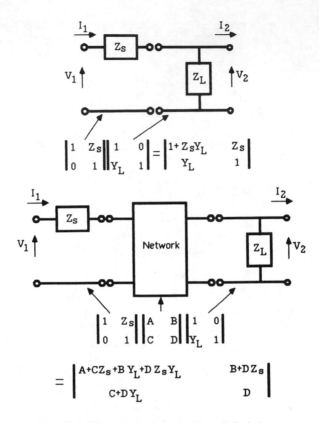

Fig. 7A.2 Networks for Insertion-Loss Calculation

7A.3 OPTIMUM BALANCE NETWORK

To derive a procedure for determining an optimal balance network for a set of arbitrary impedances, we first derive some conditions that follow from the return-loss equation.

For any two impedances Z_1 and Z_2, the return loss, RL, is defined as

$$RL = -20 \log|\rho| \text{ dB} \qquad (7A.14)$$

where

$$\rho = (Z_1 - Z_2)/(Z_1 + Z_2) \qquad (7A.15)$$

and Z_1 is said to have a return loss of RL against Z_2 or, Z_2 against Z_1. The magnitude requirement of (7A.14) makes (7A.15) symmetrical with regard to Z_1 and Z_2 for return-loss purposes. We choose Z_1 to be the constant but unknown impedance of a balance network and Z_2 to be arbitrary.

With Z_1 fixed we find the locus of Z_2 such that RL is constant. From (7A.15) we have

$$|\rho|^2 = \frac{|Z_1 - Z_2|^2}{|Z_1 + Z_2|^2} \qquad (7A.16)$$

Let

$$Z_1 = r_1 \cos \alpha + jr_1 \sin \alpha = a + jb$$
$$Z_2 = r_2 \cos \beta + jr_2 \sin \beta = x + jy$$
$$|\rho|^2 = k^2$$

then (7A.16) may be written as

$$k^2 = [(a - x)^2 + (b - y)^2]/[(a + x)^2 + (b + y)^2] \qquad (7A.17)$$

Expanding the squares, multiplying through by the denominator, collecting terms and dividing by the quantity $(1 - k^2)$, and rearranging, we have

$$x^2 - 2ax \frac{1 + k^2}{1 - k^2} + y^2 - 2by \frac{1 + k^2}{1 - k^2} = -a^2 - b^2 \qquad (7A.18)$$

Completing the squares in x and y on the left and taking all constants to the right:

$$\left(x - a\frac{1 + k^2}{1 - k^2}\right)^2 + \left(y - b\frac{1 + k^2}{1 - k^2}\right)^2$$
$$= a^2\left(-1 + \frac{1 + k^2}{1 - k^2}\right) + b^2\left(-1 + \frac{1 + k^2}{1 - k^2}\right) \qquad (7A.19)$$

and simplifying the right-hand side:

$$\left(x - a\frac{1 + k^2}{1 - k^2}\right)^2 + \left(y - b\frac{1 + k^2}{1 - k^2}\right)^2 = 4k^2 \frac{a^2 + b^2}{1 - k^2} \qquad (7A.20)$$

(7A.20) is seen to be the equation for a circle centered at the point:

$$a\left(\frac{1 + k^2}{1 - k^2}\right)^2, \; b\left(\frac{1 + k^2}{1 - k^2}\right)^2$$

with a radius, R given by

$$R = 2k \left[\frac{a^2 + b^2}{1 - k^2} \right]^{1/2}$$

and by the definitions of a, b, and k above, the circle is found to centered at the location, C, given by

$$C = Z_1 \frac{1 + |\rho|^2}{1 - |\rho|^2} \tag{7A.21}$$

with a radius of

$$R = |Z_1| \frac{2|\rho|}{1 - |\rho|^2} \tag{7A.22}$$

The locus of Z_2 for constant return loss is therefore a circle with the center given by (7A.21) lying on the extension of the vector Z_1. Note that as $|\rho|$ ranges from one to zero, R ranges from infinity to zero. With the return loss, RL, equal to $-20 \log|\rho|$, RL is inversely proportional to the radius of the circle. Thus, any impedance Z lying within the circle defined by (7A.21) and (7A.22) will have a return loss against Z_1 that is larger than that for Z_2. The circle defines the minimum return loss for all points on or within it.

Now define the quantity A as the ratio of $|C|/R$. From (7A.21) and (7A.22):

$$A = |C|/R = (1 + |\rho|^2)/(2|\rho|) \tag{7A.23}$$

and solving for $|\rho|$, we have

$$|\rho| = A - (A^2 - 1)^{1/2} \tag{7A.24}$$

Some of these quantities are illustrated in Figure 7A.3. In the figure if R is taken as unity, then, by measuring, we find C is about 1.35. Using (7A.23) and (7A.24), we find $|\rho| = 0.44$ and the return loss, by (7A.14), for all points on or within the circle to be equal or greater than 7.07 dB. The center of the circle is located at about $C = 0.74 + j1.02$. The locus of the reflection coefficient, ρ, is also illustrated.

In view of the properties of the circular locus of Z_2 and the intersections of the circle with the extension to the vector Z_1, which we chose to be the balance network, we define the procedures for finding the optimum

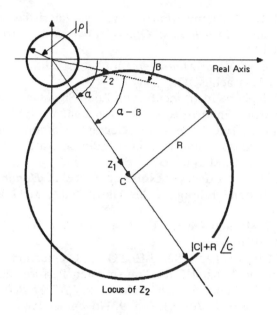

Fig. 7A.3 Parameters Associated with the Equation for RL

Z_1. Given an arbitrary set of impedances that must be balanced as best as possible against a single network, the procedure is:

1. Draw the smallest possible circle enclosing all points.
2. Draw the vector C from the origin to the center of the circle.
3. Measure $|C|$ and the radius R.
4. Use (7A.24) to find $|\rho|$.
5. Use (7A.14) to find the return loss.
6. Use (7A.22) to solve for $|Z_1|$.
7. Measure off $|Z_1|$ from the origin and read the coordinates for the balance network.

REFERENCES

1. Gresh, P.A., "Physical and Transmission Characteristics of Customer Loop Plant," *Bell System Technical Journal,* Vol. 48, No. 10, December 1969.
2. IEEE, "IEEE Standard Method for Measuring Transmission Performance of Telephone Sets," *IEEE Std. 269-1971,* February 4, 1971.

3. Webb, P.K., "Computation of the Characteristics of Telephone Connections," *British Post Office Research Department Report,* No. 630, 1977.
4. *Handbook of Noise Measurement,* Ninth Edition, GenRad, Inc. Concord, Massachusetts.
5. EIA, "Telephone Instruments with Loop Signalling," *EIA Std. RS-470,* Electronic Industries Association, January 1981.
6. International Telephone and Telegraph Corporation, *Reference Data for Radio Engineers,* Fourth Edition, American Book-Stratford Press, Inc., March 1957.
7. Webster, J.C., "Effects of Noise on Speech," *Handbook of Noise Control,* Second Edition, C.M. Harris (ed.), McGraw-Hill, New York, 1979.
8. Sessler, G.M., and West, J.E., *J. Acoust. Soc. Am.,* Vol. 46, 1969, pp. 1081–1086.
9. Carey, M.B., *et al.,* "1982/83 End Office Connection Study: Analog Voice and Voiceband Data Transmission Performance Characterization of the Public Switched Network," *AT&T Bell Laboratories Technical Journal,* Vol. 63, No. 9, November 1984.
10. CCITT, "General Characteristics for International Telephone Connections and Circuits," *Red Book, Volume III—Fascicle III.1,* ITU, Geneva, 1985.
11. CCITT, "Digital Transit Exchanges in Integrated Digital Networks and Mixed Analogue-Digital Networks," *Red Book, Volume VI—Fascicle VI.5,* Geneva, 1985.
12. Gilliman, E.A., *Synthesis of Passive Networks,* John Wiley and Sons, New York, 1957.
13. Waldhaur, F.D., *Feedback,* John Wiley and Sons, New York, 1982.

Glossary

The following definitions of telecommunication terms, acronyms, and abbreviations apply for the meanings used in this book. Some of the terms may have additional or more general meanings.

A-Law: *See* Instantaneous Compander.

ABCD Parameters: A set of descriptors used in a matrix relating input and output voltages and currents of an electrical network. The ABCD parameters of a number of cascaded networks are determined by matrix multiplication of the parameters of the individual networks. This property makes the ABCD parameters particularly useful in transmission work.

Administration: The usage here is the same as that within the CCITT. It is a short form meaning either a national telecommunication administration or a recognized private organization engaged in the provision of telecommunication services.

AEN: An abbreviation for the French name, *"Affaiblissement Équivalent pour la Netteté,"* which is translated as "articulation reference equivalent," in United Kingdom terminology, or "equivalent articulation loss," in US terms. It is a measure of speech intelligibility.

Aliasing: When an analog signal is sampled to produce PAM pulses, the spectrum of the signal is replicated an infinite number of times at multiples of the sampling frequency. Only one of these spectra is desired, and it is selectively filtered. If the filtering process admits portions of the remaining spectra, distortion results. This type of distortion is known as "aliasing."

AM: *See* Amplitude Modulation.

AML: Actual Measured Loss. The loss of a trunk as *measured* at a frequency of 0.8 or 1.0 kHz, expressed in dB.

Amplitude Jitter: Small continuous variations of the instantaneous amplitude of a signal from its expected value, which are limited by definition to causes other than noise.

Amplitude Modulation: A method of modulation in which the instantaneous amplitude of a sinusoidal carrier signal is made to depart from its rest value in proportion to the instantaneous amplitude of the information signal.

ANSI: American National Standards Institute.

Analog Communication: Communication accomplished by maintaining an analog waveform from origin (source) to destination (sink).

Analog Office: A central office using switches capable of transferring an analog waveform.

ARAEN: Abbreviation for the French name, *"Appareil de Référence pour la détermination des Affaiblissements Équivalents pour la Netteté,"* which means "reference apparatus for the determination of AEN," but is translated as "reference apparatus for the determination of transmission performance ratings." It is a subjective test facility in Geneva, Switzerland, introduced in the 1960s.

Attenuation Distortion: Distortion of a signal due to changes in amplitude that are not constant with frequency.

Balance: Term used to describe the degree of electrical symmetry from two points in a circuit or transmission line to ground. A high degree of symmetry is synonymous with high balance, and is generally desirable.

Bridging (measurement): A measurement made in a manner that has a negligible effect on the normal operation of the device or circuit being measured. For purposes of the measurement, the connection of the measuring apparatus has no effect on normal operation.

Bridging Loss:
1. Error in a loss measurement due to power absorbed by the test instrument.
2. The power absorbed from one circuit due to the connection of another circuit or device.

CCITT: Abbreviation for the French name, *Comité Consultatif International Télégraphique et Téléphonique,* known in English as the International Telegraph and Telephone Consultative Committee. The CCITT is a committee of the International Telecommunication Union (ITU).

CDF: *See* Cumulative Distribution Function.

Central Office (CO): An entity comprising a switching machine and all associated equipment. CO often includes the premises of the switching equipment.

Channel: Generic name for a communication facility. Usually implies a capacity suitable for a single user, but the application, and hence the bandwidth, is not defined.

Class N Office: A central office in the Nth level of a switching hierarchy. In the North American switching hierarchy, $N = 5$ designates the CO connecting to subscriber loops.

C-Message: A filter characteristic used in noise measuring sets to make the measurements correlate well with subjective reaction to the noise. Abbreviated as C-MESS. Also used as an identifier for noise measurements made by using such a filter. C-message weighting is similar to psophometric weighting.

C-Notched: A C-message filter with a narrow rejection band centered at 1020 Hz to suppress a tone, transmitted during the measurement, that is used to activate companders and quantizers. Abbreviated as C-NOTCH. Also used as an identifier for noise measurements made by using such a filter.

Codec: A device that converts analog signals into an encoded digital (often PCM) format in one direction, and performs the reverse operation in the other direction.

Compander: A device in which the gain is proportional to the input signal amplitude. Used to suppress noise on a channel during pauses in speech. The term is a contraction of the "compressor" action of the input portion and the "expander" action of the output portion. *See* Instantaneous Compander.

Connection: If used without a qualifier, connection refers to a specific transmission path between an origin (source) and destination (sink).

Contrast: The change in transmission loss between calls placed within a defined environment. For PBXs, contrast refers to the degree of uniformity of losses encountered, within the switch, between different pairs of ports. It is also used to describe the uniformity of loss among connections within a network.

Cradling: Term used to describe the practice of supporting a handset by resting it on the shoulder, and bending the head downward and to the side to support the handset, to have both hands free for other tasks.

Crosstalk: Name given to the phenomenon of signals in one channel being detected or heard on another.

Cumulative Distribution Function (CDF): A mathematical function describing the fraction of some population that has an attribute no greater than a value given by the independent variable. By definition, a CDF must range from zero to one.

Data Communication: Transmission of information that is in a discrete format at the source.

D_0: A correction factor used to adjust the sum of loudness losses from portions of a connection to approximate the overall connection loudness loss.

dB: Abbreviation for decibel, one tenth of a bel. Describes the ratio, R, of two powers P_1 and P_2 on a logarithmic scale:

$$R = 10 \log_{10}(P_1/P_2) \text{ dB}$$

See also dBm, dBm0, and dBV.

dBA: Refers to a sound pressure measured in decibels with a frequency weighting characteristic known as type A. Type A weighting de-emphasizes the low frequency components, and is often used in measurements of loudness of sounds. The reference pressure for the decibel scale is usually 20 micropascals, 0.000204 dynes per square centimeter.

dBm: Abbreviation for dB referred to one milliwatt of power. If P_0 is one milliwatt, and P_1 and P_0 are expressed in the same units, then the power X, in dBm, is

$$X = 10 \log_{10}(P_1/P_0)$$

dBm0: Abbreviation for a signal power P, at a point in a network, referred to the level of a standard zero dBm test tone at that point. If the expected power at a point in the network is Y dB below the power of a standard test tone at that point, and the power is expressed as R dBm0, then the absolute power at that point is P, where

$$P = R - Y \text{ (dBm)}$$

dBt: A measure of acoustic pressure. dBt = dB relative to $2.04 \cdot 10^{-5}$ newton per square meter (0.000204 dynes per square centimeter).

dBV: Abbreviation for a voltage expressed in dB relative to one volt:

$$\text{dBV} = 20 \log_{10}(V/V_0)$$

with $V_0 = 1$. The measure of voltage may be rms, peak, *etcetera*. It is understood to be rms unless otherwise specified. The multiplier is 20 because the definition is related to that for power and power is proportional to V^2.

Delay: The time required for a signal to traverse a transmission path. Delay is used most often in association with echoes. For distortionless transmission, delay must be a linear function of frequency.

Delay Distortion: Distortion due to delay being a nonlinear function of frequency. Also, it is the function or graph describing the relative delay of a transmission system as a function of frequency, normalized at the frequency of minimum delay.

DD: *See* Delay Distortion.

Digital Communication: Transmission of information by use of discrete signaling elements (pulses).

Digital Frame: *See* Frame.

Digital Office: A central office using switches capable of transferring discrete pulses between ports but not analog signals.

Distribution Frame: An assembly of connecting devices (connecting blocks) used to interconnect cables within a building to cables that interconnect among buildings, or cables between separate devices. A central office uses many frames, and the one used to terminate subscriber loops is known as the "master distribution frame" (MDF), while others are commonly referred to as "intermediate distribution frame" (IDF).

Diversity: Any means of providing more than one transmission path between two points. One path is used as back-up in case of failure or deterioration of the other.

Dropout: *See* Hit.

Drop Wire: The pair of wires used to connect an individual telephone (a subscriber location) to the distribution facility serving the area.

EARS: Abbreviation for the Bell Telephone Laboratories' Electro-Acoustic Rating System. An objective measurement system, and the associated metrics, used to describe loudness loss or electroacoustic conversion efficiencies.

Echo: The delayed reappearance of a previously sensed signal.

Echo Return Loss: A frequency-weighted measure of the loss incurred by an echo before it is sensed.

ED: *See* Envelope Delay.

EDD: *See* Envelope Delay Distortion.

EMF: Electromotive Force. Electric potential (voltage).

EML: Expected Measured Loss. The design value of the one-kilohertz transmission loss of a trunk, expressed in dB.

Encoder: A device for encoding information in a prescribed format.

Envelope Delay: The derivative of the function describing the relation between delay and radian frequency. Envelope delay must be a constant for distortionless transmission.

Envelope Delay Distortion: If the envelope delay is not zero, the function is referred to as envelope delay distortion. For proper presentation, the function must be normalized at the frequency of minimum envelope delay.

Eye Pattern: Name given to a particular oscilloscope display obtained from received data signals. The pattern typically covers the entire screen with one or more openings along the horizontal axis. The size of the opening is an indication of the quality of the received signal. Noise and distortion cause the opening to decrease in size, or "close the eye."

Exchange Area: Geographic area small enough to be served by one or a small number of central offices and not requiring carrier transmission facilities. Sometimes restricted to include only those locations with a common telephone number prefix.

Facility: A physical transmission system, such as a length of cable, a microwave transmission system, *etcetera*.

Far End: Refers to the end of a communication channel that is distant from the location being considered. The location distant from the talker's or tester's location.

FDM: *See* Frequency Division Multiplexing.

Feed Circuit: Name applied to the source of dc in a central office or PBX used to supply power to a terminal device.

Four-Wire: Used as prefix to indicate that the named device uses separate paths for the forward and reverse directions of transmission.

FM: *See* Frequency Modulation.

Frame: In synchronous digital transmission, the digital stream is segmented into units of a fixed number of bits, called frames. Each frame is an administrative and synchronous unit.

Framing Interval: The length of a frame. Usually expressed as a number of bits.

Frequency Distortion: *See* Attenuation Distortion.

Frequency Diversity: A technique used in radio transmission that provides diversity by using more than one carrier frequency. *See* Diversity.

Frequency Division Multiplexing: *See* Multiplexer.

Frequency Modulation: A method of modulation in which the instantaneous frequency of a sinusoidal carrier signal is made to depart from its rest value in proportion to the instantaneous amplitude of the information signal.

Frequency Offset: *See* Offset.

Gain Hit: *See* Hit.

Half-Duplex Transmission: A protocol that permits transmission in only one direction at a time. A return transmission must wait until the forward one is completed.

Harmonic Distortion: A measure of nonlinear distortion made by measuring harmonics, at the output of a system, of a pure tone applied at the input.

Histogram: A bar graph showing the quantized fractions of a population that have an attribute equal to the value stated on the abscissa. A histogram is the discrete variable equivalent of a probability density function.

Hit: Term used to describe a transient disturbance on a channel. Usually preceded with a descriptor: gain hit, phase hit, *etcetera*. A negative gain hit (decrease in amplitude) greater than 12 dB and lasting for at least 4 ms is called a "dropout."

Hoth Spectrum: Noise with a spectrum taken to be representative of that common in indoor environments. Spectral shape is similar to that of pink noise, but rolls off more rapidly, about 5 dB per octave to about 4 kHz, and then about 12 dB per octave to 8 kHz.

Hubbing: A method of interconnecting a number of network locations by arranging trunks from each to a single central point (switch). A star topology.

IDF: *See* Distribution Frame.

ICL: Inserted connection loss. The change in power delivered to a load when a device is inserted between the source and the load. The change is expressed in dB.

IEEE: Institute of Electrical and Electronics Engineers.

Impulse Noise: Short duration, high amplitude noise pulses, usually of a sporadic nature. In a voice-band channel, noise that exceeds the rms noise level by 12 dB or more is considered to be impulse noise. The noise pulses are typically on the order of one to a few milliseconds in duration.

Instantaneous Compander: A compander used in digital transmission systems that operates on PAM pulses. The amplitude of the output pulses of the compressor portion are a nonlinear function of the amplitudes of the input pulses. The inverse function is implemented at the expander portion. Two nonlinear functions in common use are known as the A-law and mu-law characteristics.

Intermediate Reference System (IRS): A standardized frequency response for simulating a high quality voice-grade channel. Used as the reference system for the CCITT loudness rating metrics.

Intermodulation Distortion: A measure of nonlinear distortion made by measuring cross products, at the output of a system, generated from two or more pure tones applied at the input.

Intertandem Tie Trunk: A tie trunk used between intermediate, nonterminating, switches in a tandem tie trunk network.

IRS: *See* Intermediate Reference System.

ISC: International Switching Center. Any switching location for trunks used in the international network.

ISO: The International Organization for Standardization. An international standards group concerned primarily with the specification of mechanical systems, but also active in the area of standards for telecommunications. (The initials, ISO, are derived from its former name, the International Standards Organization.)

Jitter: Term applied to small continuous variations around the expected time of occurrence of the leading edge, or other epoch, of a pulse in a digital transmission system. Jitter is cumulative through regenerators and may limit the length of such a system. *See* also Phase Jitter.

JRE: Junction Reference Equivalent. The reference equivalent (RE) for that portion of a telephone connection comprising all equipment (switches, trunks, *etcetera*) between the two local terminations.

Key Set: A telephone arranged to permit manual selection of one from a number of subscriber lines or loops.

Link: Term loosely used to indicate any kind of connection between two points. If the points are switching entities, the term may imply a trunk. In radio transmission, it refers to a single point-to-point radio connection, or "hop," as in one radio repeater link.

LIL: *See* Loudness Insertion Loss.

Listener Echo: An echo of the far-end talker's speech that is sensed at the near end.

Loading: Name for a procedure in which inductors are placed in series with cable pairs at fixed distances. The cable capacitance and the inductors form a low-pass filter structure, providing reduced loss and nearly constant frequency response over a limited bandwidth.

Load Coil: An inductor specially designed and packaged for use in loading.

Loaded Cable: A cable assembly comprising metallic pairs and load coils.

Loudness (n): A number, n, of sones, equal to the number of times that a sound is louder than one sone (*see* Loudness Level and Sone).

Loudness Insertion Loss: Insertion loss of a device expressed in terms of one of the measures of loudness.

Loudness Level: The loudness level of a sound, in phons, is numerically equal to the median value of the sound pressure, in dBt, of a 1000 Hz free progressive wave, which is judged to be equally loud as the sound. Note that this definition is identical to that of the phon, but is made for complex sounds instead of single frequencies.

Loudness Loss: The loss of a network or system, expressed in dB, but calculated by using one of the loudness weighting functions, such as IEEE or CCITT.

Loudness Rating: The basic definition for loudness rating is a ratio, expressed in dB, of the loudness of speech at the entrance to a listener's ear to the loudness of the speech as it leaves the talker's mouth. The definition is extended to include the ratio (dB) of any two signals that represent a speech waveform. The ratio may be acoustic pressure to an electrical voltage, for example. The ratio is arranged in such a manner that one expressing a decrease in loudness is a positive quantity. The numeric for loudness rating depends upon the metric system used: IEEE, CCITT, EARS. *See* Loudness Loss and Loudness Insertion Loss.

Loudness Unit (LU): Loudness unit is defined in the same manner as loudness level, but is reserved for the case in which the "sound" compared to the 1000 Hz tone is speech. This definition and usage eliminates any ambiguities or concerns over equivalent loudness of noises versus speech. Speech of n LUs has a loudness of n sones.

Loop:
1. A pair of wires connecting a customer location to a central office.
2. Any closed electrical path.

LR: *See* Loudness Rating.

MDF: *See* Distribution Frame.

Modal Distance: The distance between a talker's lips and the center of the opening to the transmitter on the mouthpiece of a handset when it is in use.

Modem: A device used to convert a signal from a digital source into an analog form for transmission, and to convert received analog signals into digital form for delivery to the sink.

Mu-Law: *See* Instantaneous Compander.

Multiplexer: A device to combine many signals using relatively small bandwidths (many channels) into a composite signal using a larger bandwidth (one channel). For example, analog signals are often combined by modulating each onto one of a set of uniformly spaced carrier frequencies, resulting in frequency division multiplexing (FDM). Or, a number N of digital channels, each of bit rate r, may be combined by dividing a single bit stream of rate R, where $R > (N \cdot r)$, into uniformly spaced segments of k bits each, resulting in $K = r/k$ blocks per second. Blocks are sequentially assigned k bits from channels 1 through N, resulting in time division multiplexing (TDM).

Multiplexing: *See* Multiplexer.

Near End: Refers to the end of a communication channel that is collocated with the talker or the testing location.

Nonlinear Distortion: Distortion of a signal due to passage through a device in which the gain or loss is a nonlinear function of amplitude. This definition restricts the term to its most common usage, meaning "nonlinear amplitude distortion." Nonlinear distortions of other signal parameters are common, but usually they have their own descriptors (i.e., envelope delay distortion).

Nonloaded Cable: A cable comprising metallic pairs in which none of the pairs have had loading coils connected to them.

NOSFER: Abbreviation for the French name "*NOuveau Système Fondamental pour la détermination des Equivalents de Référence,*" which is translated as "new standard system for the determination of reference equivalents." A CCITT test arrangement for measuring reference equivalents.

NOSFER-84: Name for a technological upgrade of NOSFER, implemented in 1984.

NPA: *See* Telephone Number.

Nyquist Rate: For pulse transmission, it has been shown that the maximum transmission rate, in pulses per second, obtainable with no inter-symbol interference, is numerically equal to twice the bandwidth of the channel, expressed in Hz. This value is known as the Nyquist rate.

Objective Measurement: A measurement made with instruments, as opposed to one made by human judgment.

Objectives: A set of goals.

Offset (frequency): A change in frequency of a signal, usually small compared to the frequency.

OLR, OOLR: Overall Loudness Rating, or Overall Objective Loudness Rating. Loudness rating for a complete telephone connection from talker to listener. OOLR is the IEEE adaptation of the EARS metric, OLR.

ONS: On-Premises Station. A station geographically close (within about 5,000 ft) to a customer's switching machine, usually a PBX.

Open Wire Transmission Line: A transmission medium comprising two parallel conductors, suspended in air with a spacing of about six to nine inches.

OPS: Off-Premises Station. A station geographically distant (beyond about 5,000 ft) from a customer's switching machine, usually a PBX. An OPS usually requires transmission considerations, such as added gain, different from those of an ONS.

ORE: Overall Reference Equivalent. The reference equivalent (RE) for an entire telephone connection from talker to listener.

OREM, A or B: Objective Reference Equivalent Measurement, A or B. An objective measuring system yielding results in close agreement with subjective reference equivalent measurements for certain types of handsets. OREMA and OREMB refer to specific instruments, and the measurements made with them. OREMA is designed to measure handsets; OREMB is designed to measure carbon transmitters. OREMA is also used with a descriptor, such as "transmitting," to indicate the type of measurement (i.e., OREMA for transmitting).

Orthotelephonic: An adjective describing the acoustical relationship between two people of equal height facing each other at a distance of one meter in a quiet, nonreverbatory room. The distance is measured as that from the talker's lips to the central point of a line drawn through the centers of the listeners' ears.

Orthotelephonic Response: The frequency response of a system, in dB, referred to that obtained in the orthotelephonic situation. If ideal band-limiting to 8 kHz is used, it is assumed that there is zero degradation. (Frequencies above 8 kHz are ignored.)

Orthotelephonic System: Any system that duplicates the acoustic characteristics obtained in the orthotelephonic situation, except that band-limiting may be included.

PABX: An automatic private branch exchange (PBX).

Pad: Name applied to an attenuator used in a transmission or testing arrangement.

PAM: *See* Pulse Amplitude Modulation.

P/AR: A measurement system designed to provide a single-number evaluation of the transmission quality of a channel. The evaluation is based upon the dispersion and amplitude compression of a transmitted pulse train. P/AR stands for peak-to-average ratio.

Pascal (Pa): A unit of acoustic pressure. One pascal is equal to $2.04 \cdot 10^{-5}$ newtons per square meter (0.000204 dynes per square centimeter), a value taken as the threshold of hearing.

Pair Integrity: Transmission cables are manufactured with a high degree of symmetry within and between the pairs of wires intended for use as individual metallic transmission lines. When two cables are spliced, care must be taken to match pairs of the two cables with regard to location within the bundle of pairs as well as matching the tip and ring conductors of the pairs. When the matching is properly done, it is referred to as maintaining pair integrity.

PBX: Private Branch Exchange. A switching device for use on a customer's premises.

PCM: *See* Pulse Code Modulation.

PDF: *See* Probability Density Function.

Pink Noise: White noise that has been shaped with a frequency characteristic having a negative slope of 3 dB per octave to attenuate the shorter wavelengths (higher frequencies).

Phase Distortion: A constant displacement from the desired phase of components of an analog signal.

Phase Jitter: Undesired, and usually sinusoidal, variations in the phase of components of a received analog signal. Phase jitter is associated with interfering tones, thus phase variations due to noise are not classified as phase jitter.

Phon: The level, in dBt, of a 1000 Hz tone that is judged to be of the same loudness as that of a tone at another frequency. This definition is in terms of single frequencies, but the same measure is used when comparing complex sounds to a 1000 Hz tone. *See* Loudness Level.

Phase Hit: *See* Hit.

Phase Modulation: Modulation accomplished by changing the instantaneous phase of a carrier in a manner proportional to the instantaneous amplitude of the modulating signal.

PM: *See* Phase Modulation.

Power Addition: A method for adding two amounts of power expressed in dB. The sum, P_{total}, in dB, of two powers, P_1 and P_2, that are expressed in dB is found by using the relation:

$$P_{total} = 10 \log_{10} (10^{P_1/10} + 10^{P_2/10})$$

(If $P_1 = P_2$, then the sum is 3.01 dB greater.)

Prefix: *See* Telephone Number.

Probability Density Function: A mathematical function describing the incremental fraction of a population that has an attribute equal to the value of the independent variable. The derivative of a cumulative distribution function. Often simply referred to as the "distribution" of some quantity.

Psophometric: A filter shape used in noise measuring sets to make the reading correlate well with subjective reaction to the noise. Also employed as an identifier for noise measurements made when using such a filter. Psophometric weighting is similar to C-message weighting.

Pulse Amplitude Modulation: A modulation technique in which the amplitudes of a train of pulses are made equal to the amplitudes of samples of the information waveform.

Pulse Code Modulation: Modulation accomplished by encoding the amplitudes of samples of an analog signal in a digital stream.

QAM: *See* Quadrature Amplitude Modulation.

Quadrature Amplitude Modulation: Modulation accomplished by altering the amplitudes of two carrier signals of the same frequency, but in quadrature. The carriers may be modulated by separate modulating signals.

Quantizer: A device used to segment an analog signal into discrete amplitude levels.

Quantizing: The process of changing an analog signal to one containing only a fixed number of discrete values of amplitude.

Quantizing Distortion: *See* Quantizing Noise.

Quantizing Noise: Distortion in an analog signal reconstructed from a digital signal, due to the approximation to the true instantaneous amplitude of the original signal by the quantizing process.

Quefrency: A term used by some authors to indicate and measure periodicity on a frequency scale. Frequency of a periodic frequency phenomenon.

RE: *See* Reference Equivalent.

Return Loss: The ratio of incident voltage or current at an impedance mismatch to the reflected voltage or current, expressed in dB.

RRE: Receive Reference Equivalent. The reference equivalent (RE) for that portion of a telephone connection comprising the handset, station, local loop, and CO feed circuit used in the receiving mode.

Reference Equivalent: A subjectively determined metric for describing electroacoustic and acoustic-acoustic losses in components of telephone systems.

Repeater: A device or collection of equipment used to amplify signals in a communication path in both directions of transmission.

RLR, ROLR: Receive Loudness Rating, or Receive Objective Loudness Rating. Loudness rating (LR) for the portion of a connection, comprising the telephone and all equipment, including the local CO's dc feed circuit, used in the receiving mode.

1. ROLR is the IEEE adaptation of the EARS system, RLR.
2. RLR is also the abbreviation for the CCITT receive loudness rating.

See United Kingdom Rating System.

Sag: Term used in descriptions of the frequency response of channels to define high-frequency roll-off. If a parabola is fit to the response at 1 kHz, the response at the frequency at which slope is defined, and the response at 1750 Hz, then sag = (actual channel response) − (value of the line defining slope), evaluated at 1750 Hz. The two numbers describing sag and slope then define the parabola, and so its approximation to the actual response. Sag may also be used to describe other characteristics such as envelope delay. *See* Slope.

Satellite PBX: A PBX in a switched private line network that has direct access to only one other PBX and has no PBX-CO trunks.

SFI: *See* Single-Frequency Interference.

Sheath: A protective coat surrounding a pair, or a number of pairs, of conductors in a cable. A sheath may be metallic or nonmetallic. Metallic sheaths can provide protection from electrical interference as well as physical protection.

Sheath Current: Current flowing in a metallic sheath on a cable. Sheath currents are usually induced by external electromagnetic fields.

SIBYL: The name of a subjective test facility that was used at Bell Telephone Laboratories.

Singing Margin: The minimum loss incurred by an echo at any frequency within the channel passband. It is a measure of stability in a transmission feedback loop.

Single-Frequency Interference: Interference in a transmission system that consists of a single sinusoid.

Slip: The phenomenon, in a digital transmission system, of one complete frame of information being repeated or deleted.

Slope: Term used in describing the frequency response of a channel. Slope is the difference between the channel response at (usually) 2800 Hz and the response at 1 kHz. Frequencies other than 2800 Hz are sometimes used to define slope. Slope may also be used to describe other characteristics such as envelope delay. *See* Sag.

SLR, SOLR: Sidetone Loudness Rating, or Sidetone Objective Loudness Rating. Loudness rating (LR) for the acoustic-to-acoustic loss between the mouthpiece and receiver of a handset.
1.   SOLR is the IEEE adaptation of the EARS system, SLR.
2.   SLR is also the abbreviation for the CCITT send loudness rating.

See United Kingdom Rating System.

Sone: The unit sone is the loudness of a 1 kHz tone 40 dB above the listener's threshold of hearing. The sone is thus the bridge between the subjective phenomenon of loudness and the objective measure, dBt, of the level of the 1 kHz tone. The sone scale is linear in loudness; N sones means that a sound is N times louder than the unit sone. N may be a fraction.

Space Diversity: A technique used in radio transmission that uses more than one receiving antenna. The paths from the transmitter to each antenna cannot be identical, so some protection against selective fading is provided by this arrangement. Selective fading is a wave interference phenomenon, and the probability of two nodes occurring simultaneously at both receiving antennas is small.

SRAEN: Abbreviation for the French name, *"Système de Référence pour la Détermination des Affaiblissements Équivalents pour la Netteté."*

Known in English as "reference system for the determination of AEN." SRAEN is an adaptation of ARAEN, expressly for use in articulation or intelligibility tests.

SRE: Send Reference Equivalent. The reference equivalent (RE) for that portion of a telephone connection comprising the handset, station, local loop, and CO feed circuit used in the talking mode.

Statistical Parameter: A parameter which has a large variation within systems or with time, or arises from a multitude of sources, such that it can only be measured or described in statistical terms.

STMR: Sidetone Masking Rating. CCITT measure of telephone sidetone path loss. *See* United Kingdom Rating System.

STRE: Sidetone Reference Equivalent. The reference equivalent (RE) for the acoustic-to-acoustic loss between the transmitter and receiver of a handset through the sidetone path.

Subscriber Line: The transmission facility or loop providing subscriber access to the local central office.

Talker Echo: An echo of the talker's speech sensed at the talker's end.

Tandem Tie Trunk: Trunk used between two switches in a tandem tie line network when one of the switches is a terminating location for that connection.

Telephone Number: A sequence of numbers assigned to, and identifying, a particular network termination point. In the US, for the 10-digit telephone number, yyy-nnn-xxxx, the area code yyy is referred to as the NPA, and the nnn as the prefix. The xxxx identifies one of the 10,000 possible numbers in the basic switching machine arrangement. It does not have a unique name, but is referenced simply as the subscriber number. The prefix, nnn, is also referred to as the NNX code or the exchange identifier.

Telephonometric: Adjective used to specify human participation in a telecommunication situation. A telephonometric measurement for example means a measurement by "voice and ear."

TEPL: Talker Echo Path Loss. TEPL is the one-kilohertz loss value.

TEPOLR: Talker Echo Path Overall Loudness Rating. TEPOLR may be expressed in terms of the weighted echo path loss (WEPL), or the IEEE loudness loss.

Terminal: Any device connecting to the customer premises end of a loop or similar facility, such as a station loop served by a PBX.

Tie Trunk: Any trunk between two switches in a private network.

Time Division Multiplexing: *See* Multiplexer.

Tip and Ring: When a pair of wires is used as a transmission line, it is important, for polarity identification in supplying dc power and signaling, to distinguish the wires within the pair. Administratively, this is accomplished by referring to one of them as the *tip* conductor and the other as the *ring* conductor. The names date back to the use of "tip and ring" jacks and plugs in manual switchboards.

TL: Transmission Level. A specification of expected signal power in a network.

TLP: A specified point in a network that is designed to have a certain transmission level (TL).

TLR, TOLR: Transmit Loudness Rating, or Transmit Objective Loudness Rating. Loudness rating (LR) for the portion of a connection, comprising the telephone and all equipment, including the local CO feed circuit, used in the transmitting or talking mode. TOLR is the IEEE adaptation of the EARS system, TLR.

Toll Area: The collection of switches and trunks in a public network providing connections between exchange areas. Includes all levels above the lowest in a switching hierarchy. (The lowest level is designated by the largest number. The numbering system is from the top down.)

Total Distortion: A measurement made with a noise meter at the output of a digital transmission device or digital system after a tone inserted at the far end is removed by means of a narrow rejection filter. The reading on the noise meter may be a composite of noise, quantizing distortion, aliasing, and harmonic distortion.

Tracking Error: A measure of the accuracy of the instantaneous amplitude of the output signal of a compander.

Trunk: The entire transmission facility between similar points within two switching machines in a switched network, or between distribution frames in a nonswitched network.

Trunk Group: A collection of any number of trunks with common end points.

TT: *See* Tie Trunk.

TTT: *See* Tandem Tie Trunk.

TTTN: Tandem Tie Trunk Network. A type of switched private line network.

Two-Wire: Used as a prefix to indicate that the named device uses the same path for forward and reverse directions of transmission.

UI: *See* Unit Interval.

Unit Interval: The ideal time between two identical, significant, and repetitive epochs in a digital transmission system. The time between leading-edge zero crossings of successive pulses, for example.

United Kingdom Rating System: A system of transmission ratings originated in the United Kingdom. The metrics are called *loudness ratings* (LR) with a preceding descriptor: *send loudness rating* (SLR); *receive loudness rating* (RLR); and *overall loudness rating* (OLR). Sidetone loss is referenced to the natural sidetone paths, acoustic and internal, which are considered to be competing with the telephone sidetone path, and is referred to as *sidetone masking rating* (STMR).

VB (weighting): Voice Band (weighting). A filter characteristic formerly used in impulse noise measuring sets.

VNL: Via Net Loss. A method for choosing the loss to assign to trunks in a network. The assigned loss is proportional to the transmission delay expected on the trunk.

Voltage Addition: The sum, V_{total}, in dB, of two voltages, V_1 and V_2, which are expressed in dB, is found by using the relation:

$$V_{total} = 20 \log_{10} (10^{V_1/20} + 10^{V_2/20})$$

(If $V_1 = V_2$, then the sum is 6.02 dB greater.)

VU: Volume Unit. A unit of measure used to describe speech power. It is the calibration unit of a VU meter. A VU meter is a voltmeter, with certain ballistic characteristics, intended for the measurement of speech waveforms. For a sine wave and a 600 Ω termination, one VU is equal to 0 dBm. For speech, it is customary to assume that power in dBm = (VU − 1.4).

White Noise: Noise that has a uniform power density spectrum over the bandwidth of interest.

Index

The Artech House Telecommunication Library

Jansky, Donald M., ed., **World Atlas of Satellites**

Kantor, L. Ya., ed., **Handbook of Satellite Telecommunications and Broadcasting**

Kremer, Hermann, **Numerical Analysis of Linear Networks and Systems**

Krommenacker, Raymond J., **World-Traded Services: The Challenge for the Eighties**

Lauria White, Rita, and Harold M. White Jr., **The Law and Regulation of International Space Communication**

Lewin, Leonard, ed., **Telecommunications: An Interdisciplinary Text**

Lewin, Leonard, ed., **Telecommunications in the U.S.: Trends and Policies**

Noll, A. Michael, **Introduction to Telecommunication Electronics**

Noll, A. Michael, **Introduction to Telephones and Telephone Systems**

Olgren, Christine H., and Lorne A. Parker, **Teleconferencing Technology and Applications**

Pipe, G. Russell, ed., **The ISDN Workshop: INTUG Proceedings**

Roda, Giovanni, **Troposcatter Radio Links**

Rutkowski, Anthony M., **Integrated Services Digital Networks**

Taylor, Henry F., ed., **Fiber Optics Communications**

Terpstra, Randall W., **Twisted Pair Network Technology**

Tröndle, K., and G. Söder, **Optimization of Digital Transmission Systems**

Whalen, Timothy, **Writing and Managing Winning Technical Proposals**